Student Solutions Manual

SEARS AND ZEMANSKY'S

UNIVERSITY PHYSICS

TENTH EDITION

VOLUMES 2 & 3

YOUNG & FREEDMAN

A. LEWIS FORD

TEXAS A&M UNIVERSITY

▲ ADDISON-WESLEY

An imprint of Addison Wesley Longman, Inc.

San Francisco • Reading, Massachusetts
New York • Harlow, England • Don Mills, Ontario
Sydney • Mexico City • Madrid • Amsterdam

ISBN 0-201-64395-2
    4 5 6 7 8 9 10—CRS—03 02

Addison Wesley Longman, Inc.
1301 Sansome Street
San Francisco, California 94111

PREFACE

This *Student Solutions Manual*, Volumes 2 & 3, contains detailed solutions for approximately one third of the Exercises and Problems in Chapters 22 through 46 of the Tenth Edition of *University Physics*. The Exercises and Problems included in this manual are selected solely from the odd-numbered Exercises and Problems in the text (for which the answers are tabulated in the back of the textbook). The Exercises and Problems included were not selected at random but rather were carefully chosen to include at least one representative example of each problem type. The remaining Exercises and Problems, for which solutions are not given here, constitute an ample set of problems for you to tackle on your own. In addition, there are the Challenge Problems in the text for which no solutions are given here.

This manual greatly expands the set of worked-out examples that accompanies the presentation of physics laws and concepts in the text. This manual was written to provide you with models to follow in working physics problems. The problems are worked out in the manner and style in which you should carry out your own problem solutions.

The author will gratefully receive comments as to style, points of physics, errors, or anything else relating to this manual. The *Student Solutions Manual* Volume 1 companion volume is also available from your college bookstore.

A. Lewis Ford
Physics Department
Texas A&M University
College Station TX 77843
Email: ford@physics.tamu.edu

CONTENTS

CHAPTER 22
ELECTRIC CHARGE AND ELECTRIC FIELD

Exercises 1, 5, 7, 9, 13, 17, 19, 21, 23, 25, 29, 33, 35, 45, 47, 49
Problems 53, 55, 57, 59, 61, 65, 67, 69, 77, 81, 83

Exercises

22-1 **a)** The charge of one electron is $-e = -1.602 \times 10^{-19}$ C.

The number of excess electrons needed to produce net charge q is

$$\frac{q}{-e} = \frac{-3.20 \times 10^{-9} \text{ C}}{-1.602 \times 10^{-19} \text{ C/electron}} = 2.00 \times 10^{10} \text{ electrons.}$$

b) Find the number of lead atoms in 8.00×10^{-3} kg of lead. The atomic mass of lead is 207×10^{-3} kg/mol, so the number of moles in 8.00×10^{-3} kg is

$$n = \frac{m_{\text{tot}}}{M} = \frac{8.00 \times 10^{-3} \text{ kg}}{207 \times 10^{-3} \text{ kg/mol}} = 0.03865 \text{ mol.}$$

N_{A} (Avogadro's number) is the number of atoms in 1 mole, so the number of lead atoms is $N = nN_{\text{A}} = (0.03865 \text{ mol})(6.022 \times 10^{23} \text{ atoms/mol}) = 2.328 \times 10^{22}$ atoms.

The number of excess electrons per lead atom is

$$\frac{2.00 \times 10^{10} \text{ electrons}}{2.328 \times 10^{22} \text{ atoms}} = 8.59 \times 10^{-13}.$$

22-5 $N = nN_{\text{A}} = (1.80 \text{ mol})(6.022 \times 10^{23} \text{ atoms/mol}) = 1.084 \times 10^{24}$ atoms

There is 1 electron per hydrogen atom and each electron has charge $-e = -1.602 \times 10^{-19}$ C, so $Q = (-1.602 \times 10^{-19} \text{ C/electron})(1.084 \times 10^{24} \text{ electrons}) = -1.74 \times 10^5$ C.

This is a large amount of charge.

22-7 **a)** $q_1 = q_2 = q$

$$F = \frac{1}{4\pi\epsilon_0} \frac{|q_1 q_2|}{r^2} = \frac{q^2}{4\pi\epsilon_0 r^2} \text{ so}$$

$$q = r\sqrt{\frac{F}{(1/4\pi\epsilon_0)}} = 0.150 \text{ m}\sqrt{\frac{0.220 \text{ N}}{8.988 \times 10^9 \text{ N} \cdot \text{m}^2/\text{C}^2}} = 7.42 \times 10^{-7} \text{ C (on each)}$$

b) $q_2 = 4q_1$

$$F = \frac{1}{4\pi\epsilon_0}\frac{|q_1 q_2|}{r^2} = \frac{4q_1^2}{4\pi\epsilon_0 r^2} \text{ so}$$

$$q_1 = r\sqrt{\frac{F}{4(1/4\pi\epsilon_0)}} = \tfrac{1}{2}r\sqrt{\frac{F}{(1/4\pi\epsilon_0)}} = \tfrac{1}{2}(7.42 \times 10^{-7} \text{ C}) = 3.71 \times 10^{-7} \text{ C}.$$

And then $q_2 = 4q_1 = 1.48 \times 10^{-6}$ C.

22-9   The weight of an electron is $m_e g$. The nucleus of a hydrogen atom is a single proton with charge $+e$. The Coulomb force between the electron and the nucleus is $\dfrac{1}{4\pi\epsilon_0}\dfrac{e^2}{r^2}$.

Equating these two forces gives $m_e g = \dfrac{1}{4\pi\epsilon_0}\dfrac{e^2}{r^2}$.

$$r = e\sqrt{\frac{(1/4\pi\epsilon_0)}{m_e g}} = 1.602 \times 10^{-19} \text{ C}\sqrt{\frac{8.988 \times 10^9 \text{ N} \cdot \text{m}^2/\text{C}^2}{(9.109 \times 10^{-31} \text{ kg})(9.80 \text{ m/s}^2)}} = 5.08 \text{ m}$$

22-13   The three charges are placed as follows:

Like charges repel and unlike attract, so the free-body diagram for q_3 is

$$F_1 = \frac{1}{4\pi\epsilon_0}\frac{|q_1 q_3|}{r_{13}^2}$$

$$F_2 = \frac{1}{4\pi\epsilon_0}\frac{|q_2 q_3|}{r_{23}^2}$$

$$F_1 = (8.988 \times 10^9 \text{ N} \cdot \text{m}^2/\text{C}^2)\frac{(1.50 \times 10^{-9} \text{ C})(5.00 \times 10^{-9} \text{ C})}{(0.200 \text{ m})^2} = 1.685 \times 10^{-6} \text{ N}$$

$$F_2 = (8.988 \times 10^9 \text{ N} \cdot \text{m}^2/\text{C}^2) \frac{(3.20 \times 10^{-9} \text{ C})(5.00 \times 10^{-9} \text{ C})}{(0.400 \text{ m})^2} = 8.988 \times 10^{-7} \text{ N}$$

The resultant force is $\vec{R} = \vec{F}_1 + \vec{F}_2$.

$R_x = 0$.

$R_y = F_1 + F_2 = 1.685 \times 10^{-6} \text{ N} + 8.988 \times 10^{-7} \text{ N} = 2.58 \times 10^{-6} \text{ N}$.

The resultant force has magnitude 2.58×10^{-6} N and is in the $-y$-direction.

22-17 a) The charges are placed as shown:

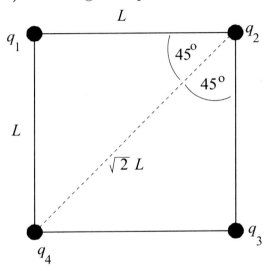

$$q_1 = q_2 = q_3 = q_4 = q$$

Consider forces on q_4. Take the y-axis to be parallel to the diagonal between q_2 and q_4 and let $+y$ be in the direction away from q_2. Then $\vec{F}_2$ is in the $+y$-direction.

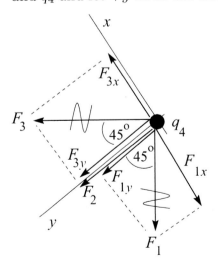

$$F_3 = F_1 = \frac{1}{4\pi\epsilon_0} \frac{q^2}{L^2}$$

$$F_2 = \frac{1}{4\pi\epsilon_0} \frac{q^2}{2L^2}$$

$F_{1x} = -F_1 \sin 45° = -F_1/\sqrt{2}$
$F_{1y} = +F_1 \cos 45° = +F_1/\sqrt{2}$
$F_{3x} = +F_3 \sin 45° = +F_3/\sqrt{2}$
$F_{3y} = +F_3 \cos 45° = +F_3/\sqrt{2}$
$F_{2x} = 0, \quad F_{2y} = F_2$

b) $R_x = F_{1x} + F_{2x} + F_{3x} = 0$

$$R_y = F_{1y} + F_{2y} + F_{3y} = (2/\sqrt{2}) \frac{1}{4\pi\epsilon_0} \frac{q^2}{L^2} + \frac{1}{4\pi\epsilon_0} \frac{q^2}{2L^2} = \frac{q^2}{8\pi\epsilon_0 L^2}(1 + 2\sqrt{2})$$

$R = \dfrac{q^2}{8\pi\epsilon_0 L^2}(1 + 2\sqrt{2})$. Same for all four charges. In general the resultant force on one of the charges is directed away from the opposite corner.

22-19 a) The gravity force is downward. For the net force to be zero the force exerted by the electric field must be upward. The electric field is downward. Since the electric field and the electric force are in opposite directions the charge of the particle is negative.

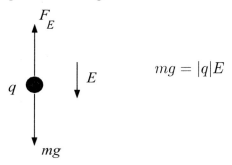

$$mg = |q|E$$

$$|q| = \frac{mg}{E} = \frac{(1.45 \times 10^{-3} \text{ kg})(9.80 \text{ m/s}^2)}{650 \text{ N/C}} = 2.19 \times 10^{-5} \text{ C and}$$

$$q = -21.9 \ \mu\text{C}$$

b) The electrical force has magnitude $F_E = |q|E = eE$.

The weight of a proton is $w = mg$.

$F_E = w$ so $eE = mg$

$$E = \frac{mg}{e} = \frac{(1.673 \times 10^{-27} \text{ kg})(9.80 \text{ m/s}^2)}{1.602 \times 10^{-19} \text{ C}} = 1.02 \times 10^{-7} \text{ N/C}.$$

This is a very small electric field.

22-21 a) For a negative charge $\vec{F}$ and field $\vec{E}$ are in opposite directions. $\vec{F}$ is downward so $\vec{E}$ is upward.

$$E = \frac{F}{|q|} = \frac{6.20 \times 10^{-9} \text{ N}}{55.0 \times 10^{-6} \text{ C}} = 1.13 \times 10^{-4} \text{ N/C}$$

b) The copper nucleus has charge $+29e$. For a positive charge the field and force are in the same direction so $\vec{F}$ is upward when $\vec{E}$ is upward.

$$F = |q|E = (29)(1.602 \times 10^{-19} \text{ C})(1.13 \times 10^{-4} \text{ N/C}) = 5.25 \times 10^{-22} \text{ N}$$

22-23 a)

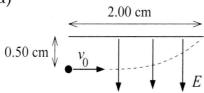

For an electron $q = -e$.

$\vec{F} = q\vec{E}$ and q negative gives that $\vec{F}$ and $\vec{E}$ are in opposite directions, so $\vec{F}$ is upward.

$$\sum F_y = ma_y$$
$$eE = ma$$

Solve the kinematics to find the acceleration of the electron:

Just misses upper plate says that $x - x_0 = 2.00$ cm when $y - y_0 = +0.500$ cm.

<u>x-component</u>

$v_{0x} = v_0 = 1.60 \times 10^6$ m/s, $a_x = 0$, $x - x_0 = 0.0200$ m, $t =?$

$x - x_0 = v_{0x}t + \frac{1}{2}a_x t^2$

$$t = \frac{x - x_0}{v_{0x}} = \frac{0.0200 \text{ m}}{1.60 \times 10^6 \text{ m/s}} = 1.25 \times 10^{-8} \text{ s}$$

In this same time t the electron travels 0.0050 m vertically:

<u>y-component</u>

$t = 1.25 \times 10^{-8}$ s, $v_{0y} = 0$, $y - y_0 = +0.0050$ m, $a_y = ?$

$y - y_0 = v_{0y}t + \frac{1}{2}a_y t^2$

$$a_y = \frac{2(y - y_0)}{t^2} = \frac{2(0.0050 \text{ m})}{(1.25 \times 10^{-8} \text{ s})^2} = 6.40 \times 10^{13} \text{ m/s}^2$$

(This analysis is very similar to that used in Chapter 3 for projectile motion, except that here the acceleration is upward rather than downward.)

This acceleration must be produced by the electric-field force:

$eE = ma$

$$E = \frac{ma}{e} = \frac{(9.109 \times 10^{-31} \text{ kg})(6.40 \times 10^{13} \text{ m/s}^2)}{1.602 \times 10^{-19} \text{ C}} = 364 \text{ N/C}$$

Note that the acceleration produced by the electric field is <u>much</u> larger than g, the acceleration produced by gravity, so it is perfectly ok to neglect the gravity force on the electron in this problem.

b) $a = \dfrac{eE}{m_{\mathrm{p}}} = \dfrac{(1.602 \times 10^{-19}\ \mathrm{C})(364\ \mathrm{N/C})}{1.673 \times 10^{-27}\ \mathrm{kg}} = 3.49 \times 10^{10}\ \mathrm{m/s}^2$

This is much less than the acceleration of the electron in part (a) so the vertical deflection is less and the proton won't hit the plates.

The proton has the same initial speed, so the proton takes the same time $t = 1.25 \times 10^{-8}$ s to travel horizontally the length of the plates. The force on the proton is downward (in the same direction as $\vec{E}$, since q is positive), so the accleration is downward and $a_y = -3.49 \times 10^{10}\ \mathrm{m/s}^2$.

$y - y_0 = v_{0y}t + \frac{1}{2}a_y t^2 = \frac{1}{2}(-3.49 \times 10^{10}\ \mathrm{m/s}^2)(1.25 \times 10^{-8}\ \mathrm{s})^2 = -2.73 \times 10^{-6}$ m.
The displacement is 2.73×10^{-6} m, downward.

c) The displacements are in opposite directions because the electron has negative charge and the proton has positive charge. The electron and proton have the same magnitude of charge, so the force the electric field exerts has the same magnitude for each charge. But the proton has a mass larger by a factor of 1836 so its acceleration and its vertical displacement are smaller by this factor.

22-25 **a)** Gravitational force exerted by the earth is $w_{\mathrm{e}} = m_{\mathrm{e}}g$.

$w_{\mathrm{e}} = (9.109 \times 10^{-31}\ \mathrm{kg})(9.80\ \mathrm{m/s}^2) = 8.93 \times 10^{-30}$ N

In Examples 22-7 and 22-8, $E = 1.00 \times 10^4$ N/C, so the electric force on the electron has magnitude

$F_E = |q|E = eE = (1.602 \times 10^{-19}\ \mathrm{C})(1.00 \times 10^4\ \mathrm{N/C}) = 1.602 \times 10^{-15}$ N.

$\dfrac{w_{\mathrm{e}}}{F_E} = \dfrac{8.93 \times 10^{-30}\ \mathrm{N}}{1.602 \times 10^{-15}\ \mathrm{N}} = 5.57 \times 10^{-15}$

The gravitational force is much smaller than the electric force and can be neglected.

b) From part (a) we know that this object must have mass much greater than an electron's mass.

$mg = |q|E$

$m = |q|E/g = (1.602 \times 10^{-19}\ \mathrm{C})(1.00 \times 10^4\ \mathrm{N/C})/(9.80\ \mathrm{m/s}^2) = 1.63 \times 10^{-16}$ kg

$\dfrac{m}{m_{\mathrm{e}}} = \dfrac{1.63 \times 10^{-16}\ \mathrm{kg}}{9.109 \times 10^{-31}\ \mathrm{kg}} = 1.79 \times 10^{14}; \quad m = 1.79 \times 10^{14} m_{\mathrm{e}}.$

Thus, m is much larger than m_{e}.

c) The electric field in the region between the plates is uniform so the force it exerts on the charged object is independent of where between the plates the object is placed.

22-29 a)

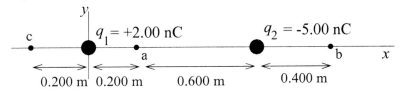

The electric field of a point charge is directed away from the point charge if the charge is positive and toward the point charge if the charge is negative.

The magnitude of the electric field is $E = \dfrac{1}{4\pi\epsilon_0}\dfrac{|q|}{r^2}$, where r is the distance between the point where the field is calculated and the point charge.

(i) At point a the fields $\vec{E}_1$ of q_1 and $\vec{E}_2$ of q_2 are:

$$E_1 = \frac{1}{4\pi\epsilon_0}\frac{|q_1|}{r_1^2} = (8.988 \times 10^9 \ \text{N} \cdot \text{m}^2/\text{C}^2)\frac{2.00 \times 10^{-9} \ \text{C}}{(0.200 \ \text{m})^2} = 449.4 \ \text{N/C}$$

$$E_2 = \frac{1}{4\pi\epsilon_0}\frac{|q_2|}{r_2^2} = (8.988 \times 10^9 \ \text{N} \cdot \text{m}^2/\text{C}^2)\frac{5.00 \times 10^{-9} \ \text{C}}{(0.600 \ \text{m})^2} = 124.8 \ \text{N/C}$$

$E_{1x} = 449.4 \ \text{N/C}, \quad E_{1y} = 0$

$E_{2x} = 124.8 \ \text{N/C}, \quad E_{2y} = 0$

$E_x = E_{1x} + E_{2x} = +449.4 \ \text{N/C} + 124.8 \ \text{N/C} = +574.2 \ \text{N/C}$

$E_y = E_{1y} + E_{2y} = 0$

The resultant field at point a has magnitude 574 N/C and is in the $+x$-direction.

(ii) At point b the fields $\vec{E}_1$ of q_1 and $\vec{E}_2$ of q_2 are:

$$E_1 = \frac{1}{4\pi\epsilon_0}\frac{|q_1|}{r_1^2} = (8.988 \times 10^9 \ \text{N} \cdot \text{m}^2/\text{C}^2)\frac{2.00 \times 10^{-9} \ \text{C}}{(1.20 \ \text{m})^2} = 12.5 \ \text{N/C}$$

$$E_2 = \frac{1}{4\pi\epsilon_0}\frac{|q_2|}{r_2^2} = (8.988 \times 10^9 \ \text{N} \cdot \text{m}^2/\text{C}^2)\frac{5.00 \times 10^{-9} \ \text{C}}{(0.400 \ \text{m})^2} = 280.9 \ \text{N/C}$$

$E_{1x} = 12.5 \text{ N/C}, \quad E_{1y} = 0$

$E_{2x} = -280.9 \text{ N/C}, \quad E_{2y} = 0$

$E_x = E_{1x} + E_{2x} = +12.5 \text{ N/C} - 280.9 \text{ N/C} = -268.4 \text{ N/C}$

$E_y = E_{1y} + E_{2y} = 0$

The resultant field at point b has magnitude 268 N/C and is in the $-x$-direction.

(iii) At point c the fields $\vec{E}_1$ of q_1 and $\vec{E}_2$ of q_2 are:

$E_1 = \dfrac{1}{4\pi\epsilon_0} \dfrac{|q_1|}{r_1^2} = (8.988 \times 10^9 \text{ N} \cdot \text{m}^2/\text{C}^2) \dfrac{2.00 \times 10^{-9} \text{ C}}{(0.200 \text{ m})^2} = 449.4 \text{ N/C}$

$E_2 = \dfrac{1}{4\pi\epsilon_0} \dfrac{|q_2|}{r_2^2} = (8.988 \times 10^9 \text{ N} \cdot \text{m}^2/\text{C}^2) \dfrac{5.00 \times 10^{-9} \text{ C}}{(1.00 \text{ m})^2} = 44.9 \text{ N/C}$

$E_{1x} = -449.4 \text{ N/C}, \quad E_{1y} = 0$

$E_{2x} = +44.9 \text{ N/C}, \quad E_{2y} = 0$

$E_x = E_{1x} + E_{2x} = -449.4 \text{ N/C} + 44.9 \text{ N/C} = -404.5 \text{ N/C}$

$E_y = E_{1y} + E_{2y} = 0$

The resultant field at point b has magnitude 404 N/C and is in the $-x$-direction.

b) Since we have calculated $\vec{E}$ at each point the simplest way to get the force is to use $\vec{F} = -e\vec{E}$.

(i) $F = (1.602 \times 10^{-19} \text{ C})(574.2 \text{ N/C}) = 9.20 \times 10^{-17} \text{ N}, \ -x$-direction

(ii) $F = (1.602 \times 10^{-19} \text{ C})(268.4 \text{ N/C}) = 4.30 \times 10^{-17} \text{ N}, \ +x$-direction

(iii) $F = (1.602 \times 10^{-19} \text{ C})(404.5 \text{ N/C}) = 6.48 \times 10^{-17} \text{ N}, \ +x$-direction

22-33 The electric field of a positive charge is directed radially outward from the charge and has magnitude $E = \dfrac{1}{4\pi\epsilon_0} \dfrac{|q|}{r^2}$. The resultant electric field is the vector sum of the fields of the individual charges.

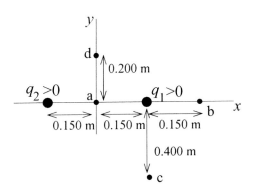

a)

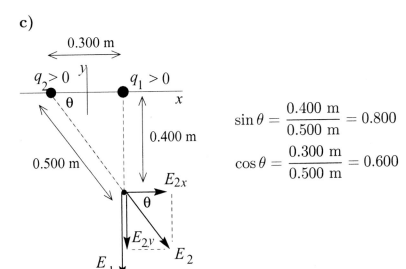

$E_1 = E_2 = \dfrac{1}{4\pi\epsilon_0}\dfrac{|q|}{r^2}$ with $r = 0.150$ m.

$E = E_2 - E_1 = 0$; $E_x = 0$, $E_y = 0$

b)

$E = E_1 + E_2$, in the $+x$-direction

$E_1 = \dfrac{1}{4\pi\epsilon_0}\dfrac{|q_1|}{r_1^2} = (8.988 \times 10^9 \text{ N} \cdot \text{m}^2/\text{C}^2)\dfrac{6.00 \times 10^{-9} \text{ C}}{(0.150 \text{ m})^2} = 2396.8 \text{ N/C}$

$E_2 = \dfrac{1}{4\pi\epsilon_0}\dfrac{|q_2|}{r_2^2} = (8.988 \times 10^9 \text{ N} \cdot \text{m}^2/\text{C}^2)\dfrac{6.00 \times 10^{-9} \text{ C}}{(0.450 \text{ m})^2} = 266.3 \text{ N/C}$

$E = E_1 + E_2 = 2396.8 \text{ N/C} + 266.3 \text{ N/C} = 2660 \text{ N/C}$; $E_x = +2260 \text{ N/C}$, $E_y = 0$

c)

$\sin\theta = \dfrac{0.400 \text{ m}}{0.500 \text{ m}} = 0.800$

$\cos\theta = \dfrac{0.300 \text{ m}}{0.500 \text{ m}} = 0.600$

$E_1 = \dfrac{1}{4\pi\epsilon_0}\dfrac{|q_1|}{r_1^2}$

$E_1 = (8.988 \times 10^9 \text{ N} \cdot \text{m}^2/\text{C}^2)\dfrac{6.00 \times 10^{-9} \text{ C}}{(0.400 \text{ m})^2} = 337.1 \text{ N/C}$

$$E_2 = \frac{1}{4\pi\epsilon_0}\frac{|q_2|}{r_2^2}$$

$$E_2 = (8.988 \times 10^9 \text{ N} \cdot \text{m}^2/\text{C}^2)\frac{6.00 \times 10^{-9} \text{ C}}{(0.500 \text{ m})^2} = 215.7 \text{ N/C}$$

$$E_{1x} = 0, \quad E_{1y} = -E_1 = -337.1 \text{ N/C}$$

$$E_{2x} = +E_2 \cos\theta = +(215.7 \text{ N/C})(0.600) = +129.4 \text{ N/C}$$

$$E_{2y} = -E_2 \sin\theta = -(215.7 \text{ N/C})(0.800) = -172.6 \text{ N/C}$$

$$E_x = E_{1x} + E_{2x} = +129 \text{ N/C}$$

$$E_y = E_{1y} + E_{2y} = -337.4 \text{ N/C} - 172.6 \text{ N/C} = -510 \text{ N/C}$$

$$E = \sqrt{E_x^2 + E_y^2} = \sqrt{(129 \text{ N/C})^2 + (-510 \text{ N/C})^2} = 526 \text{ N/C}$$

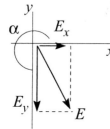

$$\tan\alpha = \frac{E_y}{E_x}$$

$$\tan\alpha = \frac{-510 \text{ N/C}}{+129 \text{ N/C}} = -3.953$$

$\alpha = 284°\text{C}$, counterclockwise from $+x$-axis

d)

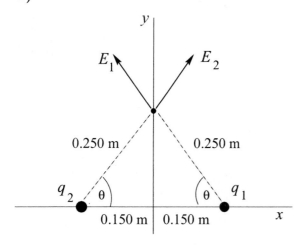

$$\sin\theta = \frac{0.200 \text{ m}}{0.250 \text{ m}} = 0.800$$

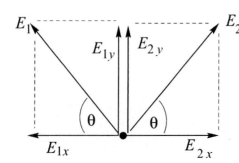

$$E_1 = E_2 = \frac{1}{4\pi\epsilon_0}\frac{|q|}{r^2}$$

$$E_1 = (8.988 \times 10^9 \text{ N} \cdot \text{m}^2/\text{C}^2)\frac{6.00 \times 10^{-9} \text{ C}}{(0.250 \text{ m})^2}$$

$$E_1 = E_2 = 862.8 \text{ N/C}$$

$$E_{1x} = -E_1 \cos\theta, \quad E_{2x} = +E_2 \cos\theta$$
$$E_x = E_{1x} + E_{2x} = 0$$
$$E_{1y} = +E_1 \sin\theta, \; E_{2y} = +E_2 \sin\theta$$
$$E_y = E_{1y} + E_{2y} = 2E_{1y} = 2E_1 \sin\theta = 2(862.8 \text{ N/C})(0.800) = 1380 \text{ N/C}$$
$$E = 1380 \text{ N/C, in the } +y\text{-direction.}$$

22-35 The resultant electric field is the vector sum of the field $\vec{E}_1$ of q_1 and $\vec{E}_2$ of q_2.

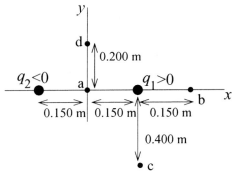

a)

$$E_1 = E_2 = \frac{1}{4\pi\epsilon_0} \frac{|q_1|}{r_1^2}$$

$$E_1 = (8.988 \times 10^9 \text{ N} \cdot \text{m}^2/\text{C}^2) \frac{6.00 \times 10^{-9} \text{ C}}{(0.150 \text{ m})^2}$$

$$E_1 = E_2 = 2397 \text{ N/C}$$

$$E_{1x} = -2397 \text{ N/C}, \; E_{1y} = 0 \quad E_{2x} = -2397 \text{ N/C}, \; E_{2y} = 0$$
$$E_x = E_{1x} + E_{2x} = 2(-2397 \text{ N/C}) = -4790 \text{ N/C}$$
$$E_y = E_{1y} + E_{2y} = 0$$

The resultant electric field at point a in the sketch has magnitude 4790 N/C and is in the $-x$-direction.

b)

$$E_1 = \frac{1}{4\pi\epsilon_0} \frac{|q_1|}{r_1^2} = (8.988 \times 10^9 \text{ N} \cdot \text{m}^2/\text{C}^2) \frac{6.00 \times 10^{-9} \text{ C}}{(0.150 \text{ m})^2} = 2397 \text{ N/C}$$

$$E_2 = \frac{1}{4\pi\epsilon_0} \frac{|q_2|}{r_2^2} = (8.988 \times 10^9 \text{ N} \cdot \text{m}^2/\text{C}^2) \frac{6.00 \times 10^{-9} \text{ C}}{(0.450 \text{ m})^2} = 266 \text{ N/C}$$

$E_{1x} = +2397$ N/C, $E_{1y} = 0$ $E_{2x} = -266$ N/C, $E_{2y} = 0$

$E_x = E_{1x} + E_{2x} = +2397$ N/C $- 266$ N/C $= +2130$ N/C

$E_y = E_{1y} + E_{2y} = 0$

The resultant electric field at point b in the sketch has magnitude 2130 N/C and is in the $+x$-direction.

c)

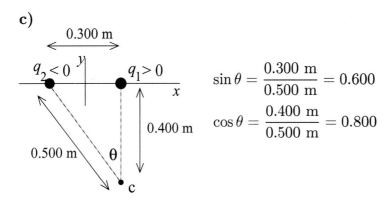

$$\sin\theta = \frac{0.300 \text{ m}}{0.500 \text{ m}} = 0.600$$

$$\cos\theta = \frac{0.400 \text{ m}}{0.500 \text{ m}} = 0.800$$

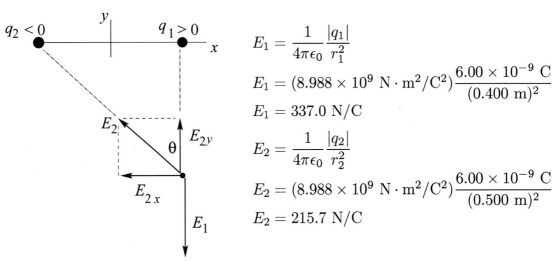

$$E_1 = \frac{1}{4\pi\epsilon_0}\frac{|q_1|}{r_1^2}$$

$$E_1 = (8.988 \times 10^9 \text{ N}\cdot\text{m}^2/\text{C}^2)\frac{6.00 \times 10^{-9} \text{ C}}{(0.400 \text{ m})^2}$$

$$E_1 = 337.0 \text{ N/C}$$

$$E_2 = \frac{1}{4\pi\epsilon_0}\frac{|q_2|}{r_2^2}$$

$$E_2 = (8.988 \times 10^9 \text{ N}\cdot\text{m}^2/\text{C}^2)\frac{6.00 \times 10^{-9} \text{ C}}{(0.500 \text{ m})^2}$$

$$E_2 = 215.7 \text{ N/C}$$

$E_{1x} = 0$, $E_{1y} = -E_1 = -337.0$ N/C

$E_{2x} = -E_2\sin\theta = -(215.7 \text{ N/C})(0.600) = -129.4$ N/C

$E_{2y} = +E_2\cos\theta = +(215.7 \text{ N/C})(0.800) = +172.6$ N/C

$E_x = E_{1x} + E_{2x} = -129$ N/C

$E_y = E_{1y} + E_{2y} = -337.4$ N/C $+ 172.6$ N/C $= -164$ N/C

$E = \sqrt{E_x^2 + E_y^2} = 209$ N/C

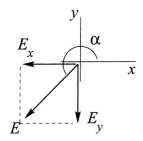

$$\tan \alpha = \frac{E_y}{E_x}$$

$$\tan \alpha = \frac{-164 \text{ N/C}}{-129 \text{ N/C}} = +1.271$$

$\alpha = 232°$, counterclockwise from $+x$-axis

d)

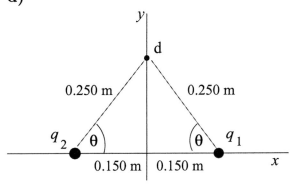

$$\sin \theta = \frac{0.200 \text{ m}}{0.250 \text{ m}} = 0.800$$

$$\cos \theta = \frac{0.150 \text{ m}}{0.250 \text{ m}} = 0.600$$

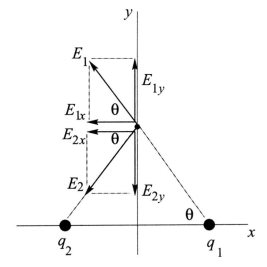

$$E_1 = E_2 = \frac{1}{4\pi\epsilon_0} \frac{|q|}{r^2}$$

$$E_1 = (8.988 \times 10^9 \text{ N} \cdot \text{m}^2/\text{C}^2) \frac{6.00 \times 10^{-9} \text{ C}}{(0.250 \text{ m})^2}$$

$$E_1 = 862.8 \text{ N/C}$$

$$E_2 = E_1 = 862.8 \text{ N/C}$$

$E_{1x} = -E_1 \cos\theta, \quad E_{2x} = -E_2 \cos\theta$

$E_x = E_{1x} + E_{2x} = -2(862.8 \text{ N/C})(0.600) = -1040 \text{ N/C}$

$E_{1y} = +E_1 \sin\theta, \ E_{2y} = -E_2 \sin\theta$

$E_y = E_{1y} + E_{2y} = 0$

$E = 1040 \text{ N/C}$, in the $-x$-direction.

22-45 **a)** The electric field lines are perpendicular to the line of charge.

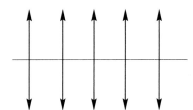

b) The magnitude of the electric field is inversely proportional to the spacing of the field lines. Consider a circle of radius r with the line of charge passing through the center.

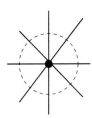

The spacing of field lines is the same all around the circle, and in the direction perpendicular to the plane of the circle the lines are equally spaced, so E depends only on the distance r. The number of field lines passing out through the circle is independent of the radius of the circle, so the spacing of the field lines is proportional to the reciprocal of the circumference $2\pi r$ of the circle. Hence E is proportional to $1/r$.

22-47 a) Use Eq.(22-14): $p = qd = (4.5 \times 10^{-9} \text{ C})(3.1 \times 10^{-3} \text{ m}) = 1.4 \times 10^{-11} \text{ C·m}$; The direction of $\vec{p}$ is from q_1 toward q_2.

b) $\tau = pE \sin \phi$ so

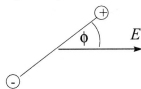

$$E = \frac{\tau}{p \sin \phi}$$

$$E = \frac{7.2 \times 10^{-9} \text{ N/m}}{(1.4 \times 10^{-11} \text{ C} \cdot \text{m}) \sin 36.9°} = 860 \text{ N/C}$$

22-49 a) $U(\phi) = -\vec{p} \cdot \vec{E} = -pE \cos \phi$, where ϕ is the angle between $\vec{p}$ and $\vec{E}$.

parallel: $\phi = 0$ and $U(0°) = -pE$

perpendicular: $\phi = 90°$ and $U(90°) = 0$

$\Delta U = U(90°) - U(0°)) = pE = (5.0 \times 10^{-30} \text{ C} \cdot \text{m})(1.6 \times 10^{6} \text{ N/C}) = 8.0 \times 10^{-24}$ J.

b) $\frac{3}{2}kT = \Delta U$

$$T = \frac{2\,\Delta U}{3k} = \frac{2(8.0 \times 10^{-24} \text{ J})}{3(1.381 \times 10^{-23} \text{ J/K})} = 0.39 \text{ K}$$

Problems

22-53 a)

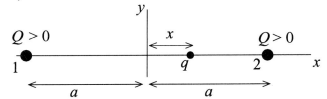

Find the net force on q:

$F_x = F_{1x} + F_{2x}$ and
$F_{1x} = +F_1$, $F_{2x} = -F_2$

$$F_1 = \frac{1}{4\pi\epsilon_0}\frac{qQ}{(a+x)^2}, \quad F_2 = \frac{1}{4\pi\epsilon_0}\frac{qQ}{(a-x)^2}$$

$$F_x = F_1 - F_2 = \frac{qQ}{4\pi\epsilon_0}\left[\frac{1}{(a+x)^2} - \frac{1}{(a-x)^2}\right]$$

$$F_x = \frac{qQ}{4\pi\epsilon_0 a^2}\left[+\left(1+\frac{x}{a}\right)^{-2} - \left(1-\frac{x}{a}\right)^{-2}\right]$$

Since $x << a$ we can use the binomial expansion for $(1-x/a)^{-2}$ and $(1+x/a)^{-2}$ and keep only the first two terms: $(1+z)^n \approx 1 + nz$.

For $(1-x/a)^{-2}$, $z = -x/a$ and $n = -2$ so $(1-x/a)^{-2} \approx 1 + 2x/a$.

For $(1+x/a)^{-2}$, $z = +x/a$ and $n = -2$ so $(1+x/a)^{-2} \approx 1 - 2x/a$.

Then $F \approx \dfrac{qQ}{4\pi\epsilon_0 a^2}\left[\left(1-\dfrac{2x}{a}\right) - \left(1+\dfrac{2x}{a}\right)\right] = -\left(\dfrac{qQ}{\pi\epsilon_0 a^3}\right)x.$

For simple harmonic motion $F = -kx$ and the frequency of oscillation is $f = (1/2\pi)\sqrt{k/m}$. The net force here is of this form, with $k = qQ/\pi\epsilon_0 a^3$.

Thus $f = \dfrac{1}{2\pi}\sqrt{\dfrac{qQ}{\pi\epsilon_0 m a^3}}.$

b)

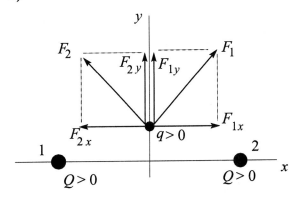

The x-components of the forces exerted by the two charges cancel, the y-components add, and the net force is in the $+y$-direction when $y > 0$ and in the $-y$-direction when $y < 0$. The charge moves away from the origin on the y-axis and never returns.

22-55 a)

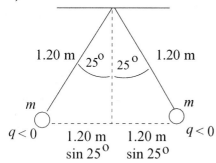

sphere on the left: sphere on the right:

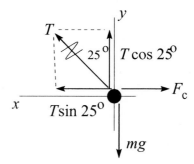

F_c is the repulsive Coulomb force exerted by one sphere on the other.

b) From either force diagram in part (a):

$\sum F_y = ma_y$

$T\cos 25.0° - mg = 0$ and $T = \dfrac{mg}{\cos 25.0°}$

$\sum F_x = ma_x$

$T\sin 25.0° - F_c = 0$ and $F_c = T\sin 25.0°$

Use the first equation to eliminate T in the second:

$F_c = (mg/\cos 25.0°)(\sin 25.0°) = mg\tan 25.0°$

$F_c = \dfrac{1}{4\pi\epsilon_0}\dfrac{|q_1 q_2|}{r^2} = \dfrac{1}{4\pi\epsilon_0}\dfrac{q^2}{r^2} = \dfrac{1}{4\pi\epsilon_0}\dfrac{q^2}{[2(1.20 \text{ m})\sin 25.0°]^2}$

Combine this with $F_c = mg\tan 25.0°$ and get

$$mg \tan 25.0° = \frac{1}{4\pi\epsilon_0} \frac{q^2}{[2(1.20 \text{ m}) \sin 25.0°]^2}$$

$$q = (2.40 \text{ m}) \sin 25.0° \sqrt{\frac{mg \tan 25.0°}{(1/4\pi\epsilon_0)}}$$

$$q = (2.40 \text{ m}) \sin 25.0° \sqrt{\frac{(15.0 \times 10^{-3} \text{ kg})(9.80 \text{ m/s}^2) \tan 25.0°}{8.988 \times 10^9 \text{ N} \cdot \text{m}^2/\text{C}^2}} = 2.80 \times 10^{-6} \text{ C}$$

c) The separation between the two spheres is given by $2L\sin\theta$. $q = 2.80 \ \mu C$ as found in part (b). $F_c = (1/4\pi\epsilon_0)q^2/(2L\sin\theta)^2$ and $F_c = mg\tan\theta$. Thus $(1/4\pi\epsilon_0)q^2/(2L\sin\theta)^2 = mg\tan\theta$.

$$(\sin\theta)^2 \tan\theta = \frac{1}{4\pi\epsilon_0} \frac{q^2}{4L^2mg} =$$

$$(8.988 \times 10^9 \text{ N} \cdot \text{m}^2/\text{C}^2) \frac{(2.80 \times 10^{-6})^2}{4(0.600 \text{ m})^2(15.0 \times 10^{-3} \text{ kg})(9.80 \text{ m/s}^2)} = 0.3328.$$

Solve this equation by trial and error. This will go quicker if we can make a good estimate of the value of θ that solves the equation. For θ small, $\tan\theta \approx \sin\theta$. With this approximation the equation becomes $\sin^3\theta = 0.3328$ and $\sin\theta = 0.6930$, so $\theta = 43.9°$.

Now refine this guess:

θ	$\sin^2\theta \tan\theta$	
45.0°	0.5000	
40.0°	0.3467	
39.6°	0.3361	
39.5°	0.3335	
39.4°	0.3309	so $\theta = 39.5°$

22-57 **a)** The number of Na^+ ions in 0.100 mol of NaCl is $N = nN_A$.

The charge of one ion is $+e$, so the total charge is $q_1 = nN_Ae =$
(0.100 mol)(6.022 × 10²³ ions/mol)(1.602 × 10⁻¹⁹C/ion) $= 9.647 \times 10^3$ C

There are the same number of Cl^- ions and each has charge $-e$, so
$q_2 = -9.647 \times 10^3$ C.

$$F = \frac{1}{4\pi\epsilon_0} \frac{|q_1 q_2|}{r^2} = (8.988 \times 10^9 \text{ N} \cdot \text{m}^2/\text{C}^2) \frac{(9.647 \times 10^3 \text{ C})^2}{(0.0200 \text{ m})^2} = 2.09 \times 10^{21} \text{ N}$$

b) $a = F/m$

Need the mass of 0.100 mol of Cl^- ions. For Cl, $M = 35.453 \times 10^{-3}$ kg/mol, so
$m = (0.100 \text{ mol})(35.453 \times 10^{-3} \text{ kg/mol}) = 35.45 \times 10^{-4}$ kg.

Then $a = \dfrac{F}{m} = \dfrac{2.09 \times 10^{21} \text{ N}}{35.45 \times 10^{-4} \text{ kg}} = 5.90 \times 10^{23} \text{ m/s}^2$.

c) It is not reasonable to have such a huge force.

22-59 a)

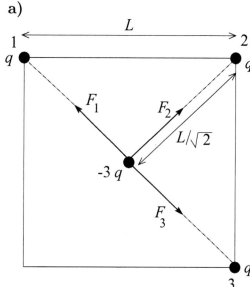

$\vec{F}_1 + \vec{F}_3 = \mathbf{0}$, so the net force is $\vec{F} = \vec{F}_2$.

$$F = \frac{1}{4\pi\epsilon_0} \frac{q(3q)}{(L/\sqrt{2})^2} = \frac{6q^2}{4\pi\epsilon_0 L^2},$$
away from the vacant corner.

b)

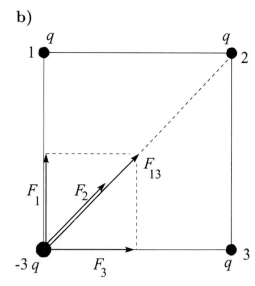

$$F_2 = \frac{1}{4\pi\epsilon_0} \frac{q(3q)}{(\sqrt{2}L)^2} = \frac{3q^2}{4\pi\epsilon_0(2L^2)}$$

$$F_1 = F_3 = \frac{1}{4\pi\epsilon_0} \frac{q(3q)}{L^2} = \frac{3q^2}{4\pi\epsilon_0 L^2}$$

The vector sum of F_1 and F_3 is
$F_{13} = \sqrt{F_1^2 + F_3^2}$.

$F_{13} = \sqrt{2}F_1 = \dfrac{3\sqrt{2}q^2}{4\pi\epsilon_0 L^2}$; $\vec{F}_{13}$ and $\vec{F}_2$ are in the same direction.

$F = F_{13} + F_2 = \dfrac{3q^2}{4\pi\epsilon_0 L^2}\left(\sqrt{2} + \dfrac{1}{2}\right)$, and is directed toward the center of the square.

22-61 a)

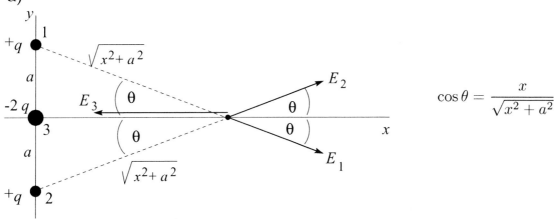

$$\cos\theta = \frac{x}{\sqrt{x^2 + a^2}}$$

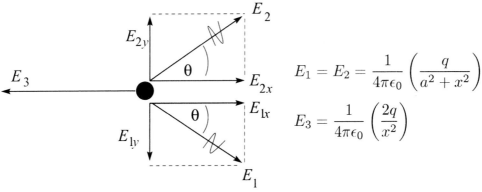

$$E_1 = E_2 = \frac{1}{4\pi\epsilon_0}\left(\frac{q}{a^2 + x^2}\right)$$

$$E_3 = \frac{1}{4\pi\epsilon_0}\left(\frac{2q}{x^2}\right)$$

$E_{1y} = -E_1\sin\theta,\ E_{2y} = +E_2\sin\theta$ so $E_y = E_{1y} + E_{2y} = 0$.

$$E_{1x} = E_{2x} = +E_1\cos\theta = \frac{1}{4\pi\epsilon_0}\left(\frac{q}{a^2 + x^2}\right)\left(\frac{x}{\sqrt{x^2 + a^2}}\right), \quad E_{3x} = -E_3$$

$$E_x = E_{1x} + E_{2x} + E_{3x} = 2\left(\frac{1}{4\pi\epsilon_0}\left(\frac{q}{a^2 + x^2}\right)\left(\frac{x}{\sqrt{x^2 + a^2}}\right)\right) - \frac{2q}{4\pi\epsilon_0 x^2}$$

$$E_x = -\frac{2q}{4\pi\epsilon_0}\left(\frac{1}{x^2} - \frac{x}{(a^2 + x^2)^{3/2}}\right) = -\frac{2q}{4\pi\epsilon_0 x^2}\left(1 - \frac{1}{(1 + a^2/x^2)^{3/2}}\right)$$

Thus $E = \dfrac{2q}{4\pi\epsilon_0 x^2}\left(1 - \dfrac{1}{(1 + a^2/x^2)^{3/2}}\right)$, in the $-x$-direction.

b) $x >> a$ implies $a^2/x^2 << 1$ and $(1 + a^2/x^2)^{-3/2} \approx 1 - 3a^2/2x^2$.

Thus $E \approx \dfrac{2q}{4\pi\epsilon_0 x^2}\left(1 - \left(1 - \dfrac{3a^2}{2x^2}\right)\right) = \dfrac{3qa^2}{4\pi\epsilon_0 x^4}$.

$E \sim 1/x^4$. For a point charge $E \sim 1/x^2$ and for a dipole $E \sim 1/x^3$.

22-65 **a)** The electric field is upward so the electric force on the positively charged proton is upward and has magnitude $F = eE$. Use coordinates where positive y is downward. Then applying $\sum \vec{F} = m\vec{a}$ to the proton gives that $a_x = 0$ and $a_y = -eE/m$. In these coordinates the initial velocity has components $v_x = +v_0 \cos \alpha$ and $v_y = +v_0 \sin \alpha$.

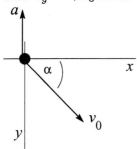

Finding h_{max}: At $y = h_{max}$ the y-component of the velocity is zero.

$v_y = 0$, $\quad v_{0y} = v_0 \sin \alpha$, $\quad a_y = -eE/m$, $\quad y - y_0 = h_{max} = ?$

$$v_y^2 = v_{0y}^2 + 2a_y(y - y_0)$$

$$y - y_0 = \frac{v_y^2 - v_{0y}^2}{2a_y}$$

$$h_{max} = \frac{-v_0^2 \sin^2 \alpha}{2(-eE/m)} = \frac{mv_0^2 \sin^2 \alpha}{2eE}$$

b) Use the vertical motion to find the time t:

$y - y_0 = 0$, $\quad v_{0y} = v_0 \sin \alpha$, $\quad a_y = -eE/m$, $\quad t = ?$

$$y - y_0 = v_{0y}t + \tfrac{1}{2}a_y t^2$$

With $y - y_0 = 0$ this gives $t = -\dfrac{2v_{0y}}{a_y} = -\dfrac{2(v_0 \sin \alpha)}{-eE/m} = \dfrac{2mv_0 \sin \alpha}{eE}$

Then use the x-component motion to find d:

$a_x = 0$, $\quad v_{0x} = v_0 \cos \alpha$, $\quad t = 2mv_0 \sin \alpha / eE$, $\quad x - x_0 = d = ?$

$x - x_0 = v_{0x}t + \tfrac{1}{2}a_x t^2$ gives

$$d = v_0 \cos \alpha \left(\frac{2mv_0 \sin \alpha}{eE} \right) = \frac{mv_0^2 2 \sin \alpha \cos \alpha}{eE} = \frac{mv_0^2 \sin 2\alpha}{eE}$$

c)

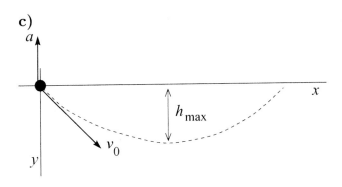

d) Use the expression in part (a):

$$h_{max} = \frac{[(4.00 \times 10^5 \text{ m/s})(\sin 30.0^\circ)]^2(1.673 \times 10^{-27} \text{ kg})}{2(1.602 \times 10^{-19} \text{ C})(500 \text{ N/C})} = 0.418 \text{ m}$$

Use the expression in part (b):

$$d = \frac{(1.673 \times 10^{-27} \text{ kg})(4.00 \times 10^5 \text{ m/s})^2 \sin 60.0^\circ}{(1.602 \times 10^{-19} \text{ C})(500 \text{ N/C})} = 2.89 \text{ m}$$

22-67

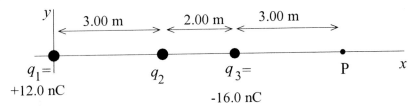

At point P the resultant field has magnitude 12.0 N/C and is in the $+x$-direction, so $E_x = +12.0$ N/C.

$\vec{E} = \vec{E}_1 + \vec{E}_2 + \vec{E}_3$ so $\vec{E}_2 = \vec{E} - \vec{E}_1 - \vec{E}_3$

$E_{2x} = E_x - E_{1x} - E_{3x}$; calculate E_{2x} and from it deduce the sign and magnitude of q_2.

The electric field $\vec{E}_1$ of q_1 at point P has magnitude

$$E_1 = \frac{1}{4\pi\epsilon_0} \frac{|q_1|}{r_1^2} = (8.988 \times 10^9 \text{ N} \cdot \text{m}^2/\text{C}^2)\frac{12.0 \times 10^{-9} \text{ C}}{(8.00 \text{ m})^2} = 1.685 \text{ N/C}$$

and is in the $+x$-direction. Thus $E_{1x} = +1.685$ N/C.

The electric field $\vec{E}_3$ of q_3 at point P has magnitude

$$E_3 = \frac{1}{4\pi\epsilon_0} \frac{|q_3|}{r_3^2} = (8.988 \times 10^9 \text{ N} \cdot \text{m}^2/\text{C}^2)\frac{16.0 \times 10^{-9} \text{ C}}{(3.00 \text{ m})^2} = 15.98 \text{ N/C}$$

and is in the $-x$-direction. Thus $E_{3x} = -15.98$ N/C.

Thus $E_{2x} = E_x - E_{1x} - E_{3x} = +12.0 \text{ N/C} - 1.68 \text{ N/C} - (-15.98 \text{ N/C}) = +26.30 \text{ N/C}$.

Since E_{2x} is positive, $\vec{E}_2$ is in the $+x$-direction at P and q_2 is positive.

$$E_2 = \frac{1}{4\pi\epsilon_0} \frac{|q_2|}{r_2^2} \text{ so}$$

$$|q_2| = \frac{r_2^2 E_2}{(1/4\pi\epsilon_0)} = \frac{(5.00 \text{ m})^2 (26.30 \text{ N/C})}{(8.988 \times 10^9 \text{ N} \cdot \text{m}^2/\text{C}^2)} = 7.32 \times 10^{-8} \text{ C}$$

$$q_2 = +73.2 \text{ nC}$$

22-69 a)

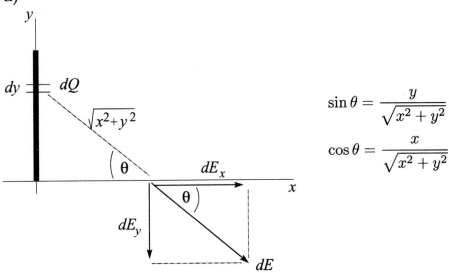

$$\sin\theta = \frac{y}{\sqrt{x^2 + y^2}}$$

$$\cos\theta = \frac{x}{\sqrt{x^2 + y^2}}$$

Slice the charge distribution up into small pieces of length dy. The charge dQ in each slice is $dQ = Q(dy/a)$. The electric field this produces at a distance x along the x-axis is dE. Calculate the components of $d\vec{E}$ and then integrate over the charge distribution to find the components of the total field.

$$dE = \frac{1}{4\pi\epsilon_0} \left(\frac{dQ}{x^2 + y^2} \right) = \frac{Q}{4\pi\epsilon_0 a} \left(\frac{dy}{x^2 + y^2} \right)$$

$$dE_x = dE \cos\theta = \frac{Qx}{4\pi\epsilon_0 a} \left(\frac{dy}{(x^2 + y^2)^{3/2}} \right)$$

$$dE_y = -dE \sin\theta = -\frac{Q}{4\pi\epsilon_0 a} \left(\frac{y \, dy}{(x^2 + y^2)^{3/2}} \right)$$

$$E_x = \int dE_x = \frac{Qx}{4\pi\epsilon_0 a} \int_0^a \frac{dy}{(x^2 + y^2)^{3/2}} =$$

$$\frac{Qx}{4\pi\epsilon_0 a} \left[\frac{1}{x^2} \frac{y}{\sqrt{x^2 + y^2}} \right]_0^a = \frac{Q}{4\pi\epsilon_0 x} \frac{1}{\sqrt{x^2 + a^2}}$$

$$E_y = \int dE_y = -\frac{Q}{4\pi\epsilon_0 a} \int_0^a \frac{y\, dy}{(x^2 + y^2)^{3/2}} =$$

$$-\frac{Q}{4\pi\epsilon_0 a} \left[-\frac{1}{\sqrt{x^2 + y^2}} \right]_0^a = -\frac{Q}{4\pi\epsilon_0 a} \left(\frac{1}{x} - \frac{1}{\sqrt{x^2 + a^2}} \right)$$

b) $\vec{F} = q_0 \vec{E}$

$$F_x = -qE_x = \frac{-qQ}{4\pi\epsilon_0 x} \frac{1}{\sqrt{x^2 + a^2}}; \quad F_y = -qE_y = \frac{qQ}{4\pi\epsilon_0 a} \left(\frac{1}{x} - \frac{1}{\sqrt{x^2 + a^2}} \right)$$

c) For $x \gg a$, $\dfrac{1}{\sqrt{x^2 + a^2}} = \dfrac{1}{x} \left(1 + \dfrac{a^2}{x^2} \right)^{-1/2} = \dfrac{1}{x} \left(1 - \dfrac{a^2}{2x^2} \right) = \dfrac{1}{x} - \dfrac{a^2}{2x^3}$

$$F_x \approx -\frac{qQ}{4\pi\epsilon_0 x^2}, \quad F_y \approx \frac{qQ}{4\pi\epsilon_0 a} \left(\frac{1}{x} - \frac{1}{x} + \frac{a^2}{2x^3} \right) = \frac{qQ}{8\pi\epsilon_0 a x^3}$$

For $x \gg a$, $F_y \ll F_x$ and $F \approx |F_x| = \dfrac{qQ}{4\pi\epsilon_0 x^2}$ and $\vec{F}$ is in the $-x$-direction. For $x \gg a$ the charge distribution Q acts like a point charge.

22-77   Divide the charge distribution into small segments, use the point charge formula for the electric field due to each small segment and integrate over the charge distribution to find the x and y components of the total field.

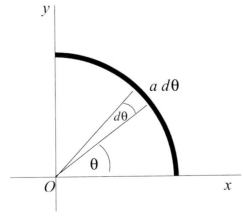

A small segment that subtends angle $d\theta$ has length $a\, d\theta$ and contains charge

$$dQ = \left(\frac{a\, d\theta}{\frac{1}{2}\pi a} \right) Q = \frac{2Q}{\pi}\, d\theta.$$

($\frac{1}{2}\pi a$ is the total length of the charge distribution.)

The charge is negative, so the field at the origin is directed toward the small segment. The small segment is located at angle θ as shown in the sketch.

$$dE = \frac{1}{4\pi\epsilon_0}\frac{|dQ|}{a^2}$$

$$dE = \frac{Q}{2\pi^2\epsilon_0 a^2}\,d\theta$$

$$dE_x = dE\cos\theta = (Q/2\pi^2\epsilon_0 a^2)\cos\theta\,d\theta$$

$$E_x = \int dE_x = \frac{Q}{2\pi^2\epsilon_0 a^2}\int_0^{\pi/2}\cos\theta\,d\theta = \frac{Q}{2\pi^2\epsilon_0 a^2}\left(\sin\theta\Big|_0^{\pi/2}\right) = \frac{Q}{2\pi^2\epsilon_0 a^2}$$

$$dE_y = dE\sin\theta = (Q/2\pi^2\epsilon_0 a^2)\sin\theta\,d\theta$$

$$E_y = \int dE_y = \frac{Q}{2\pi^2\epsilon_0 a^2}\int_0^{\pi/2}\sin\theta\,d\theta = \frac{Q}{2\pi^2\epsilon_0 a^2}\left(-\cos\theta\Big|_0^{\pi/2}\right) = \frac{Q}{2\pi^2\epsilon_0 a^2}$$

Note that $E_x = E_y$, as expected from symmetry.

22-81 For an infinite charge sheet $E = |\sigma|/2\epsilon_0$, the same at all points, and is directed away from the sheet if the sheet has positive charge and toward the sheet if it has negative charge.

Obtain the net electric field as the vector sum of the fields due to each sheet.

a)

$$E = E_I + E_{II} - E_{III}$$

$$E = (|\sigma_I| + |\sigma_{II}| - |\sigma_{III}|)/2\epsilon_0$$

$E = (+0.0200\ \text{C/m}^2 + 0.0100\ \text{C/m}^2 - 0.0200\ \text{C/m}^2)/2\epsilon_0 =$
$(+0.0100\ \text{C/m}^2)/2(8.854\times10^{-12}\ \text{C}^2/\text{N}\cdot\text{m}^2) = 5.65\times10^8\ \text{N/C, to left.}$

b)

$$E = E_I + E_{III} - E_{II}$$

$$E = (|\sigma_I| + |\sigma_{III}| - |\sigma_{II}|)/2\epsilon_0$$

$E = (+0.0200\ \text{C/m}^2 + 0.0200\ \text{C/m}^2 - 0.0100\ \text{C/m}^2)/2\epsilon_0 =$
$(+0.0300\ \text{C/m}^2)/2(8.854\times10^{-12}\ \text{C}^2/\text{N}\cdot\text{m}^2) = 1.69\times10^9\ \text{N/C, to right.}$

c)

$E = E_I + E_{II} + E_{III}$

$E = (|\sigma_I| + |\sigma_{II}| + |\sigma_{III}|)/2\epsilon_0$

$E = (+0.0200 \text{ C/m}^2 + 0.0100 \text{ C/m}^2 + 0.0200 \text{ C/m}^2)/2\epsilon_0 =$
$(+0.0500 \text{ C/m}^2)/2(8.854 \times 10^{-12} \text{ C}^2/\text{N} \cdot \text{m}^2) = 2.82 \times 10^9 \text{ N/C, to right.}$

d)

$E = E_I + E_{II} - E_{III}$,

the same as at point P.

$E = 5.65 \times 10^8 \text{ N/C, to right.}$

22-83 The electric field due to the first sheet, which is in the xy-plane, is
$\vec{E}_1 = (\sigma/2\epsilon_0)\hat{k}$ for $z > 0$ and $\vec{E}_1 = -(\sigma/2\epsilon_0)\hat{k}$ for $z < 0$.

We can write this as $\vec{E}_1 = (\sigma/2\epsilon_0)(z/|z|)\hat{k}$, since $z/|z| = +1$ for $z > 0$ and $z/|z| = -z/z = -1$ for $z < 0$.

Similarly, we can write the electric field due to the second sheet as
$\vec{E}_2 = -(\sigma/2\epsilon_0)(x/|x|)\hat{i}$, since its charge density is $-\sigma$.

The net field is $\vec{E} = \vec{E}_1 + \vec{E}_2 = (\sigma/2\epsilon_0)(-(x/|x|)\hat{i} + (z/|z|)\hat{k})$.

CHAPTER 23
GAUSS'S LAW

Exercises

23-1 a) $\Phi_E = \int E \cos\phi \, dA$, where ϕ is the angle between the normal to the sheet $\hat{n}$ and the electric field $\vec{E}$. In this problem E and $\cos\phi$ are constant over the surface so

$$\Phi_E = E\cos\phi \int dA = E\cos\phi A = (14 \text{ N/C})(\cos 60°)(0.250 \text{ m}^2) = 1.8 \text{ N} \cdot \text{m}^2/\text{C}.$$

b) Φ_E is independent of the shape of the sheet as long as ϕ and E are constant at all points on the sheet.

c) (i) $\Phi_E = E\cos\phi \, A$. Φ_E is largest for $\phi = 0°$, so $\cos\phi = 1$ and $\Phi_E = EA$.

(ii) Φ_E is smallest for $\phi = 90°$, so $\cos\phi = 0$ and $\Phi_E = 0$.

23-3 a) $\vec{E} = -B\hat{i} + C\hat{j} - D\hat{k}; \quad A = L^2$

<u>face S_1:</u>

$\hat{n} = -\hat{j}$

$\Phi_E = \vec{E} \cdot \vec{A} = \vec{E} \cdot (A\hat{n}) = (-B\hat{i} + C\hat{j} - D\hat{k}) \cdot (-A\hat{j}) = -CL^2.$

<u>face S_2:</u>

$\hat{n} = +\hat{k}$

$\Phi_E = \vec{E} \cdot \vec{A} = \vec{E} \cdot (A\hat{n}) = (-B\hat{i} + C\hat{j} - D\hat{k}) \cdot (A\hat{k}) = -DL^2.$

<u>face S_3:</u>

$\hat{n} = +\hat{j}$

$\Phi_E = \vec{E} \cdot \vec{A} = \vec{E} \cdot (A\hat{n}) = (-B\hat{i} + C\hat{j} - D\hat{k}) \cdot (A\hat{j}) = +CL^2.$

<u>face S_4:</u>

$\hat{n} = -\hat{k}$

$\Phi_E = \vec{E} \cdot \vec{A} = \vec{E} \cdot (A\hat{n}) = (-B\hat{i} + C\hat{j} - D\hat{k}) \cdot (-A\hat{k}) = +DL^2.$

<u>face S_5:</u>

$\hat{n} = +\hat{i}$

$\Phi_E = \vec{E} \cdot \vec{A} = \vec{E} \cdot (A\hat{n}) = (-B\hat{i} + C\hat{j} - D\hat{k}) \cdot (A\hat{i}) = -BL^2.$

face S_6:

$\hat{n} = -\hat{i}$

$\Phi_E = \vec{E} \cdot \vec{A} = \vec{E} \cdot (A\hat{n}) = (-B\hat{i} + C\hat{j} - D\hat{k}) \cdot (-A\hat{i}) = +BL^2.$

b) Add the flux through each of the six faces:

$\Phi_E = -CL^2 - DL^2 + CL^2 + DL^2 - BL^2 + BL^2 = 0$

The total electric flux through all sides is zero.

23-5 a)

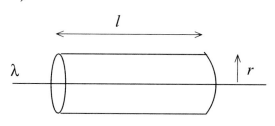
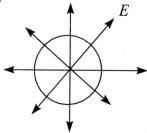

The area of the curved part of the cylinder is $A = 2\pi r l$.

The electric field is parallel to the end caps of the cylinder, so $\vec{E} \cdot \vec{A} = 0$ for the ends and the flux through the cylinder end caps is zero.

The electric field is normal to the curved surface of the cylinder and has the same magnitude $E = \lambda/2\pi\epsilon_0 r$ at all points on this surface. Thus $\phi = 0°$ and

$\Phi_E = EA \cos\phi = EA = (\lambda/2\pi\epsilon_0 r)(2\pi r l) = \dfrac{\lambda l}{\epsilon_0} = \dfrac{(6.00 \times 10^{-6} \text{ C/m})(0.400 \text{ m})}{8.854 \times 10^{-12} \text{ C}^2/\text{N} \cdot \text{m}^2} =$

$2.71 \times 10^5 \text{ N} \cdot \text{m}^2/\text{C}$

b) In the calculation in part (a) the radius r of the cylinder divided out, so the flux remains the same, $\Phi_E = 2.71 \times 10^5 \text{ N} \cdot \text{m}^2/\text{C}$.

c) $\Phi_E = \dfrac{\lambda l}{\epsilon_0} = \dfrac{(6.00 \times 10^{-6} \text{ C/m})(0.800 \text{ m})}{8.854 \times 10^{-12} \text{ C}^2/\text{N} \cdot \text{m}^2} = 5.42 \times 10^5 \text{ N} \cdot \text{m}^2/\text{C}$ (twice the flux calculated in parts (b) and (c)).

23-11 a) It is rather difficult to calculate the flux directly from $\Phi = \oint \vec{E} \cdot d\vec{A}$ since the magnitude of $\vec{E}$ and its angle with $d\vec{A}$ varies over the surface of the cube. A much easier approach is to use Gauss's law to calculate the total flux through the cube:

$\Phi_E = Q_{encl}/\epsilon_0 = \dfrac{9.60 \times 10^{-6} \text{ C}}{8.854 \times 10^{-12} \text{ C}^2/\text{N} \cdot \text{m}^2} = 1.084 \times 10^6 \text{ N} \cdot \text{m}^2/\text{C}.$

By symmetry the flux is the same through each of the six faces, so the flux through one face is $\frac{1}{6}(1.084 \times 10^6 \text{ N} \cdot \text{m}^2/\text{C}) = 1.81 \times 10^5 \text{ N} \cdot \text{m}^2/\text{C}$.

b) In part (a) the size of the cube did not enter into the calculations. The flux through one face depends only on the amount of charge at the center of the cube. So the answer to (a) would not change if the size of the cube were changed.

23-15 Example 23-5 derived that the electric field just outside the surface of a spherical conductor that has net charge q is $E = \dfrac{1}{4\pi\epsilon_0}\dfrac{q}{R^2}$.

$$q = \frac{R^2 E}{(1/4\pi\epsilon_0)} = \frac{(0.160 \text{ m})^2(1150 \text{ N/C})}{8.988 \times 10^9 \text{ N} \cdot \text{m}^2/\text{C}^2} = 3.275 \times 10^{-9} \text{ C}.$$

Each electron has a charge of magnitude $e = 1.602 \times 10^{-19}$ C, so the number of excess electrons needed is

$$\frac{3.275 \times 10^{-9} \text{ C}}{1.602 \times 10^{-19} \text{ C}} = 2.04 \times 10^{10}.$$

23-19 a) Consider the charge on a length l of the cylinder.

This can be expressed as $q = \lambda l$. But since the surface area is $2\pi Rl$ it can also be expressed as $q = \sigma 2\pi Rl$. These two expressions must be equal, so $\lambda l = \sigma 2\pi Rl$ and $\lambda = 2\pi R\sigma$.

b) Apply Gauss's law to a Gaussian surface that is a cylinder of length l, radius r, and whose axis coincides with the axis of the charge distribution.

$$Q_{\text{encl}} = \sigma(2\pi Rl)$$
$$\Phi_E = 2\pi rlE$$

$\Phi_E = \dfrac{Q_{\text{encl}}}{\epsilon_0}$ gives $2\pi rlE = \dfrac{\sigma(2\pi Rl)}{\epsilon_0}$

$$E = \frac{\sigma R}{\epsilon_0 r}$$

c) Example 23-6 shows that the electric field of an infinite line of charge is $E = \lambda/2\pi\epsilon_0 r$.

$\sigma = \dfrac{\lambda}{2\pi R}$, so $E = \dfrac{\sigma R}{\epsilon_0 r} = \dfrac{R}{\epsilon_0 r}\left(\dfrac{\lambda}{2\pi R}\right) = \dfrac{\lambda}{2\pi\epsilon_0 r}$, the same as for an infinite line of charge that is along the axis of the cylinder.

Problems

23-23 a) Find the net flux through the parallelepiped surface and then use that in Gauss's law to find the net charge within. Flux out of the surface is positive and flux into the surface is negative.

$\vec{E}_1$ gives flux out of the surface:

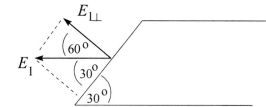

$\Phi_1 = +E_{1\perp}A$
$A = (0.0600\text{ m})(0.0500\text{ m}) = 3.00\times10^{-3}\text{ m}^2$
$E_{1\perp} = E_1\cos60° = (2.50\times10^4\text{ N/C})\cos60°$
$E_{1\perp} = 1.25\times10^4\text{ N/C}$

$\Phi_{E_1} = +E_{1\perp}A = +(1.25\times10^4\text{ N/C})(3.00\times10^{-3}\text{ m}^2) = 37.5\text{ N}\cdot\text{m}^2/\text{C}$

$\vec{E}_2$ gives flux into the surface:

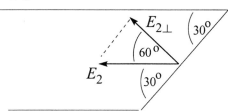

$\Phi_2 = -E_{2\perp}A$
$A = (0.0600\text{ m})(0.0500\text{ m}) = 3.00\times10^{-3}\text{ m}^2$
$E_{2\perp} = E_2\cos60° = (7.00\times10^4\text{ N/C})\cos60°$
$E_{1\perp} = 3.50\times10^4\text{ N/C}$

$\Phi_{E_2} = -E_{2\perp}A = -(3.50\times10^4\text{ N/C})(3.00\times10^{-3}\text{ m}^2) = -105.0\text{ N}\cdot\text{m}^2/\text{C}$

The net flux is $\Phi_E = \Phi_{E_1} + \Phi_{E_2} = +37.5\text{ N·m}^2/\text{C} - 105.0\text{ N·m}^2/\text{C} = -67.5\text{ N·m}^2/\text{C}$. The net flux is negative (inward), so the net charge enclosed is negative.

Apply Gauss's law: $\Phi_E = \dfrac{Q_{\text{encl}}}{\epsilon_0}$

$Q_{\text{encl}} = \Phi_E\epsilon_0 = (-67.5\text{ N}\cdot\text{m}^2/\text{C})(8.854\times10^{-12}\text{ C}^2/\text{N}\cdot\text{m}^2) = -5.98\times10^{-10}\text{ C}.$

b) If there were no charge within the parallelpiped the net flux would be zero. This is not the case, so there is charge inside. The electric field lines that pass out through the surface of the parallelpiped must terminate on charges, so there also must be charges outside the parallelpiped.

23-27 a) (i) $r < a$

Apply Gauss's law to a spherical Gaussian surface with radius $r < a$.

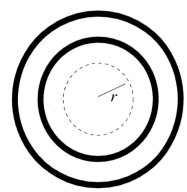

$$\Phi_E = EA = E(4\pi r^2)$$

$Q_{encl} = 0$; no charge is enclosed

$$\Phi_E = \frac{Q_{encl}}{\epsilon_0} \text{ says}$$

$E(4\pi r^2) = 0$ and $E = 0$.

(ii) $a < r < b$

Points in this region are in the conductor of the small shell, so $E = 0$.

(iii) $b < r < c$

Apply Gauss's law to a spherical Gaussian surface with radius $b < r < c$.

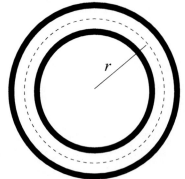

$$\Phi_E = EA = E(4\pi r^2)$$

The Gaussian surface encloses all of the small shell and none of the large shell, so $Q_{encl} = +2q$.

$\Phi_E = \dfrac{Q_{encl}}{\epsilon_0}$ gives $E(4\pi r^2) = \dfrac{2q}{\epsilon_0}$ so $E = \dfrac{2q}{4\pi\epsilon_0 r^2}$. Since the enclosed charge is positive the electric field is radially outward.

(iv) $c < r < d$

Points in this region are in the conductor of the large shell, so $E = 0$.

(v) $r > d$

Apply Gauss's law to a spherical Gaussian surface with radius $r > d$.

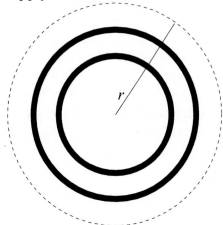

$$\Phi_E = EA = E(4\pi r^2)$$

The Gaussian surface encloses all of the small shell and all of the large shell, so $Q_{encl} = +2q + 4q = 6q$.

$$\Phi_E = \frac{Q_{\text{encl}}}{\epsilon_0} \text{ gives } E(4\pi r^2) = \frac{6q}{\epsilon_0}$$

$E = \frac{6q}{4\pi\epsilon_0 r^2}$.   Since the enclosed charge is positive the electric field is radially outward.

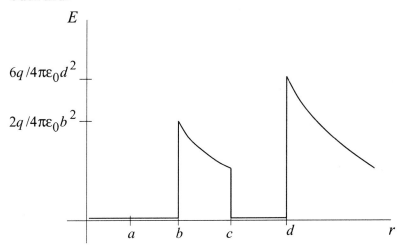

b) (i) charge on inner surface of the small shell:

Apply Gauss's law to a spherical Gaussian surface with radius $a < r < b$. This surface lies within the conductor of the small shell, where $E = 0$, so $\Phi_E = 0$. Thus by Gauss's law $Q_{\text{encl}} = 0$, so there is zero charge on the inner surface of the small shell.

(ii) charge on outer surface of the small shell:

The total charge on the small shell is $+2q$. We found in part (i) that there is zero charge on the inner surface of the shell, so all $+2q$ must reside on the outer surface.

(iii) charge on inner surface of large shell:

Apply Guass's law to a spherical Gaussian surface with radius $c < r < d$. The surface lies within the conductor of the large shell, where $E = 0$, so $\Phi_E = 0$. Thus by Gauss's law $Q_{\text{encl}} = 0$. The surface encloses the $+2q$ on the small shell so there must be charge $-2q$ on the inner surface of the large shell to make the total enclosed charge zero.

(iv) charge on outer surface of large shell

The total charge on the large shell is $+4q$. We showed in part (iii) that the charge on the inner surface is $-2q$, so there must be $+6q$ on the outer surface.

23-31 Find the electric field $\vec{E}$ produced by the shell

and then use $\vec{F} = q\vec{E}$ to find the force the shell exerts on the point charge.

a) Apply Gauss's law to a spherical Gaussian surface that has radius $r > R$ and that is concentric with the shell.

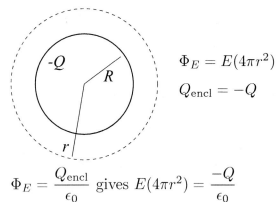

$$\Phi_E = E(4\pi r^2)$$

$$Q_{\text{encl}} = -Q$$

$$\Phi_E = \frac{Q_{\text{encl}}}{\epsilon_0} \text{ gives } E(4\pi r^2) = \frac{-Q}{\epsilon_0}$$

The magnitude of the field is $E = \dfrac{Q}{4\pi\epsilon_0 r^2}$ and it is directed toward the center of the shell.

Then $F = qE = \dfrac{qQ}{4\pi\epsilon_0 r^2}$, directed toward then center of the shell. (Since q is positive, $\vec{E}$ and $\vec{F}$ are in the same direction.)

b) Apply Gauss's law to a spherical Gaussian surface that has radius $r < R$ and that is concentric with the shell.

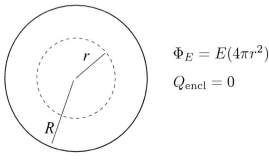

$$\Phi_E = E(4\pi r^2)$$

$$Q_{\text{encl}} = 0$$

$$\Phi_E = \frac{Q_{\text{encl}}}{\epsilon_0} \text{ gives } E(4\pi r^2) = 0$$

Then $E = 0$ so $F = 0$.

23-33 **a)** Apply Gauss's law to a Gaussian cylinder of length l and radius r, where $a < r < b$.

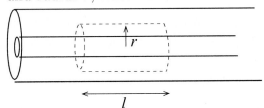

$$\Phi_E = E(2\pi r l)$$

$Q_{\text{encl}} = \lambda l$ (the charge on the length l of the inner conductor that is inside the Gaussian surface.

$$\Phi_E = \frac{Q_{encl}}{\epsilon_0} \text{ gives } E(2\pi rl) = \frac{\lambda l}{\epsilon_0}$$

$E = \dfrac{\lambda}{2\pi\epsilon_0 r}$. The enclosed charge is positive so the direction of $\vec{E}$ is radially outward.

b) Apply Gauss's law to a Gaussian cylinder of length l and radius r, where $r > c$:

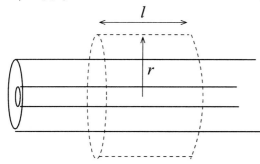

$\Phi_E = E(2\pi rl)$

$Q_{encl} = \lambda l$ (the charge on the length l of the inner conductor that is inside the Gaussian surface; the outer conductor carries no net charge.

$$\Phi_E = \frac{Q_{encl}}{\epsilon_0} \text{ gives } E(2\pi rl) = \frac{\lambda l}{\epsilon_0}$$

$E = \dfrac{\lambda}{2\pi\epsilon_0 r}$. The enclosed charge is positive so the direction of $\vec{E}$ is radially outward.

c) $E = 0$ within a conductor. Thus $E = 0$ for $r < a$;

$E = \dfrac{\lambda}{2\pi\epsilon_0 r}$ for $a < r < b$; $E = 0$ for $b < r < c$;

$E = \dfrac{\lambda}{2\pi\epsilon_0 r}$ for $r > c$.

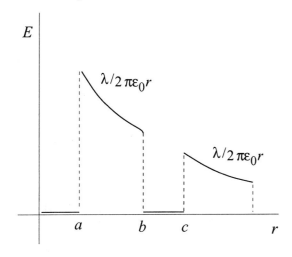

d) <u>inner surface:</u>

Apply Gauss's law to a Gaussian cylinder with radius r, where $b < r < c$. This surface lies within the conductor of the outer cylinder, where $E = 0$, so $\Phi_E = 0$. Thus by Gauss's law $Q_{\text{encl}} = 0$. The surface encloses charge λl on the inner conductor, so it must enclose charge $-\lambda l$ on the inner surface of the outer conductor. The charge per unit length on the inner surface of the outer cylinder is $-\lambda$.

<u>outer surface:</u>

The outer cylinder carries no net charge. So if there is charge per unit length $-\lambda$ on its inner surface there must be charge per unit length $+\lambda$ on the outer surface.

23-35 a) (i) <u>$r < a$</u>: Apply Gauss's law to a cylindrical Gaussian surface of length l and radius r, where $r < a$.

$\Phi_E = E(2\pi r l)$

$Q_{\text{encl}} = \alpha l$ (the charge on the length l of the line of charge)

$\Phi_E = \dfrac{Q_{\text{encl}}}{\epsilon_0}$ gives $E(2\pi r l) = \dfrac{\alpha l}{\epsilon_0}$

$E = \dfrac{\alpha}{2\pi\epsilon_0 r}$. The enclosed charge is positive so the direction of $\vec{E}$ is radially outward.

(ii) <u>$a < r < b$</u>: Points in this region are within the conducting tube, so $E = 0$.

(iii) <u>$r > b$</u>: Apply Gauss's law to a cylindrical Gaussian surface of length l and radius r, where $r > b$.

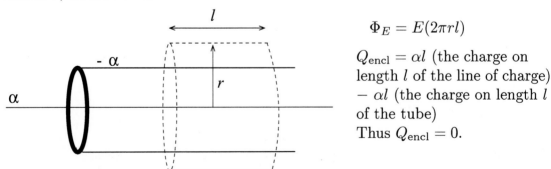

$\Phi_E = E(2\pi r l)$

$Q_{\text{encl}} = \alpha l$ (the charge on length l of the line of charge) $- \alpha l$ (the charge on length l of the tube) Thus $Q_{\text{encl}} = 0$.

$\Phi_E = \dfrac{Q_{\text{encl}}}{\epsilon_0}$ gives $E(2\pi r l) = 0$ and $E = 0$.

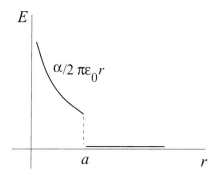

b) (i) <u>inner surface</u>

Apply Gauss's law to a cylindrical Gaussian surface of length l and radius r, where $a < r < b$. This surface lies within the conductor of the tube, where $E = 0$, so $\Phi_E = 0$. Then by Gauss's law $Q_{\text{encl}} = 0$. The surface encloses charge αl on the line of charge so must enclose charge αl on the inner surface of the tube. The charge per unit length on the inner surface of the tube is $-\alpha$.

(ii) <u>outer surface</u>

The net charge per unit length on the tube is $-\alpha$. We have shown in part (i) that this must all reside on the inner surface, so there is no net charge on the outer surface of the tube.

23-39

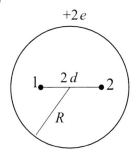

If the electrons are in equilibrium the net force on each one is zero.

Consider the forces on electron 2. There is a repulsive force F_1 due to the other electron, electron 1.:

$$F_1 = \frac{1}{4\pi\epsilon_0}\frac{e^2}{(2d)^2}$$

The electric field inside the uniform distribution of positive charge is

$$E = \frac{Qr}{4\pi\epsilon_0 R^3} \text{ (Example 23-9), where } Q = +2e. \text{ At the position of electron 2, } r = d.$$

The force F_{cd} exerted by the positive charge distribution is

$F_{cd} = eE = \dfrac{e(2e)d}{4\pi\epsilon_0 R^3}$ and is attractive.

The force diagram for electron 2 is

$\overset{\displaystyle F_{cd}}{\longleftarrow} \bullet \overset{\displaystyle F_1}{\longrightarrow}$

Net force equals zero implies $F_1 = F_{cd}$ and $\dfrac{1}{4\pi\epsilon_0}\dfrac{e^2}{4d^2} = \dfrac{2e^2 d}{4\pi\epsilon_0 R^3}$

Thus $(1/4d^2) = 2d/R^3$, so $d^3 = R^3/8$ and $d = R/2$.

23-41 Example 23-9 shows that the electric field of a uniformly charged

sphere is $E = \dfrac{Qr}{4\pi\epsilon_0 R^3}$ for $r < R$ and $E = \dfrac{Q}{4\pi\epsilon_0 r^2}$ for $r > R$.

The force on the electron has magnitude $F = qE = eE$. The acceleration has magnitude $a = F/m = eE/m$, where $e/m = (1.60 \times 10^{-19}\text{ C})/(9.109 \times 10^{-31}\text{ kg}) = 1.757 \times 10^{11}$ C/kg.

a) $\underline{r = 2R}$

$E = \dfrac{Q}{4\pi\epsilon_0 r^2} = \dfrac{Q}{4\pi\epsilon_0 (2R)^2} = \dfrac{1}{4}\left(\dfrac{Q}{4\pi\epsilon_0 R^2}\right)$

$E = \dfrac{1}{4}\left(\dfrac{82e}{4\pi\epsilon_0 R^2}\right) = \dfrac{1}{4}\dfrac{(8.988 \times 10^9\text{ N}\cdot\text{m}^2/\text{C}^2)(82)(1.602 \times 10^{-19}\text{ C})}{(7.1 \times 10^{-15}\text{ m})^2} =$

$\tfrac{1}{4}(2.342 \times 10^{21}\text{ N/C}) = 5.855 \times 10^{20}\text{ N/C}$

$a = (e/m)E = (1.757 \times 10^{11}\text{ C/kg})(5.855 \times 10^{20}\text{ N/C}) = 1.0 \times 10^{32}\text{ m/s}^2$

b) $\underline{r = R}$

$E = \dfrac{Q}{4\pi\epsilon_0 r^2} = \dfrac{Q}{4\pi\epsilon_0 R^2} = 2.342 \times 10^{21}\text{ N/C}.$

$a = (e/m)E = (1.757 \times 10^{11}\text{ C/kg})(2.342 \times 10^{21}\text{ N/C}) = 4.1 \times 10^{32}\text{ m/s}^2$ (4 times larger than for $r = 2R$)

c) $\underline{r = R/2}$

$E = \dfrac{Qr}{4\pi\epsilon_0 R^3} = \dfrac{Q(R/2)}{4\pi\epsilon_0 R^3} = \dfrac{1}{2}\left(\dfrac{Q}{4\pi\epsilon_0 R^2}\right) = \dfrac{1}{2}(2.342 \times 10^{21}\text{ N/C}) =$ $1.171 \times 10^{21}\text{ N/C}.$

$a = (e/m)E = (1.757 \times 10^{11}\text{ C/kg})(1.171 \times 10^{21}\text{ N/C}) = 2.1 \times 10^{32}\text{ m/s}^2$ ($\tfrac{1}{2}$ the value at $r = R$)

d) At $r = 0$, $E = 0$ so $F = 0$ and $a = 0$.

23-43 a)

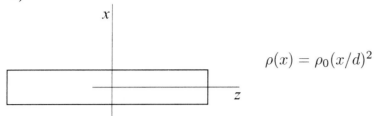

$$\rho(x) = \rho_0 (x/d)^2$$

The charge distribution is symmetric about $x = 0$, so by symmetry $E(x) = E(-x)$. But for $x > 0$ the field is in the $+x$ direction and for $x < 0$ the field is in the $-x$ direction. At $x = 0$ the field can't be both in the $+x$ and $-x$ directions so must be zero. That is, $E_x(x) = -E_x(-x)$. At point $x = 0$ this gives $E_x(0) = -E_x(0)$ and this equation is satisfied only for $E_x(0) = 0$.

b) $\underline{|x| > d}$ (outside the slab)

Apply Gauss's law to a cylindrical Gaussian surface whose axis is perpendicular to the slab and whose end caps have area A and are the same distance $|x| > d$ from $x = 0$.

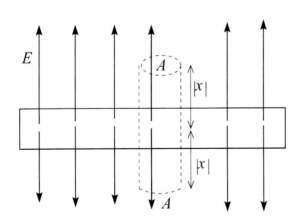

$$\Phi_E = 2EA$$

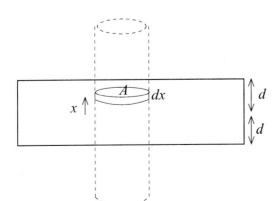

To find Q_{encl} consider a thin disk at coordinate x and with thickness dx. The charge within this disk is
$$dq = \rho\, dV = \rho A\, dx = (\rho_0 A/d^2) x^2\, dx.$$

The total charge enclosed by the Gaussian cylinder is

$Q_{\text{encl}} = 2 \int_0^d dq = (2\rho_0 A/d^2) \int_0^d x^2 \, dx = (2\rho_0 A/d^2)(d^3/3) = \frac{2}{3}\rho_0 Ad.$

Then $\Phi_E = \dfrac{Q_{\text{encl}}}{\epsilon_0}$ gives $2EA = 2\rho_0 Ad/3\epsilon_0.$

$E = \rho_0 d/3\epsilon_0$

$\vec{E}$ is directed away from $x = 0$, so $\vec{E} = (\rho_0 d/3\epsilon_0)(x/|x|)\hat{i}.$

<u>$|x| < d$</u> (inside the slab)

Apply Gauss's law to a cylindrical Gaussian surface whose axis is perpendicular to the slab and whose end caps have area A and are the same distance $|x| < d$ from $x = 0$.

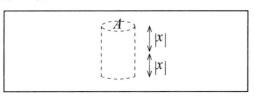

$\Phi_E = 2EA$

Q_{encl} is found as above, but now the integral on dx is only from 0 to x instead of 0 to d.

$Q_{\text{encl}} = 2 \int_0^x dq = (2\rho_0 A/d^2) \int_0^x x^2 \, dx = (2\rho_0 A/d^2)(x^3/3).$

Then $\Phi_E = \dfrac{Q_{\text{encl}}}{\epsilon_0}$ gives $2EA = 2\rho_0 Ax^3/3\epsilon_0 d^2.$

$E = \rho_0 x^3/3\epsilon_0 d^2$

$\vec{E}$ is directed away from $x = 0$, so $\vec{E} = (\rho_0 x^3/3\epsilon_0 d^2)(x/|x|)\hat{i}.$

Note that $E = 0$ at $x = 0$ as stated in part (a). Note also that the expressions for $|x| > d$ and $|x| < d$ agree for $x = d$.

23-45 $\rho(r) = \rho_0(1 - r/R)$ for $r \leq R$ where $\rho_0 = 3Q/\pi R^3$.

$\rho(r) = 0$ for $r \geq R$

a) The charge density varies with r inside the spherical volume. Divide the volume up into thin concentric shells, of radius r and thickness dr.

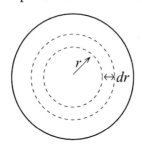

The volume of such a shell is
$dV = 4\pi r^2 \, dr$

The charge contained within the shell is
$dq = \rho(r) \, dV = 4\pi r^2 \rho_0 (1 - r/R) \, dr$

The total charge Q in the charge distribution is obtained by integrating dq over all such shells into which the sphere can be subdivided:

$$Q = \int dq = \int_0^R 4\pi r^2 \rho_0 (1 - r/R)\, dr = 4\pi \rho_0 \int_0^R (r^2 - r^3/R)\, dr$$

$$Q = 4\pi \rho_0 \left[\frac{r^3}{3} - \frac{r^4}{4R} \right]_0^R = 4\pi \rho_0 \left(\frac{R^3}{3} - \frac{R^4}{4R} \right) = 4\pi \rho_0 (R^3/12) =$$

$4\pi (3Q/\pi R^3)(R^3/12) = Q$, as was to be shown.

b) Apply Gauss's law to a spherical surface of radius r, where $r > R$:

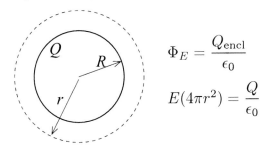

$$\Phi_E = \frac{Q_{encl}}{\epsilon_0}$$

$$E(4\pi r^2) = \frac{Q}{\epsilon_0}$$

$E = \dfrac{Q}{4\pi \epsilon_0 r^2}$; same as for point charge of charge Q.

c) Apply Gauss's law to a spherical surface of radius r, where $r < R$:

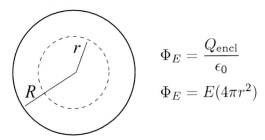

$$\Phi_E = \frac{Q_{encl}}{\epsilon_0}$$

$$\Phi_E = E(4\pi r^2)$$

To calculate the enclosed charge Q_{encl} use the same technique as in part (a), except integrate dq out to r rather than R. (We want the charge that is inside radius r.)

$$Q_{encl} = \int_0^r 4\pi r'^2 \rho_0 \left(1 - \frac{r'}{R} \right) dr' = 4\pi \rho_0 \int_0^r \left(r'^2 - \frac{r'^3}{R} \right) dr'$$

$$Q_{encl} = 4\pi \rho_0 \left[\frac{r'^3}{3} - \frac{r'^4}{4R} \right]_0^r = 4\pi \rho_0 \left(\frac{r^3}{3} - \frac{r^4}{4R} \right) = 4\pi \rho_0 r^3 \left(\frac{1}{3} - \frac{r}{4R} \right)$$

$$\rho_0 = \frac{3Q}{\pi R^3} \text{ so } Q_{encl} = 12Q \frac{r^3}{R^3} \left(\frac{1}{3} - \frac{r}{4R} \right) = Q \left(\frac{r^3}{R^3} \right) \left(4 - 3\frac{r}{R} \right).$$

Thus Gauss's law gives $E(4\pi r^2) = \dfrac{Q}{\epsilon_0} \left(\dfrac{r^3}{R^3} \right) \left(4 - 3\dfrac{r}{R} \right)$

$$E = \frac{Qr}{4\pi\epsilon_0 R^3}\left(4 - \frac{3r}{R}\right), \ r \leq R$$

d)

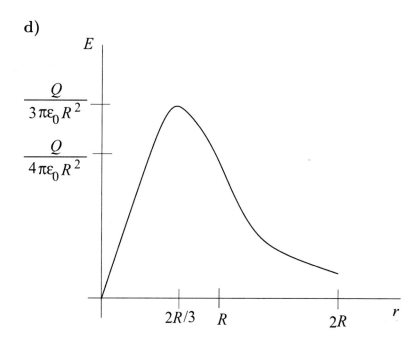

e) Where the electric field is a maximum, $\dfrac{dE}{dr} = 0$. Thus

$$\frac{d}{dr}\left(4r - \frac{3r^2}{R}\right) = 0 \text{ so } 4 - 6r/R = 0 \text{ and } r = 2R/3.$$

At this value of r, $E = \dfrac{Q}{4\pi\epsilon_0 R^3}\left(\dfrac{2R}{3}\right)\left(4 - \dfrac{3}{R}\dfrac{2R}{3}\right) = \dfrac{Q}{3\pi\epsilon_0 R^2}$

23-49 a) For an insulating sphere of uniform charge density ρ and

centered at the origin, the electric field inside the sphere is given by $E = Qr'/4\pi\epsilon_0 R^3$ (Example 23-9), where $\vec{r}'$ is the vector from the center of the sphere to the point where E is calculated.

But $\rho = 3Q/4\pi R^3$ so this may be written as $E = \rho r/3\epsilon_0$. And $\vec{E}$ is radially outward, in the direction of $\vec{r}'$, so $\vec{E} = \rho\vec{r}'/3\epsilon_0$.

For a sphere whose center is located by vector $\vec{b}$, a point inside the sphere and located by $\vec{r}$ is located by the vector $\vec{r}' = \vec{r} - \vec{b}$ relative to the center of the sphere.

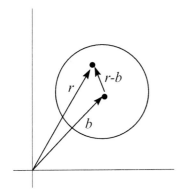

Thus $\vec{E} = \dfrac{\rho(\vec{r} - \vec{b})}{3\epsilon_0}$

b) $\vec{E} = \vec{E}_{\text{uniform}} + \vec{E}_{\text{hole}}$, where $\vec{E}_{\text{uniform}}$ is the field of a uniformly charged sphere with charge density ρ and $\vec{E}_{\text{hole}}$ is the field of a sphere located at the hole and with charge density $-\rho$. (Within the spherical hole the net charge density is $+\rho-\rho = 0$.)

$\vec{E}_{\text{uniform}} = \dfrac{\rho\vec{r}}{3\epsilon_0}$, where $\vec{r}$ is a vector from the center of the sphere.

$\vec{E}_{\text{hole}} = \dfrac{-\rho(\vec{r} - \vec{b})}{3\epsilon_0}$, at points inside the hole.

Then $\vec{E} = \dfrac{\rho\vec{r}}{3\epsilon_0} + \left(\dfrac{-\rho(\vec{r} - \vec{b})}{3\epsilon_0} \right) = \dfrac{\rho\vec{b}}{3\epsilon_0}$.

$\vec{E}$ is independent of $\vec{r}$ so is uniform inside the hole. The direction of $\vec{E}$ inside the hole is in the direction of the vector $\vec{b}$, the direction from the center of the insulating sphere to the center of the hole.

23-51   The electric field at each point is the vector sum of the fields of the two charge distributions.

Inside a sphere of uniform positive charge, $E = \dfrac{\rho r}{3\epsilon_0}$.

$\rho = \dfrac{Q}{\frac{4}{3}\pi R^3} = \dfrac{3Q}{4\pi R^3}$ so $E = \dfrac{Qr}{4\pi\epsilon_0 R^3}$, directed away from the center of the sphere.

Outside a sphere of uniform positive charge, $E = \dfrac{Q}{4\pi\epsilon_0 r^2}$, directed away from the center of the sphere.

a) $x = 0$ This point is inside sphere 1 and outside sphere 2.

$E_1 = \dfrac{Qr}{4\pi\epsilon_0 R^3} = 0$, since $r = 0$.

$E_2 = \dfrac{Q}{4\pi\epsilon_0 r^2}$ with $r = 2R$ so $E_2 = \dfrac{Q}{16\pi\epsilon_0 R^2}$, in the $-x$-direction.

Thus $\vec{E} = \vec{E}_1 + \vec{E}_2 = -\dfrac{Q}{16\pi\epsilon_0 R^2}\hat{i}$.

b) $x = R/2$. This point is inside sphere 1 and outside sphere 2. Each field is directed away from the center of the sphere that produces it.

$E_1 = \dfrac{Qr}{4\pi\epsilon_0 R^3}$ with $r = R/2$ so

$E_1 = \dfrac{Q}{8\pi\epsilon_0 R^2}$

$E_2 = \dfrac{Q}{4\pi\epsilon_0 r^2}$ with $r = 3R/2$ so $E_2 = \dfrac{Q}{9\pi\epsilon_0 R^2}$

$E = E_1 - E_2 = \dfrac{Q}{72\pi\epsilon_0 R^2}$, in the $+x$-direction.

$$\vec{E} = \dfrac{Q}{72\pi\epsilon_0 R^2}\hat{i}$$

c) $x = R$. This point is at the surface of each sphere. The fields have equal magnitudes and opposite directions, so $E = 0$.

d) $x = 3R$. This point is outside both spheres. Each field is directed away from the center of the sphere that produces it.

$E_1 = \dfrac{Q}{4\pi\epsilon_0 r^2}$ with $r = 3R$ so

$E_1 = \dfrac{Q}{36\pi\epsilon_0 R^2}$

$E_2 = \dfrac{Q}{4\pi\epsilon_0 r^2}$ with $r = R$ so $E_2 = \dfrac{Q}{4\pi\epsilon_0 R^2}$

$E = E_1 + E_2 = \dfrac{5Q}{18\pi\epsilon_0 R^2}$, in the $+x$-direction.

$$\vec{E} = \dfrac{5Q}{18\pi\epsilon_0 R^2}\hat{i}$$

CHAPTER 24
ELECTRIC POTENTIAL

Exercises

24-1

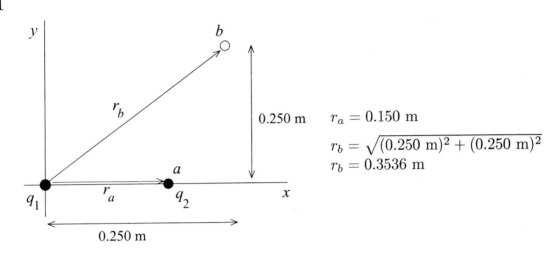

$r_a = 0.150$ m

$r_b = \sqrt{(0.250 \text{ m})^2 + (0.250 \text{ m})^2}$

$r_b = 0.3536$ m

$W_{a \to b} = U_a - U_b$

$$U_a = \frac{1}{4\pi\epsilon_0} \frac{q_1 q_2}{r_a} = (8.988 \times 10^9 \text{ N} \cdot \text{m}^2/\text{C}^2)\frac{(+2.40 \times 10^{-6} \text{ C})(-4.30 \times 10^{-6} \text{ C})}{0.150 \text{ m}}$$

$U_a = -0.6184$ J

$$U_b = \frac{1}{4\pi\epsilon_0} \frac{q_1 q_2}{r_b} = (8.988 \times 10^9 \text{ N} \cdot \text{m}^2/\text{C}^2)\frac{(+2.40 \times 10^{-6} \text{ C})(-4.30 \times 10^{-6} \text{ C})}{0.3536 \text{ m}}$$

$U_b = -0.2623$ J

$W_{a \to b} = U_a - U_b = -0.6184 \text{ J} - (-0.2623 \text{ J}) = -0.356 \text{ J}$

The attractive force on q_2 is toward the origin, so it does negative work on q_2 when q_2 moves to larger r.

24-3   **a)** Use conservation of energy:

$K_a + U_a + W_{\text{other}} = K_b + U_b$

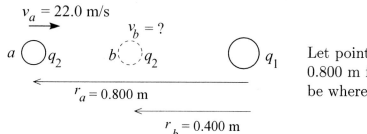

Let point a be where q_2 is 0.800 m from q_1 and point b be where q_2 is 0.400 m from q_1

Only the electric force does work, so $W_{\text{other}} = 0$ and $U = \dfrac{1}{4\pi\epsilon_0}\dfrac{q_1 q_2}{r}$.

$$K_a = \tfrac{1}{2}mv_a^2 = \tfrac{1}{2}(1.50 \times 10^{-3}\text{ kg})(22.0\text{ m/s})^2 = 0.3630\text{ J}$$

$$U_a = \frac{1}{4\pi\epsilon_0}\frac{q_1 q_2}{r_a} =$$

$$(8.988 \times 10^9\text{ N}\cdot\text{m}^2/\text{C}^2)\frac{(-2.80 \times 10^{-6}\text{ C})(-7.80 \times 10^{-6}\text{ C})}{0.800\text{ m}} = +0.2454\text{ J}$$

$$K_b = \tfrac{1}{2}mv_b^2$$

$$U_b = \frac{1}{4\pi\epsilon_0}\frac{q_1 q_2}{r_b} =$$

$$(8.988 \times 10^9\text{ N}\cdot\text{m}^2/\text{C}^2)\frac{(-2.80 \times 10^{-6}\text{ C})(-7.80 \times 10^{-6}\text{ C})}{0.400\text{ m}} = +0.4907\text{ J}$$

The conservation of energy equation then gives $K_b = K_a + (U_a - U_b)$

$$\tfrac{1}{2}mv_b^2 = +0.3630\text{ J} + (0.2454\text{ J} - 0.4907\text{ J}) = 0.1177\text{ J}$$

$$v_b = \sqrt{\frac{2(0.1177\text{ J})}{1.50 \times 10^{-3}\text{ kg}}} = 12.5\text{ m/s}$$

The potential energy increases when the two positively charged spheres get closer together, so the kinetic energy and speed decrease.

b) Let point c be where q_2 has its speed momentarily reduced to zero. Apply conservation of energy to points a and c: $K_a + U_a + W_{\text{other}} = K_c + U_c$.

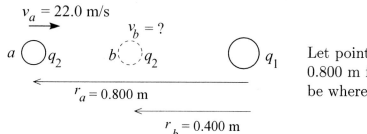

$K_a = +0.3630$ J (from part (a))
$U_a = +0.2454$ J (from part (a))

$K_c = 0$ (at distance of closest approach the speed is zero)

$$U_c = \frac{1}{4\pi\epsilon_0}\frac{q_1 q_2}{r_c}$$

Thus conservation of energy $K_a + U_a = U_c$ gives

$$\frac{1}{4\pi\epsilon_0}\frac{q_1 q_2}{r_c} = +0.3630 \text{ J} + 0.2454 \text{ J} = 0.6084 \text{ J}$$

$$r_c = \frac{1}{4\pi\epsilon_0}\frac{q_1 q_2}{0.6084 \text{ J}} =$$

$$(8.988 \times 10^9 \text{ N}\cdot\text{m}^2/\text{C}^2)\frac{(-2.80 \times 10^{-6} \text{ C})(-7.80 \times 10^{-6} \text{ C})}{+0.6084 \text{ J}} = 0.323 \text{ m}.$$

24-5   **a)** $U = \frac{1}{4\pi\epsilon_0}\frac{qq'}{r}$

$$U = (8.988 \times 10^9 \text{ N}\cdot\text{m}^2/\text{C}^2)\frac{(+4.60 \times 10^{-6} \text{ C})(+1.20 \times 10^{-6} \text{ C})}{0.250 \text{ m}} = +0.198 \text{ J}$$

b) Use conservation of energy: $K_a + U_a + W_{\text{other}} = K_b + U_b$

Only the electric force does work, so $W_{\text{other}} = 0$ and $U = \frac{1}{4\pi\epsilon_0}\frac{qQ}{r}$.

$K_a = 0$ (released from rest)
$U_a = +0.198$ J (from part (a))
$K_b = \frac{1}{2}mv_b^2$

(i) $r_b = 0.500$ m

$$U_b = \frac{1}{4\pi\epsilon_0}\frac{qQ}{r} =$$

$$(8.988 \times 10^9 \text{ N}\cdot\text{m}^2/\text{C}^2)\frac{(+4.60 \times 10^{-6} \text{ C})(+1.20 \times 10^{-6} \text{ C})}{0.500 \text{ m}} = +0.0992 \text{ J}$$

Then $K_a + U_a + W_{\text{other}} = K_b + U_b$ gives $K_b = U_a - U_b$ and

$$\tfrac{1}{2}mv_b^2 = U_a - U_b \text{ and } v_b = \sqrt{\frac{2(U_a - U_b)}{m}} = \sqrt{\frac{2(+0.198 \text{ J} - 0.0992 \text{ J})}{2.80 \times 10^{-4} \text{ kg}}} = 26.6 \text{ m/s}.$$

(ii) $r_b = 5.00$ m

r_b is now ten times larger than in (i) so U_b is ten times smaller:

$U_b = +0.0992 \text{ J}/10 = +0.00992 \text{ J}.$

$$v_b = \sqrt{\frac{2(U_a - U_b)}{m}} = \sqrt{\frac{2(+0.198 \text{ J} - 0.00992 \text{ J})}{2.80 \times 10^{-4} \text{ kg}}} = 36.7 \text{ m/s}.$$

(iii) $r_b = 50.0$ m

r_b is now ten times larger than in (ii) so U_b is ten times smaller:

$U_b = +0.00992 \text{ J}/10 = +0.000992$ J.

$$v_b = \sqrt{\frac{2(U_a - U_b)}{m}} = \sqrt{\frac{2(+0.198 \text{ J} - 0.000992 \text{ J})}{2.80 \times 10^{-4} \text{ kg}}} = 37.5 \text{ m/s}.$$

24-9 $W = -\Delta U = -(U_2 - U_1)$

Let 1 be where all the charges are infinitely far apart. Let 2 be where the charges are at the corners of the triangle.

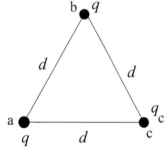

Let q_c be the third, unknown charge.

$U_1 = 0$

$$U_2 = U_{ab} + U_{ac} + U_{bc} = \frac{1}{4\pi\epsilon_0 d}(q^2 + 2qq_c)$$

Want $W = 0$, so $W = -(U_2 - U_1)$ gives $0 = -U_2$

$$0 = \frac{1}{4\pi\epsilon_0 d}(q^2 + 2qq_c)$$

$q^2 + 2qq_c = 0$ and $q_c = -q/2$.

24-11 $W_{a \to b} = q' \int_a^b \vec{E} \cdot d\vec{l}$

Use coordinates where $+y$ is upward and $+x$ is to the right. Then $\vec{E} = E\hat{j}$ with $E = 4.00 \times 10^4$ N/C.

a)

$d\vec{l} = dx\hat{i}$

$\vec{E} \cdot d\vec{l} = (E\hat{j}) \cdot (dx\hat{i}) = 0$ so $W_{a \to b} = q' \int_a^b \vec{E} \cdot d\vec{l} = 0$.

The electric force on the positive charge is upward (in the direction of the electric field) and does no work for a horizontal displacement of the charge.

b)

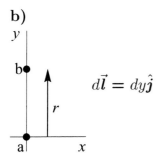

$$\vec{E} \cdot d\vec{l} = (E\hat{j}) \cdot (dy\hat{j}) = E\,dy$$
$$W_{a \to b} = q' \int_a^b \vec{E} \cdot d\vec{l} = q'E \int_a^b dy = q'E(y_b - y_a)$$

$y_b - y_a = +0.670$ m, positive since the displacement is upward and we have taken $+y$ to be upward.

$W_{a \to b} = q'E(y_b - y_a) = (+28.0 \times 10^{-9}$ C$)(4.00 \times 10^4$ N/C$)(+0.670$ m$) = +7.50 \times 10^{-4}$ J.

The electric force on the positive charge is upward so it does positive work for an upward displacement of the charge.

c) $d\vec{l} = dx\hat{i} + dy\hat{j}$ (The displacement has both horizontal and vertical components.)

$\vec{E} \cdot d\vec{l} = (E\hat{j}) \cdot (dx\hat{i} + dy\hat{j}) = E\,dy$ (Only the vertical component of the displacement contributes to the work.)

$$W_{a \to b} = q' \int_a^b \vec{E} \cdot d\vec{l} = q'E \int_a^b dy = q'E(y_b - y_a)$$

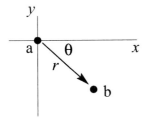

$y_a = 0$
$y_b = -r\sin\theta = -(2.60$ m$)\sin 45° = -1.838$ m
The vertical component of the 2.60 m displacement is 1.838 m downward.

$W_{a \to b} = q'E(y_b - y_a) = (+28.0 \times 10^{-9}$ C$)(4.00 \times 10^4$ N/C$)(-1.838$ m$) = -2.06 \times 10^{-3}$ J.

The electric force on the positive charge is upward so it does negative work for a displacement of the charge that has a downward component.

24-15

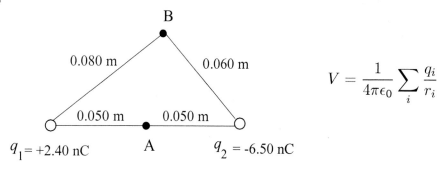

$$V = \frac{1}{4\pi\epsilon_0} \sum_i \frac{q_i}{r_i}$$

$q_1 = +2.40$ nC   A $q_2 = -6.50$ nC

a) $V_A = \dfrac{1}{4\pi\epsilon_0}\left(\dfrac{q_1}{r_{A1}} + \dfrac{q_2}{r_{A2}}\right)$

$$V_A = (8.988 \times 10^9 \text{ N} \cdot \text{m}^2/\text{C}^2)\left(\frac{+2.40 \times 10^{-9} \text{ C}}{0.050 \text{ m}} + \frac{-6.50 \times 10^{-9} \text{ C}}{0.050 \text{ m}}\right) = -737 \text{ V}$$

b) $V_B = \dfrac{1}{4\pi\epsilon_0}\left(\dfrac{q_1}{r_{B1}} + \dfrac{q_2}{r_{B2}}\right)$

$$V_B = (8.988 \times 10^9 \text{ N} \cdot \text{m}^2/\text{C}^2)\left(\frac{+2.40 \times 10^{-9} \text{ C}}{0.080 \text{ m}} + \frac{-6.50 \times 10^{-9} \text{ C}}{0.060 \text{ m}}\right) = -704 \text{ V}$$

c) $W_{B \to A} = q'(V_B - V_A) = (2.50 \times 10^{-9} \text{ C})(-704 \text{ V} - (-737 \text{ V})) = +8.2 \times 10^{-8} \text{ J}$

The electric force does positive work on the positive charge when it moves from higher potential (point B) to lower potential (point A).

24-23 **a)** The direction of $\vec{E}$ is always from high potential to low potential so point b is at higher potential.

b) $V_b - V_a = -\int_a^b \vec{E} \cdot d\vec{l} = \int_a^b E \, dx = E(x_b - x_a)$.

$$E = \frac{V_b - V_a}{x_b - x_a} = \frac{+240 \text{ V}}{0.90 \text{ m} - 0.60 \text{ m}} = 800 \text{ V/m}$$

c) $W_{b \to a} = q(V_b - V_a) = (-0.200 \times 10^{-6} \text{ C})(+240 \text{ V}) = -4.80 \times 10^{-5} \text{ J}$.

The electric force does negative work on a negative charge when the negative charge moves from high potential (point b) to low potential (point a).

24-25 **a)** When the electron is on either side of the center of the ring, the ring exerts an attractive force directed toward the center of the ring. This restoring force produces oscillatory motion of the electron along the axis of the ring, with amplitude

30.0 cm. The force on the electron is <u>not</u> of the form $F = -kx$ so the oscillatory motion is not simple harmonic motion.

b) $K_a + U_a = K_b + U_b$ with a at the initial position of the electron and b at the center of the ring. From Example 24-11, $V = \dfrac{1}{4\pi\epsilon_0} \dfrac{Q}{\sqrt{x^2 + R^2}}$, where R is the radius of the ring.

$x_a = 30.0$ cm, $x_b = 0$.

$K_a = 0$ (released from rest), $K_b = \frac{1}{2}mv^2$

Thus $\frac{1}{2}mv^2 = U_a - U_b$

And $U = qV = -eV$ so $v = \sqrt{\dfrac{2e(V_b - V_a)}{m}}$.

$V_a = \dfrac{1}{4\pi\epsilon_0} \dfrac{Q}{\sqrt{x_a^2 + R^2}} = (8.988 \times 10^9 \text{ N} \cdot \text{m}^2/\text{C}^2) \dfrac{24.0 \times 10^{-9} \text{ C}}{\sqrt{(0.300 \text{ m})^2 + (0.150 \text{ m})^2}}$

$V_a = 643$ V

$V_b = \dfrac{1}{4\pi\epsilon_0} \dfrac{Q}{\sqrt{x_b^2 + R^2}} = (8.988 \times 10^9 \text{ N} \cdot \text{m}^2/\text{C}^2) \dfrac{24.0 \times 10^{-9} \text{ C}}{0.150 \text{ m}} = 1438$ V

$v = \sqrt{\dfrac{2e(V_b - V_a)}{m}} = \sqrt{\dfrac{2(1.602 \times 10^{-19} \text{ C})(1438 \text{ V} - 643 \text{ V})}{9.109 \times 10^{-31} \text{ kg}}} = 1.67 \times 10^7$ m/s

24-27 **a)** From Example 24-9, $E = \dfrac{V_{ab}}{d} = \dfrac{360 \text{ V}}{0.0450 \text{ m}} = 8000$ V/m

b) $F = |q|E = (2.40 \times 10^{-9} \text{ C})(8000 \text{ V/m}) = +1.92 \times 10^{-5}$ N

c)

The plate with positive charge (plate a) is at higher potential. The electric field is directed from high potential toward low potential (or, $\vec{E}$ is from + charge toward − charge), so $\vec{E}$ points from a to b. Hence the force that $\vec{E}$ exerts on the positive charge is from a to b, so it does positive work.

$W = \int_a^b \vec{F} \cdot d\vec{l} = Fd$, where d is the separation between the plates.

$W = Fd = (1.92 \times 10^{-5} \text{ N})(0.0450 \text{ m}) = +8.64 \times 10^{-7}$ J

d) $V_a - V_b = +360$ V (plate a is at higher potential)

$\Delta U = U_a - U_b = q(V_a - V_b) = (2.40 \times 10^{-9} \text{ C})(360 \text{ V}) = +8.64 \times 10^{-7}$ J.

We see that $W_{a \to b} = U_a - U_b$.

24-29 a) $E = \dfrac{V_{ab}}{d}$ gives that

$$d = \frac{V_{ab}}{E} = \frac{4.75 \times 10^3 \text{ V}}{3.00 \times 10^6 \text{ N/C}} = 1.58 \times 10^{-3} \text{ m} = 1.58 \text{ mm}.$$

b) $E = \sigma/\epsilon_o$

$\sigma = E\epsilon_0 = (3.00 \times 10^6 \text{ V/m})(8.854 \times 10^{-12} \text{ C}^2/\text{N} \cdot \text{m}^2) = 2.66 \times 10^{-5} \text{ C/m}^2.$

24-31 $V = Axy - Bx^2 + Cy$

a) $E_x = -\dfrac{\partial V}{\partial x} = -Ay + 2Bx$

$E_y = -\dfrac{\partial V}{\partial y} = -Ax - C$

$E_z = -\dfrac{\partial V}{\partial z} = 0$

b) $E = 0$ requires that $E_x = E_y = E_z = 0$.

$E_z = 0$ everywhere.

$E_y = 0$ at $x = -C/A$.

And E_x is also equal zero for this x, any value of z, and $y = 2Bx/A = (2B/A)(-C/A) = -2BC/A^2$.

24-33 a) V is independent of x and y so

$$E_x = -\frac{\partial V}{\partial x} = 0 \text{ and } E_y = -\frac{\partial V}{\partial y} = 0.$$

$E_z = -\dfrac{\partial V}{\partial z}$

For $z < 0$, $V = 0$ so $E_z = 0$. $E_x = E_y = E_z = 0$ so $E = 0$.

For $0 < z < d$, $V = Cz$ so $E_z = -C$. $\vec{E} = -C\hat{k}$.

For $z > d$, $V = Cd$ which is constant and has no z-dependence, so $E_z = 0$. $E_x = E_y = E_z = 0$ so $E = 0$.

b) There is a uniform electric field in the $-z$-direction for z between 0 and d and zero field elsewhere. This is the field of two infinite sheets of charge parallel to the xy-plane, one with charge $+\sigma$ at $z = d$ and one with charge $-\sigma$ at $z = 0$.

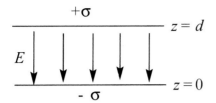

$+\sigma$
$z = d$

E

$z = 0$
$-\sigma$

For this charge distribution $E = \sigma/\epsilon_0$ between the plates, with direction from the positive plate toward the negative plate, and $E = 0$ outside the plates.

To have $E = C$ between the plates need $C = \sigma/\epsilon_0$, so $\sigma = C\epsilon_0$.

24-39 **a)** $\vec{F}_E = q\vec{E}$. Since $q = -e$ is negative $\vec{F}_E$ and $\vec{E}$ are in opposite directions; $\vec{E}$ is upward so $\vec{F}_E$ is downward. The magnitude of F_E is $F_E = |q|E = eE = (1.602 \times 10^{-19} \text{ C})(1.10 \times 10^3 \text{ N/C}) = 1.76 \times 10^{-16} \text{ N}$.

b) Calculate the acceleration of the electron produced by the electric force:
$$a = \frac{F}{m} = \frac{1.76 \times 10^{-16} \text{ N}}{9.109 \times 10^{-31} \text{ kg}} = 1.93 \times 10^{14} \text{ m/s}^2$$

This is much larger than $g = 9.80 \text{ m/s}^2$, so the gravity force on the electron can be neglected.

$\vec{F}_E$ is downward, so $\vec{a}$ is downward.

c) The acceleration is constant and downward, so the motion is like that of a projectile. Use the horizontal motion to find the time and then use the time to find the vertical displacement.

x-component
$v_{0x} = 6.50 \times 10^6 \text{ m/s};\quad a_x = 0;\quad x - x_0 = 0.060 \text{ m};\quad t = ?$
$x - x_0 = v_{0x}t + \frac{1}{2}a_x t^2$ and the a_x term is zero, so
$$t = \frac{x - x_0}{v_{0x}} = \frac{0.060 \text{ m}}{6.50 \times 10^6 \text{ s}} = 9.231 \times 10^{-9} \text{ s}$$

y-component
$v_{0y} = 0;\quad a_y = 1.93 \times 10^{14} \text{ m/s}^2;\quad t = 9.231 \times 10^{-9} \text{ m/s};\quad y - y_0 = ?$
$y - y_0 = v_{0y}t + \frac{1}{2}a_y t^2$
$y - y_0 == \frac{1}{2}(1.93 \times 10^{14} \text{ m/s}^2)(9.231 \times 10^{-9} \text{ s})^2 = 0.00822 \text{ m} = 0.822 \text{ cm}$

d)

v_x

α

v_y

v

$v_x = v_{0x} = 6.50 \times 10^6 \text{ m/s}$ (since $a_x = 0$)

$v_y = v_{0y} + a_y t$
$v_y = 0 + (1.93 \times 10^{14} \text{ m/s}^2)(9.231 \times 10^{-9} \text{ s})$
$v_y = 1.782 \times 10^6 \text{ m/s}$

$$\tan \alpha = \frac{v_y}{v_x} = \frac{1.782 \times 10^6 \text{ m/s}}{6.50 \times 10^6 \text{ m/s}} = 0.2742 \text{ so } \alpha = 15.3°.$$

e) Consider the motion of the electron after it leaves the region between the plates. Outside the plates there is no electric field, so $a = 0$. (Gravity can still be neglected since the electron is traveling at such high speed and the times are small.) Use the horizontal motion to find the time it takes the electron to travel 0.120 m horizontally to the screen. From this time find the distance downward that the electron travels.

x-component

$v_{0x} = 6.50 \times 10^6$ m/s; $a_x = 0$; $x - x_0 = 0.120$ m; $t = ?$

$x - x_0 = v_{0x}t + \frac{1}{2}a_x t^2$ and the a_x term is zero, so

$$t = \frac{x - x_0}{v_{0x}} = \frac{0.120 \text{ m}}{6.50 \times 10^6 \text{ s}} = 1.846 \times 10^{-8} \text{ s}$$

y-component

$v_{0y} = 1.782 \times 10^6$ m/s (from part (b)); $a_y = 0$; $t = 1.846 \times 10^{-8}$ m/s;

$y - y_0 = ?$

$y - y_0 = v_{0y}t + \frac{1}{2}a_y t^2 = (1.782 \times 10^6 \text{ m/s})(1.846 \times 10^{-8} \text{ s}) = 0.0329 \text{ m} = 3.29 \text{ cm}$

The electron travels downward a distance 0.822 cm while it is between the plates and a distance 3.29 cm while traveling from the edge of the plates to the screen. The total downward deflection is 0.822 cm + 3.29 cm = 4.11 cm.

Problems

24-49

a) $W_{\text{tot}} = \Delta K = K_b - K_a = K_b = 4.35 \times 10^{-5}$ J

The electric force F_E and the additional force F both do work, so that $W_{\text{tot}} = W_{F_E} + W_F$.

$W_{F_E} = W_{\text{tot}} - W_F = 4.35 \times 10^{-5} \text{ J} - 6.50 \times 10^{-5} \text{ J} = -2.15 \times 10^{-5} \text{ J}$

The electric force is to the left (in the direction of the electric field since the particle has positive charge). The displacement is to the right, so the electric force does negative work. The additional force F is in the direction of the displacement, so it does positive work.

b) For the work done by the electric force,

$$W_{a \to b} = q(V_a - V_b) \text{ and } V_a - V_b = \frac{W_{a \to b}}{q} = \frac{-2.15 \times 10^{-5} \text{ J}}{7.60 \times 10^{-9} \text{ C}} = -2.83 \times 10^3 \text{ V}.$$

The starting point (point a) is at 2.83×10^3 V lower potential than the ending point (point b). We know that $V_b > V_a$ because the electric field always points from high potential toward low potential.

c) Since the electric field is uniform and directed opposite to the displacement $W_{a \to b} = -F_E d = -qEd$, where $d = 8.00$ cm is the displacement of the particle.

$$E = -\frac{W_{a \to b}}{qd} = -\frac{V_a - V_b}{d} = -\frac{-2.83 \times 10^3 \text{ V}}{0.0800 \text{ m}} = 3.54 \times 10^4 \text{ V/m}.$$

24-51 a) $V = Cx^{4/3}$

$C = V/x^{4/3} = 240 \text{ V}/(13.0 \times 10^{-3} \text{ m})^{4/3} = 7.85 \times 10^4 \text{ V/m}^{4/3}$

b) $E_x = -\dfrac{\partial V}{\partial x} = -\dfrac{4}{3}Cx^{1/3} = -(1.05 \times 10^5 \text{ V/m}^{4/3})x^{1/3}$

The minus sign means that E_x is in the $-x$-direction, which says that $\vec{E}$ points from the positive anode toward the negative cathode.

c) $\vec{F} = q\vec{E}$ so $F_x = -eE_x = \frac{4}{3}eCx^{1/3}$

Halfway between the electrodes means $x = 6.50 \times 10^{-3}$ m.

$F_x = \frac{4}{3}(1.602 \times 10^{-19} \text{ V})(7.85 \times 10^4 \text{ V/m}^{4/3})(6.50 \times 10^{-3} \text{ m})^{1/3} = 3.13 \times 10^{-15} \text{ N}$

F_x is positive, so the force is directed toward the positive anode.

24-55 a)

$+e \qquad\qquad -e \qquad\qquad +e$

$r = (1.07 \times 10^{-10} \text{ m})/2 = 0.535 \times 10^{-10} \text{ m}$

The potential energy of interaction of the electron with each proton is

$$U = \frac{1}{4\pi\epsilon_0}\frac{(-e^2)}{r}, \text{ so the total potential energy is}$$

$$U = -\frac{2e^2}{4\pi\epsilon_0 r} = -\frac{2(8.988 \times 10^9 \text{ N} \cdot \text{m}^2/\text{C}^2)(1.60 \times 10^{-19} \text{ C})^2}{0.535 \times 10^{-10} \text{ m}} = -8.60 \times 10^{-18} \text{ J}$$

$$U = -8.60 \times 10^{-18} \text{ J}(1 \text{ eV}/1.602 \times 10^{-19} \text{ J}) = -53.7 \text{ eV}$$

b)

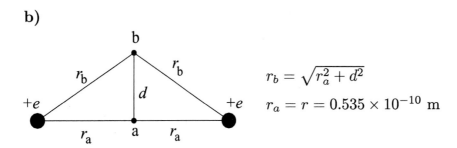

$$r_b = \sqrt{r_a^2 + d^2}$$

$$r_a = r = 0.535 \times 10^{-10} \text{ m}$$

Apply $K_a + U_a + W_{\text{other}} = K_b + U_b$ with point a midway between the protons and point b where the electron instantaneously has $v = 0$ (at its maximum displacement d from point a). Only the Coulomb force does work, so $W_{\text{other}} = 0$.

$U_a = -8.60 \times 10^{-18}$ J (from part (a)

$K_a = \frac{1}{2}mv^2 = \frac{1}{2}(9.109 \times 10^{-31} \text{ kg})(1.50 \times 10^6 \text{ m/s})^2 = 1.025 \times 10^{-18}$ J

$K_b = 0$

$U_b = -2ke^2/r_b$

Then $U_b = K_a + U_a - K_b = 1.025 \times 10^{-18}$ J $- 8.60 \times 10^{-18}$ J $=$
-7.575×10^{-18} J.

$$r_b = -\frac{2ke^2}{U_b} = -\frac{2(8.988 \times 10^9 \text{ N} \cdot \text{m}^2/\text{C}^2)(1.60 \times 10^{-19} \text{ C})^2}{-7.575 \times 10^{-18} \text{ J}} = 6.075 \times 10^{-11} \text{ m}$$

Then $d = \sqrt{r_b^2 - r_a^2} = \sqrt{(6.075 \times 10^{-11} \text{ m})^2 - (5.35 \times 10^{-11} \text{ m})^2} = 2.88 \times 10^{-11}$ m.

24-57 a) From Example 24-10, for a conducting cylinder with charge per unit length λ the potential for points outside the cylinder is given by $V = (\lambda/2\pi\epsilon_0) \ln(r_0/r)$ where r is the distance from the cylinder axis and r_0 is the distance from the axis for which we take $V = 0$. Inside the cylinder the potential has the same value as on the cylinder surface. The electric field is the same for a solid conducting cylinder or for a hollow conducting tube so this expression for V applies to both. This problem says to take $r_0 = b$.

For the hollow tube of radius b and charge per unit length $-\lambda$:

outside $V = -(\lambda/2\pi\epsilon_0) \ln(b/r)$;   inside $V = 0$ since $V = 0$ at $r = b$.

For the metal cylinder of radius a and charge per unit length λ:

outside $V = (\lambda/2\pi\epsilon_0) \ln(b/r)$, inside $V = (\lambda/2\pi\epsilon_0) \ln(b/a)$, the value at $r = a$.

(i) $r < a$; inside both $V = (\lambda/2\pi\epsilon_0) \ln(b/a)$

(ii) $a < r < b$; outside cylinder, inside tube $V = (\lambda/2\pi\epsilon_0) \ln(b/r)$

(iii) $r > b$; outside both The potentials are equal in magnitude and opposite in sign so $V = 0$.

b) For $r = a$, $V_a = (\lambda/2\pi\epsilon_0)\ln(b/a)$.

For $r = b$, $V_b = 0$.

Thus $V_{ab} = V_a - V_b = (\lambda/2\pi\epsilon_0)\ln(b/a)$.

c) $E = -\dfrac{\partial V}{\partial r} = -\dfrac{\lambda}{2\pi\epsilon_0}\dfrac{\partial}{\partial r}\ln\left(\dfrac{b}{r}\right) = -\dfrac{\lambda}{2\pi\epsilon_0}\left(\dfrac{r}{b}\right)\left(-\dfrac{b}{r^2}\right) = \dfrac{V_{ab}}{\ln(b/a)}\dfrac{1}{r}.$

d) The electric field between the cylinders is due only to the inner cylinder, so V_{ab} is not changed, $V_{ab} = (\lambda/2\pi\epsilon_0)\ln(b/a)$.

24-59 a) Problem 24-57 derived that $E = \dfrac{V_{ab}}{\ln(b/a)}\dfrac{1}{r}$, where a is the radius of the inner cylinder (wire) and b is the radius of the outer hollow cylinder. The potential difference between the two cylinders is V_{ab}. Midway between the wire and the cylinder wall is at a radius of $r = (a + b)/2 = (90.0 \times 10^{-6}\text{ m} + 0.140\text{ m})/2 = 0.07004\text{ m}$.

$$E = \dfrac{V_{ab}}{\ln(b/a)}\dfrac{1}{r} = \dfrac{50.0 \times 10^3\text{ V}}{\ln(0.140\text{ m}/90.0 \times 10^{-6}\text{ m})(0.07004\text{ m})} = 9.71 \times 10^4\text{ V/m}$$

b) $F_E = 10mg$

$$|q|E = 10mg \text{ and } |q| = \dfrac{10mg}{E} = \dfrac{10(30.0 \times 10^{-9}\text{ kg})(9.80\text{ m/s}^2)}{9.71 \times 10^4\text{ V/m}} = 3.03 \times 10^{-11}\text{ C}$$

24-61 a) Consider a thin ring of radius y and width dy. The ring has area $2\pi y\,dy$ so the charge on the ring is $dq = \sigma(2\pi y\,dy)$.

The result of Example 24-11 then says that the potential due to this thin ring at the point on the axis at a distance x from the ring is

$$dV = \dfrac{1}{4\pi\epsilon_0}\dfrac{dq}{\sqrt{x^2 + y^2}} = \dfrac{2\pi\sigma}{4\pi\epsilon_0}\dfrac{y\,dy}{\sqrt{x^2 + y^2}}$$

$$V = \int dV = \dfrac{\sigma}{2\epsilon_0}\int_0^R \dfrac{y\,dy}{\sqrt{x^2 + y^2}} = \dfrac{\sigma}{2\epsilon_0}\left[\sqrt{x^2 + y^2}\right]_0^R = \dfrac{\sigma}{2\epsilon_0}(\sqrt{x^2 + R^2} - x)$$

Note: For $x \gg R$ this result should reduce to the potential of a point charge with $Q = \sigma\pi R^2$.

$\sqrt{x^2 + R^2} = x(1 + R^2/x^2)^{1/2} \approx x(1 + R^2/2x^2)$ so $\sqrt{x^2 + R^2} - x \approx R^2/2x$

Then $V \approx \dfrac{\sigma}{2\epsilon_o}\dfrac{R^2}{2x} = \dfrac{\sigma\pi R^2}{4\pi\epsilon_0 x} = \dfrac{Q}{4\pi\epsilon_0 x}$, as expected.

b) $E_x = -\dfrac{\partial V}{\partial x} = -\dfrac{\sigma}{2\epsilon_0}\left(\dfrac{x}{\sqrt{x^2 + R^2}} - 1\right) = \dfrac{\sigma x}{2\epsilon_0}\left(\dfrac{1}{x} - \dfrac{1}{\sqrt{x^2 + R^2}}\right)$, which agrees with Eq.(22-11).

24-63 **a)** From Problem 23-36, $E(r) = \dfrac{\lambda r}{2\pi\epsilon_0 R^2}$ for $r \leq R$

(inside the cylindrical charge distribution) and

$E(r) = \dfrac{\lambda}{2\pi\epsilon_0 r}$ for $r \geq R$. Let $V = 0$ at $r = R$ (at the surface of the cylinder).

Use $V_a - V_b = \int_a^b \vec{E} \cdot d\vec{l}$.

<u>$r > R$</u>

Take point a to be at R and point b to be at r, where $r > R$. Let $d\vec{l} = d\vec{r}$. $\vec{E}$ and $d\vec{r}$ are both radially outward, so $\vec{E} \cdot d\vec{r} = E\,dr$. Thus $V_R - V_r = \int_R^r E\,dr$. Then $V_R = 0$ gives $V_r = -\int_R^r E\,dr$. In this interval $(r > R)$, $E(r) = \lambda/2\pi\epsilon_0 r$, so

$$V_r = -\int_R^r \frac{\lambda}{2\pi\epsilon_0 r}dr = -\frac{\lambda}{2\pi\epsilon_o}\int_R^r \frac{dr}{r} = -\frac{\lambda}{2\pi\epsilon_0}\ln\left(\frac{r}{R}\right).$$

Note: This expression gives $V_r = 0$ when $r = R$ and the potential decreases (becomes a negative number of larger magnitude) with increasing distance from the cylinder.

<u>$r \leq R$</u>

Take point a at r, where $r < R$, and point b at R. $\vec{E} \cdot d\vec{r} = E\,dr$ as before. Thus $V_r - V_R = \int_r^R E\,dr$. Then $V_R = 0$ then gives $V_r = \int_r^R E\,dr$. In this interval $(r < R)$, $E(r) = \lambda r/2\pi\epsilon_0 R^2$, so

$$V_r = \int_r^R \frac{\lambda r}{2\pi\epsilon_0 R^2}dr = \frac{\lambda}{2\pi\epsilon_o R^2}\int_r^R r\,dr = \frac{\lambda}{2\pi\epsilon_0 R^2}\left(\frac{R^2}{2} - \frac{r^2}{2}\right).$$

$$V_r = \frac{\lambda}{4\pi\epsilon_0}\left(1 - \left(\frac{r}{R}\right)^2\right).$$

Note: This expression also gives $V_r = 0$ when $r = R$. The potential is $\lambda/4\pi\epsilon_0$ at $r = 0$ and decreases with increasing r.

b)

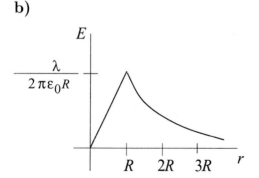
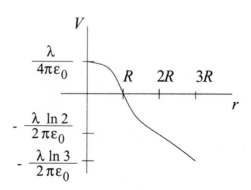

24-67 $\vec{E} = (-5.00 \text{ N/C} \cdot \text{m})x\hat{i} + (3.00 \text{ N/C} \cdot \text{m})z\hat{k}$

a) S_1 and S_3 are parallel to the xz-plane; S_1 is at $y_1 = 0$ and S_3 is at $y_3 = L$.

$$V_1 - V_3 = \int_1^3 \vec{E} \cdot d\vec{l}; \quad dl = dy\hat{j} \text{ so } \vec{E} \cdot d\vec{l} = 0 \text{ so } V_1 - V_3 = 0.$$

b) S_2 and S_4 are parallel to the xy-plane; S_4 is at $z_4 = 0$ and S_2 is at $z_2 = L$.

$$V_4 - V_2 = \int_4^2 \vec{E} \cdot d\vec{l}; \quad dl = dz\hat{k} \text{ so } \vec{E} \cdot d\vec{l} = (3.00 \text{ N/C} \cdot \text{m})z\, dz.$$

$$V_4 - V_2 = (3.00 \text{ N/C} \cdot \text{m}) \int_{z_4}^{z_2} z\, dz = (3.00 \text{ N/C} \cdot \text{m})(\frac{1}{2}z^2|_0^L) = (1.50 \text{ N/C} \cdot \text{m})L^2$$

$V_4 - V_2 = (1.50 \text{ N/C} \cdot \text{m})(0.300 \text{ m})^2 = 0.135 \text{ V}; V_4 - V_2$ is positive.

Thus S_4 is at higher potential. ($\vec{E}$ points from high potential toward low potential.)

c) S_5 and S_6 are parallel to the yz-plane; S_6 is at $x_6 = 0$ and S_5 is at $x_5 = L$.

$$V_6 - V_5 = \int_6^5 \vec{E} \cdot d\vec{l}; \quad dl = dx\hat{i} \text{ so } \vec{E} \cdot d\vec{l} = (-5.00 \text{ N/C} \cdot \text{m})x\, dx.$$

$$V_6 - V_5 = (-5.00 \text{ N/C·m}) \int_{x_6}^{x_5} x\, dx = (-5.00 \text{ N/C·m})(\frac{1}{2}x^2|_0^L) = (-2.50 \text{ N/C·m})L^2$$

$V_6 - V_5 = (-2.50 \text{ N/C} \cdot \text{m})(0.300 \text{ m})^2 = -0.225 \text{ V}; V_6 - V_5$ is negative so S_5 is at higher potential. ($\vec{E}$ points from high potential toward low potential.)

24-69 From Problem 23-25, for $R \leq r \leq 2R$ (between the sphere and the shell)
$E = Q/4\pi\epsilon_0 r^2$

Take a at R and b at $2R$. Then

$$V_{ab} = V_a - V_b = \int_R^{2R} E\, dr = \frac{Q}{4\pi\epsilon_0} \int_R^{2R} \frac{dr}{r^2} = \frac{Q}{4\pi\epsilon_0}\left[-\frac{1}{r}\right]_R^{2R} = \frac{Q}{4\pi\epsilon_0}\left(\frac{1}{R} - \frac{1}{2R}\right)$$

$$V_{ab} = \frac{Q}{8\pi\epsilon_0 R}$$

24-73 a)

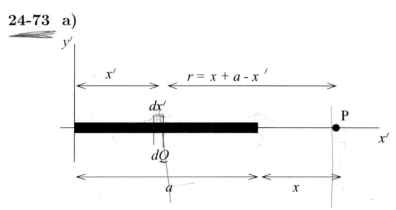

Use coordinates with the origin at the left-hand end of the rod and one axis along the rod. Call the axes x' and y' so as not to confuse them with the distance x given in the problem.

Slice the charged rod up into thin slices of width dx'. Each slice has charge $dQ = Q(dx'/a)$ and a distance $r = x + a - x'$ from point P. The potential at P due to the small slice dQ is

$$dV = \frac{1}{4\pi\epsilon_0}\left(\frac{dQ}{r}\right) = \frac{1}{4\pi\epsilon_0}\frac{Q}{a}\left(\frac{dx'}{x + a - x'}\right).$$

Compute the total V at P due to the entire rod by integrating dV over the length of the rod ($x' = 0$ to $x' = a$):

$$V = \int dV = \frac{Q}{4\pi\epsilon_0 a}\int_0^a \frac{dx'}{(x + a - x')} = \frac{Q}{4\pi\epsilon_0 a}\left[-\ln(x + a - x')\right]_0^a =$$

$$\frac{Q}{4\pi\epsilon_0 a}\ln\left(\frac{x + a}{x}\right).$$

Note: As $x \to \infty$, $V \to \dfrac{Q}{4\pi\epsilon_0 a}\ln\left(\dfrac{x}{x}\right) = 0.$

b)

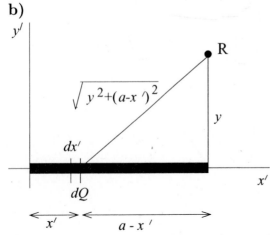

$dQ = (Q/a)dx'$ as in part (a)

Each slice dQ is a distance $r = \sqrt{y^2 + (a - x')^2}$ from point R.

The potential dV at R due to the small slice dQ is

$$dV = \frac{1}{4\pi\epsilon_0}\left(\frac{dQ}{r}\right) = \frac{1}{4\pi\epsilon_0}\frac{Q}{a}\frac{dx'}{\sqrt{y^2 + (a - x')^2}}.$$

$$V = \int dV = \frac{Q}{4\pi\epsilon_0 a}\int_0^a \frac{dx'}{\sqrt{y^2 + (a - x')^2}}.$$

In the integral make the change of variable $u = a - x'$; $du = -dx'$

$$V = -\frac{Q}{4\pi\epsilon_0 a}\int_a^0 \frac{du}{\sqrt{y^2 + u^2}} = -\frac{Q}{4\pi\epsilon_0 a}\left[\ln(u + \sqrt{y^2 + u^2})\right]_a^0$$

$$V = -\frac{Q}{4\pi\epsilon_0 a}[\ln y - \ln(a + \sqrt{y^2 + a^2})] = \frac{Q}{4\pi\epsilon_0 a}\left[\ln\left(\frac{a + \sqrt{a^2 + y^2}}{y}\right)\right].$$

(The expression for the integral was found in appendix B.)

Note: As $y \to \infty$, $V \to \frac{Q}{4\pi\epsilon_0 a}\ln\left(\frac{y}{y}\right) = 0.$

c) <u>part (a)</u>: $V = \frac{Q}{4\pi\epsilon_0 a}\ln\left(\frac{x + a}{x}\right) = \frac{Q}{4\pi\epsilon_0 a}\ln\left(1 + \frac{a}{x}\right).$

From Appendix B, $\ln(1 + u) = u - u^2/2 \ldots$, so $\ln(1 + a/x) = a/x - a^2/2x^2$ and this becomes a/x when x is large.

Thus $V \to \frac{Q}{4\pi\epsilon_0 a}\left(\frac{a}{x}\right) = \frac{Q}{4\pi\epsilon_0 x}$. For large x, V becomes the potential of a point charge.

<u>part (b)</u>: $V = \frac{Q}{4\pi\epsilon_0 a}\left[\ln\left(\frac{a + \sqrt{a^2 + y^2}}{y}\right)\right] = \frac{Q}{4\pi\epsilon_0 a}\ln\left(\frac{a}{y} + \sqrt{1 + \frac{a^2}{y^2}}\right).$

From the binomial theorem (Appendix B), $\sqrt{1 + a^2/y^2} = (1 + a^2/y^2)^{1/2} = 1 + a^2/2y^2 + \ldots$
Thus $a/y + \sqrt{1 + a^2/y^2} \to 1 + a/y + a^2/2y^2 + \ldots \to 1 + a/y$. And then using $\ln(1 + u) \approx u$ gives

$$V \to \frac{Q}{4\pi\epsilon_0 a}\ln(1 + a/y) \to \frac{Q}{4\pi\epsilon_0 a}\left(\frac{a}{y}\right) = \frac{Q}{4\pi\epsilon_0 y}.$$

For large y, V becomes the potential of a point charge.

24-75 **a)** The potential at the surface of a charged conducting sphere

is given by Example 24-8: $V = \frac{1}{4\pi\epsilon_0}\frac{q}{R}$. For spheres A and B this gives

$$V_A = \frac{Q_A}{4\pi\epsilon_0 R_A} \text{ and } V_B = \frac{Q_B}{4\pi\epsilon_0 R_B}.$$

$V_A = V_B$ gives $Q_A/4\pi\epsilon_0 R_A = Q_B/4\pi\epsilon_0 R_B$ and $Q_B/Q_A = R_B/R_A$.

And then $R_A = 3R_B$ implies $Q_B/Q_A = 1/3$.

b) The electric field at the surface of a charged conducting sphere is given in Example 23-5:

$$E = \frac{1}{4\pi\epsilon_0} \frac{|q|}{R^2}.$$

For spheres A and B this gives

$$E_A = \frac{|Q_A|}{4\pi\epsilon_0 R_A^2} \text{ and } E_B = \frac{|Q_B|}{4\pi\epsilon_0 R_B^2}.$$

$$\frac{E_B}{E_A} = \left(\frac{|Q_B|}{4\pi\epsilon_0 R_B^2}\right)\left(\frac{4\pi\epsilon_0 R_A^2}{|Q_A|}\right) = |Q_B/Q_A|(R_A/R_B)^2 = (1/3)(3)^2 = 3.$$

24-77 a) $E_1 = \dfrac{1}{4\pi\epsilon_0} \dfrac{Q_1}{R_1^2}, \quad V_1 = \dfrac{1}{4\pi\epsilon_0} \dfrac{Q_1}{R_1} = R_1 E_1$

b) Two conditions must be met:

1) Let q_1 and q_2 be the final charges on each sphere. Then $q_1 + q_2 = Q_1$ (charge conservation)

2) Let V_1 and V_2 be the final potentials of each sphere. All points of a conductor are at the same potential, so $V_1 = V_2$.

$V_1 = V_2$ requires that $\dfrac{1}{4\pi\epsilon_0} \dfrac{q_1}{R_1} = \dfrac{1}{4\pi\epsilon_0} \dfrac{q_2}{R_2}$ and then $q_1/R_1 = q_2/R_2$

$q_1 R_2 = q_2 R_1 = (Q_1 - q_1)R_1$

This gives $q_1 = (R_1/[R_1 + R_2])Q_1$ and $q_2 = Q_1 - q_1 = Q_1(1 - R_1/[R_1 + R_2]) = Q_1(R_2/[R_1 + R_2])$

c) $V_1 = \dfrac{1}{4\pi\epsilon_0} \dfrac{q_1}{R_1} = \dfrac{Q_1}{4\pi\epsilon_0(R_1 + R_2)}$ and

$V_2 = \dfrac{1}{4\pi\epsilon_0} \dfrac{q_2}{R_2} = \dfrac{Q_1}{4\pi\epsilon_0(R_1 + R_2)}$, which equals V_1 as it should.

d) $E_1 = \dfrac{V_1}{R_1} = \dfrac{Q_1}{4\pi\epsilon_0 R_1(R_1 + R_2)}$

$E_2 = \dfrac{V_2}{R_2} = \dfrac{Q_1}{4\pi\epsilon_0 R_2(R_1 + R_2)}$

CHAPTER 25
CAPACITANCE AND DIELECTRICS

Exercises 3, 5, 7, 9, 11, 17, 21, 25, 27, 29, 33, 35
Problems 37, 43, 45, 47, 49, 51, 53, 55

Exercises

25-3   **a)** $C = \dfrac{Q}{V_{ab}}$ so $V_{ab} = \dfrac{Q}{C} = \dfrac{0.148 \times 10^{-6} \text{ C}}{245 \times 10^{-12} \text{ F}} = 604 \text{ V}$

b) $C = \dfrac{\epsilon_0 A}{d}$ so

$$A = \frac{Cd}{\epsilon_0} = \frac{(245 \times 10^{-12} \text{ F})(0.328 \times 10^{-3} \text{ m})}{8.854 \times 10^{-12} \text{ C}^2/\text{N} \cdot \text{m}^2} = 9.08 \times 10^{-3} \text{ m}^2 = 90.8 \text{ cm}^2$$

c) $V_{ab} = Ed$ so $E = \dfrac{V_{ab}}{d} = \dfrac{604 \text{ V}}{0.328 \times 10^{-3} \text{ m}} = 1.84 \times 10^6 \text{ V/m}$

d) $E = \dfrac{\sigma}{\epsilon_0}$ so

$$\sigma = E\epsilon_0 = (1.84 \times 10^6 \text{ V/m})(8.854 \times 10^{-12} \text{ C}^2/\text{N} \cdot \text{m}^2) = 1.63 \times 10^{-5} \text{ C/m}^2$$

or

$$\sigma = \frac{Q}{A} = \frac{0.148 \times 10^{-6} \text{ C}}{9.08 \times 10^{-3} \text{ m}^2} = 1.63 \times 10^{-5} \text{ C/m}^2, \text{ which checks.}$$

25-5   **a)** From Example 25-4, $\dfrac{C}{L} = \dfrac{2\pi\epsilon_0}{\ln(r_b/r_a)}$

$$\frac{C}{L} = \frac{2\pi(8.854 \times 10^{-12} \text{ C}^2/\text{N} \cdot \text{m}^2)}{\ln(3.5 \text{ mm}/1.5 \text{ mm})} = 6.57 \times 10^{-11} \text{ F/m} = 66 \text{ pF/m}$$

b) $C = (6.57 \times 10^{-11} \text{ F/m})(2.8 \text{ m}) = 1.84 \times 10^{-10} \text{ F}.$

$Q = CV = (1.84 \times 10^{-10} \text{ F})(350 \times 10^{-3} \text{ V}) = 6.4 \times 10^{-11} \text{ C} = 64 \text{ pC}$

The conductor at higher potential has the positive charge, so there is +64 pC on the inner conductor and −64 pC on the outer conductor.

25-7   **a)** From Example 25-3, $C = 4\pi\epsilon_0 \dfrac{r_a r_b}{r_b - r_a}$

$$C = 4\pi(8.854 \times 10^{-12} \text{ C}^2/\text{N} \cdot \text{m}^2) \left(\frac{(0.148 \text{ m})(0.125 \text{ m})}{0.148 \text{ m} - 0.125 \text{ m}} \right) = 8.95 \times 10^{-11} \text{ F} =$$
89.5 pF

b) From Example 25-3, $E = Q/4\pi\epsilon_0 r^2$ between the spheres.
$C = Q/V$ so $Q = CV = (89.5 \times 10^{-12} \text{ F})(120 \text{ V}) = 1.074 \times 10^{-8} \text{ C}$
$E = Q/4\pi\epsilon_0 r^2 = (8.988 \times 10^9 \text{ N} \cdot \text{m}^2/\text{C}^2)(1.074 \times 10^{-8} \text{ C})/r^2 = (96.53 \text{ N} \cdot \text{m}^2/\text{C})/r^2$

For $r = 0.126$ m, $E = \dfrac{96.53 \text{ N} \cdot \text{m}^2/\text{C}}{(0.126 \text{ m})^2} = 6.08 \times 10^3 \text{ V/m}$

c) For $r = 0.147$ m, $E = \dfrac{96.53 \text{ N} \cdot \text{m}^2/\text{C}}{(0.147 \text{ m})^2} = 4.47 \times 10^3 \text{ V/m}$

d) No, the results of parts (b) and (c) show that E is not uniform in the region between the spheres.

25-9   Do parts (a) and (b) together:

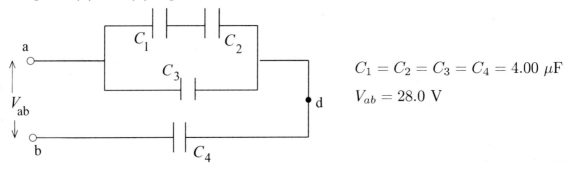

$C_1 = C_2 = C_3 = C_4 = 4.00 \ \mu\text{F}$

$V_{ab} = 28.0$ V

Simplify the circuit by replacing the capacitor combinations by their equivalents:

$$\frac{1}{C_{12}} = \frac{1}{C_1} + \frac{1}{C_2}$$

$$C_{12} = \frac{C_1 C_2}{C_1 + C_2} = \frac{(4.00 \times 10^{-6} \text{ F})(4.00 \times 10^{-6} \text{ F})}{4.00 \times 10^{-6} \text{ F} + 4.00 \times 10^{-6} \text{ F}} = 2.00 \times 10^{-6} \text{ F}$$

$C_{123} = C_{12} + C_3$
$C_{123} = 2.00 \times 10^{-6} \text{ F} + 4.00 \times 10^{-6} \text{ F}$
$C_{123} = 6.00 \times 10^{-6} \text{ F}$

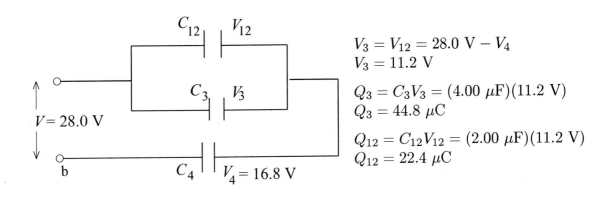

$$\frac{1}{C_{1234}} = \frac{1}{C_{123}} + \frac{1}{C_4}$$

$$C_{1234} = \frac{C_{123}C_4}{C_{123} + C_4} = \frac{(6.00 \times 10^{-6}\ \text{F})(4.00 \times 10^{-6}\ \text{F})}{6.00 \times 10^{-6}\ \text{F} + 4.00 \times 10^{-6}\ \text{F}} = 2.40 \times 10^{-6}\ \text{F}$$

The circuit is equivalent to

$$V_{1234} = V = 28.0\ \text{V}$$

$$Q_{1234} = C_{1234}V = (2.40 \times 10^{-6}\ \text{F})(28.0\ \text{V}) = 67.2\ \mu\text{C}$$

Now build back up the original circuit, step by step:

$$Q_{123} = Q_4 = Q_{1234} = 67.2\ \mu\text{C}$$
(charge same for capacitors in series)

Then $V_{123} = \dfrac{Q_{123}}{C_{123}} = \dfrac{67.2\ \mu\text{C}}{6.00\ \mu\text{F}} = 11.2\ \text{V}$

$V_4 = \dfrac{Q_4}{C_4} = \dfrac{67.2\ \mu\text{C}}{4.00\ \mu\text{F}} = 16.8\ \text{V}$

Note that $V_4 + V_{123} = 16.8\ \text{V} + 11.2\ \text{V} = 28.0\ \text{V}$, as it should.

$$V_3 = V_{12} = 28.0\ \text{V} - V_4$$
$$V_3 = 11.2\ \text{V}$$

$$Q_3 = C_3 V_3 = (4.00\ \mu\text{F})(11.2\ \text{V})$$
$$Q_3 = 44.8\ \mu\text{C}$$

$$Q_{12} = C_{12}V_{12} = (2.00\ \mu\text{F})(11.2\ \text{V})$$
$$Q_{12} = 22.4\ \mu\text{C}$$

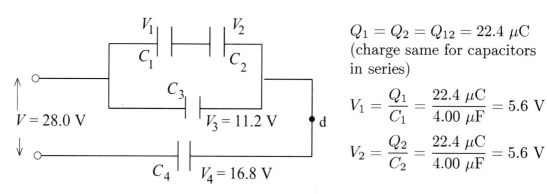

$Q_1 = Q_2 = Q_{12} = 22.4 \ \mu C$
(charge same for capacitors in series)

$$V_1 = \frac{Q_1}{C_1} = \frac{22.4 \ \mu C}{4.00 \ \mu F} = 5.6 \ V$$

$$V_2 = \frac{Q_2}{C_2} = \frac{22.4 \ \mu C}{4.00 \ \mu F} = 5.6 \ V$$

Note that $V_1 + V_2 = 11.2$ V, which equals V_3 as it should.

Summary:

$Q_1 = 22.4 \ \mu C, \quad V_1 = 5.6 \ V$

$Q_2 = 22.4 \ \mu C, \quad V_2 = 5.6 \ V$

$Q_3 = 44.8 \ \mu C, \quad V_3 = 11.2 \ V$

$Q_4 = 67.2 \ \mu C, \quad V_4 = 16.8 \ V$

c) $V_{ad} = V_3 = 11.2$ V

25-11 Do parts (a) and (b) together.

$V_1 = V_2 = V$
$V_1 = 52.0$ V
$V_2 = 52.0$ V

$C = Q/V$ so $Q = CV$

$Q_1 = C_1 V_1 = (3.00 \ \mu F)(52.0 \ V) = 156 \ \mu C$

$Q_2 = C_2 V_2 = (5.00 \ \mu F)(52.0 \ V) = 260 \ \mu C$

25-17 The energy density is given by Eq.(25-11): $u = \frac{1}{2}\epsilon_0 E^2$

Calculate E: $E = \dfrac{V}{d} = \dfrac{400 \ V}{5.00 \times 10^{-3} \ m} = 8.00 \times 10^4 \ V/m.$

Then $u = \frac{1}{2}\epsilon_0 E^2 = \frac{1}{2}(8.854 \times 10^{-12} \ C^2/N \cdot m^2)(8.00 \times 10^4 \ V/m)^2 = 0.0283 \ J/m^3$

25-21 **a)** $U = \dfrac{Q^2}{2C}; \quad C = \dfrac{\epsilon_0 A}{x} \quad \text{so} \quad U = \dfrac{xQ^2}{2\epsilon_0 A}$

b) $x \to x + dx$ gives $U = \dfrac{(x + dx)Q^2}{2\epsilon_0 A}$

$$dU = \frac{(x+dx)Q^2}{2\epsilon_0 A} - \frac{xQ^2}{2\epsilon_0 a} = \left(\frac{Q^2}{2\epsilon_0 A}\right) dx$$

c) $dW = F\,dx = dU$, so $F = \dfrac{Q^2}{2\epsilon_0 A}$

d)

$E = \dfrac{\sigma}{\epsilon_0} = \dfrac{Q}{\epsilon_0 A}$

$F = \frac{1}{2}QE$, not QE

The reason for the difference is that E is the field due to <u>both</u> plates. If we consider the positive plate only and calculate its electric field using Gauss's law:

$$\oint \vec{E} \cdot d\vec{A} = \frac{Q_{encl}}{\epsilon_0}$$

$$2EA = \frac{\sigma A}{\epsilon_0}$$

$$E = \frac{\sigma}{2\epsilon_0} = \frac{Q}{2\epsilon_0 A}$$

The force this field exerts on the other plate, that has charge $-Q$, is $F = \dfrac{Q^2}{2\epsilon_0 A}$.

25-25 **a)** From Exercise 25-7, $E = 6.08 \times 10^3$ V/m at $r = 12.6$ cm.

$u = \frac{1}{2}\epsilon_0 E^2 = \frac{1}{2}(8.854 \times 10^{-12}\ \mathrm{C^2/N \cdot m^2})(6.08 \times 10^3\ \mathrm{V/m})^2 = 1.64 \times 10^{-4}\ \mathrm{J/m^3}$

b) From Exercise 25-7, $E = 4.47 \times 10^3$ V/m at $r = 14.7$ cm.

$u = \frac{1}{2}\epsilon_0 E^2 = \frac{1}{2}(8.854 \times 10^{-12}\ \mathrm{C^2/N \cdot m^2})(4.47 \times 10^3\ \mathrm{V/m})^2 = 8.85 \times 10^{-5}\ \mathrm{J/m^3}$

c) No, the results of parts (a) and (b) show that the energy density is not uniform in the region between the spheres.

25-27 $E = \dfrac{E_0}{K}$ so $K = \dfrac{E_0}{E} = \dfrac{3.20 \times 10^5\ \mathrm{V/m}}{2.50 \times 10^5\ \mathrm{V/m}} = 1.28$

a) $\sigma_i = \sigma(1 - 1/K)$

$\sigma = \epsilon_0 E_0 = (8.854 \times 10^{-12}\ \mathrm{C^2/N \cdot m^2})(3.20 \times 10^5\ \mathrm{N/C}) = 2.833 \times 10^{-6}\ \mathrm{C/m^2}$

$\sigma_i = (2.833 \times 10^{-6}\ \mathrm{C/m^2})(1 - 1/1.28) = 6.20 \times 10^{-7}\ \mathrm{C/m^2}$

b) As calculated above, $K = 1.28$.

25-29 $C = K\epsilon_0 A/d$. Minimum A means smallest possible d.

d is limited by the requirement that E be less than 1.60×10^7 V/m when V is as large as 5500 V.

$$V = Ed \text{ so } d = \frac{V}{E} = \frac{5500 \text{ V}}{1.60 \times 10^7 \text{ V/m}} = 3.44 \times 10^{-4} \text{ m}$$

$$\text{Then } A = \frac{Cd}{K\epsilon_0} = \frac{(1.25 \times 10^{-9} \text{ F})(3.44 \times 10^{-4} \text{ m})}{(3.60)(8.854 \times 10^{-12} \text{ C}^2/\text{N} \cdot \text{m}^2)} = 0.0135 \text{ m}^2.$$

25-33 a) Since the capacitor remains connected to the power supply the potential difference doesn't change when the dielectric is inserted. Before the dielectric is inserted $U_0 = \frac{1}{2}C_0V^2$ so

$$V = \sqrt{\frac{2U_0}{C_0}} = \sqrt{\frac{2(1.85 \times 10^{-5} \text{ J})}{360 \times 10^{-9} \text{ F}}} = 10.1 \text{ V}$$

b) $K = C/C_0$
$U_0 = \frac{1}{2}C_0V^2$, $U = \frac{1}{2}CV^2$ so $C/C_0 = U/U_0$

$$K = \frac{U}{U_0} = \frac{1.85 \times 10^{-5} \text{ J} + 2.32 \times 10^{-5} \text{ J}}{1.85 \times 10^{-5} \text{ J}} = 2.25$$

25-35 a) Apply Eq.(25-22) to the dashed surface:

$$\oint K\vec{E} \cdot d\vec{A} = \frac{Q_{encl-free}}{\epsilon_0}$$

$$\oint K\vec{E} \cdot d\vec{A} = KEA'$$
since $E = 0$ outside the plates

$$Q_{encl-free} = \sigma A' = (Q/A)A'$$

Thus $KEA' = \dfrac{(Q/A)A'}{\epsilon_0}$ and $E = \dfrac{Q}{\epsilon_0 AK}$

b) $V = Ed = \dfrac{Qd}{\epsilon_0 AK}$

c) $C = \dfrac{Q}{V} = \dfrac{Q}{(Qd/\epsilon_0 AK)} = K\dfrac{\epsilon_0 A}{d} = KC_0$, so $K = C/C_0$, which is Eq.(25-12).

Problems

25-37 a) $C = \dfrac{\epsilon_0 A}{d}$

$$C = \frac{(8.854 \times 10^{-12} \text{ C}^2/\text{N} \cdot \text{m}^2)(0.16 \text{ m})^2}{9.4 \times 10^{-3} \text{ m}} = 2.4 \times 10^{-11} \text{ F} = 24 \text{ pF}$$

b) Remains connected to the battery says that V stays 12 V.

$Q = CV = (2.4 \times 10^{-11} \text{ F})(12 \text{ V}) = 2.9 \times 10^{-10} \text{ C}$

c) $E = \dfrac{V}{d} = \dfrac{12 \text{ V}}{9.4 \times 10^{-3} \text{ m}} = 1.3 \times 10^3 \text{ V/m}$

d) $U = \frac{1}{2}QV = \frac{1}{2}(2.9 \times 10^{-10} \text{ C})(12.0 \text{ V}) = 1.7 \times 10^{-9} \text{ J}$

25-43 a)

$V_1 = V_2 = 660 \text{ V}$

$Q_1 = C_1 V_1 = (4.00 \times 10^{-6} \text{ F})(660 \text{ V})$

$Q_1 = 2.64 \times 10^{-3} \text{ C}$

$Q_2 = C_2 V_2 = (6.00 \times 10^{-6} \text{ F})(660 \text{ V}) = 3.96 \times 10^{-3} \text{ C}$

b)

$q_1 + q_2 = Q_2 - Q_1$ (since terminals of unlike sign are connected)

$q_1 + q_2 = 3.96 \times 10^{-3} \text{ C} - 2.64 \times 10^{-3} \text{ C} = 1.32 \times 10^{-3} \text{ C}$

From the circuit diagram we can see that $v_1 = v_2$.

$v_1 = v_2$ says that $q_1/C_1 = q_2/C_2$ and $q_2 = (C_2/C_1)q_1 = (6.00 \ \mu\text{F}/4.00 \ \mu\text{F})q_1 = 1.50 q_1$

Use this result in the equation $q_1 + q_2 = 1.32 \times 10^{-3}$ C and get $2.50 q_1 = 1.32 \times 10^{-3}$ C

$q_1 = 5.28 \times 10^{-4} \text{ C}, \ q_2 = 7.92 \times 10^{-4} \text{ C}$

$v_1 = \dfrac{q_1}{C_1} = \dfrac{5.28 \times 10^{-4} \text{ C}}{4.00 \times 10^{-6} \text{ F}} = 132 \text{ V} \qquad v_2 = \dfrac{q_2}{C_2} = \dfrac{7.92 \times 10^{-4} \text{ C}}{6.00 \times 10^{-6} \text{ F}} = 132 \text{ V}$

So we do have $v_1 = v_2$.

25-45 a)

$C_1 = C_5 = 8.4 \ \mu\text{F}$

$C_2 = C_3 = C_4 = 4.2 \ \mu\text{F}$

Simplify the circuit by replacing the capacitor combinations by their equivalents:

$$\frac{1}{C_{34}} = \frac{1}{C_3} + \frac{1}{C_4}$$

$$\frac{1}{C_{34}} = \frac{C_3 + C_4}{C_3 C_4}$$

$$C_{34} = \frac{C_3 C_4}{C_3 + C_4} = \frac{(4.2 \ \mu\text{F})(4.2 \ \mu\text{F})}{4.2 \ \mu\text{F} + 4.2 \ \mu\text{F}} = 2.1 \ \mu\text{F}$$

$$C_{234} = C_2 + C_{34}$$

$$C_{234} = 4.2 \ \mu\text{F} + 2.1 \ \mu\text{F}$$

$$C_{234} = 6.3 \ \mu\text{F}$$

$$\frac{1}{C_{eq}} = \frac{1}{C_1} + \frac{1}{C_5} + \frac{1}{C_{234}}$$

$$\frac{1}{C_{eq}} = \frac{2}{8.4 \ \mu\text{F}} + \frac{1}{6.3 \ \mu\text{F}}$$

$$C_{eq} = 2.5 \ \mu\text{F}$$

b) Now build the original circuit back up, piece by piece:

$$Q_{eq} = C_{eq} V$$

$$Q_{eq} = (2.5 \ \mu\text{F})(220 \ \text{V}) = 550 \ \mu\text{C}$$

$Q_1 = Q_5 = Q_{234} = 550 \ \mu\text{C}$ (capacitors in series have same charge)

$$V_1 = \frac{Q_1}{C_1} = \frac{550 \ \mu\text{C}}{8.4 \ \mu\text{F}} = 65 \ \text{V}$$

$$V_5 = \frac{Q_5}{C_5} = \frac{550 \ \mu\text{C}}{8.4 \ \mu\text{F}} = 65 \ \text{V}$$

$$V_{234} = \frac{Q_{234}}{C_{234}} = \frac{550 \ \mu\text{C}}{6.3 \ \mu\text{F}} = 87 \ \text{V}$$

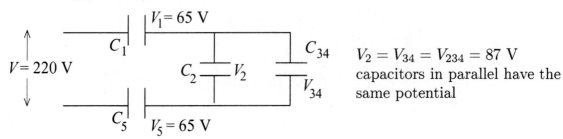

$V_2 = V_{34} = V_{234} = 87 \ \text{V}$
capacitors in parallel have the same potential

$Q_2 = C_2 V_2 = (4.2 \ \mu F)(87 \ V) = 370 \ \mu C$

$Q_{34} = C_{34} V_{34} = (2.1 \ \mu F)(87 \ V) = 180 \ \mu C$

$Q_3 = Q_4 = Q_{34} = 180 \ \mu C$

capacitors in series have the same charge

$V_3 = \dfrac{Q_3}{C_3} = \dfrac{180 \ \mu C}{4.2 \ \mu F} = 43 \ V$

$V_4 = \dfrac{Q_4}{C_4} = \dfrac{180 \ \mu C}{4.2 \ \mu F} = 43 \ V$

Summary:

$Q_1 = 550 \ \mu C, \quad V_1 = 65 \ V$

$Q_2 = 370 \ \mu C, \quad V_2 = 87 \ V$

$Q_3 = 180 \ \mu C, \quad V_3 = 43 \ V$

$Q_4 = 180 \ \mu C, \quad V_4 = 43 \ V$

$Q_5 = 550 \ \mu C, \quad V_5 = 65 \ V$

Note: $V_3 + V_4 = V_2$ and $V_1 + V_2 + V_5 = 220 \ V$ (apart from some small rounding error)

$Q_1 = Q_2 + Q_3$ and $Q_5 = Q_2 + Q_4$

25-47 a)

$C_3 = C_1/2$ so $\dfrac{1}{C_{eq}} = \dfrac{1}{C_1} + \dfrac{1}{C_2} + \dfrac{1}{C_3} = \dfrac{4}{8.4 \ \mu F}$ and $C_{eq} = 8.4 \ \mu F/4 = 2.1 \ \mu F$

$Q = C_{eq} V = (2.1 \ \mu F)(36 \ V) = 76 \ \mu C$

The three capacitors are in series so they each have the same charge:

$Q_1 = Q_2 = Q_3 = 76 \ \mu C$

b) $U = \frac{1}{2}Q_1 V_1 + \frac{1}{2}Q_2 V_2 + \frac{1}{2}Q_3 V_3$

But $Q_1 = Q_2 = Q_3 = Q$ so $U = \frac{1}{2}Q(V_1 + V_2 + V_3)$

But also $V_1 + V_2 + V_3 = V = 36$ V, so $U = \frac{1}{2}QV = \frac{1}{2}(76 \ \mu C)(36 \text{ V}) = 1.4 \times 10^{-3}$ J.

c)

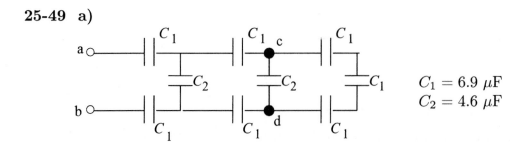

The total positive charge that is available to be distributed on the upper plates of the three capacitors is $Q_0 = Q_{01} + Q_{02} + Q_{03} = 3(76 \ \mu C) = 228 \ \mu C$.

Thus $Q_1 + Q_2 + Q_3 = 228 \ \mu C$.

After the circuit is completed the charge distributes to make $V_1 = V_2 = V_3$.

$V = Q/C$ and $V_1 = V_2$ so $Q_1/C_1 = Q_2/C_2$ and then $C_1 = C_2$ says $Q_1 = Q_2$.

$V_1 = V_3$ says $Q_1/C_1 = Q_3/C_3$ and $Q_1 = Q_3(C_1/C_3) = Q_3(8.4 \ \mu F/4.2 \ \mu F) = 2Q_3$

Using $Q_2 = Q_1$ and $Q_1 = 2Q_3$ in the above equation gives $2Q_3 + 2Q_3 + Q_3 = 228 \ \mu C$.

$5Q_3 = 228 \ \mu C$ and $Q_3 = 45.6 \ \mu C$, $Q_1 = Q_2 = 91.2 \ \mu C$

Then $V_1 = \dfrac{Q_1}{C_1} = \dfrac{91.2 \ \mu C}{8.4 \ \mu F} = 11$ V, $V_2 = \dfrac{Q_2}{C_2} = \dfrac{91.2 \ \mu C}{8.4 \ \mu F} = 11$ V, and

$V_3 = \dfrac{Q_3}{C_3} = \dfrac{45.6 \ \mu C}{4.2 \ \mu F} = 11$ V.

The voltage across each capacitor in the parallel combination is 11 V.

d) $U = \frac{1}{2}Q_1 V_1 + \frac{1}{2}Q_2 V_2 + \frac{1}{2}Q_3 V_3$.

But $V_1 = V_2 = V_3$ so $U = \frac{1}{2}V_1(Q_1 + Q_2 + Q_3) = \frac{1}{2}(11 \text{ V})(228 \ \mu C) = 1.3 \times 10^{-3}$ J.

This is less than the original energy of 1.4×10^{-3} J.

25-49 a)

$C_1 = 6.9 \ \mu F$
$C_2 = 4.6 \ \mu F$

Simplify the network by replacing the capacitor combinations by their equivalents:

$$\frac{1}{C_{eq}} = \frac{3}{C_1}$$

$$C_{eq} = \frac{C_1}{3} = \frac{6.9\ \mu F}{3} = 2.3\ \mu F$$

$$C_{eq} = 2.3\ \mu F + C_2$$
$$C_{eq} = 2.3\ \mu F + 4.6\ \mu F = 6.9\ \mu F$$

$$\frac{1}{C_{eq}} = \frac{2}{C_1} + \frac{1}{6.9\ \mu F} = \frac{3}{6.9\ \mu F}$$

$$C_{eq} = 2.3\ \mu F$$

$$C_{eq} = C_2 + 2.3\ \mu F = 4.6\ \mu F + 2.3\ \mu F$$
$$C_{eq} = 6.9\ \mu F$$

$$\frac{1}{C_{eq}} = \frac{2}{C_1} + \frac{1}{6.9\ \mu F} = \frac{3}{6.9\ \mu F}$$

$$C_{eq} = 2.3\ \mu F$$

b)

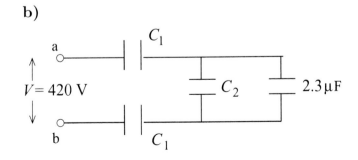

From part (a) 2.3 μF is the equivalent capacitance of the rest of the network.

The equivalent network is

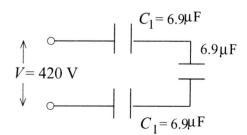

series, so all three capacitors have the same Q.

But here all three have the same C, so by $V = Q/C$ all three must have the same V. The three voltages must add to 420 V, so each capacitor has $V = 140$ V. The 6.9 μF to the right is the equivalent of C_2 and the 2.3 μF capacitor in parallel, so $V_2 = 140$ V. (Capacitors in parallel have the same potential difference.)

Hence $Q_1 = C_1 V_1 = (6.9 \ \mu\text{F})(140 \ \text{V}) = 9.7 \times 10^{-4}$ C

and $Q_2 = C_2 V_2 = (4.6 \ \mu\text{F})(140 \ \text{V}) = 6.4 \times 10^{-4}$ C.

c) From the potentials deduced in part (b) we have

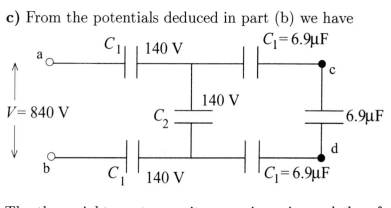

From part (a) 9.3 μF is the equivalent capacitance of the rest of the network.

The three right-most capacitors are in series and therefore have the same charge. But their capacitances are also equal, so by $V = Q/C$ they each have the same potential difference. Their potentials must sum to 140 V, so the potential across each is 47 V and $V_{cd} = 47$ V.

25-51 **a)** The conductor can be at some potential V, where $V = 0$ far from the conductor. This potential depends on the charge Q on the conductor so we can define $C = Q/V$ where C will not depend on V or Q.

b) For any point on a solid conducting sphere $V = Q/4\pi\epsilon_0 R$ if $V = 0$ at $r \to \infty$.

$$C = \frac{Q}{V} = Q\left(\frac{4\pi\epsilon_0 R}{Q}\right) = 4\pi\epsilon_0 R$$

c) $C = 4\pi\epsilon_0 R = 4\pi(8.854 \times 10^{-12} \ \text{F/m})(6.38 \times 10^6 \ \text{m}) = 7.10 \times 10^{-4} \ \text{F} = 710 \ \mu\text{F}$. The capacitance of the earth is about seven times larger than the largest capacitances in this range.

25-53 From Example 23-9, $E = \dfrac{1}{4\pi\epsilon_0}\dfrac{Qr}{R^3}$ for $r < R$

and $E = \dfrac{1}{4\pi\epsilon_0}\dfrac{Q}{r^2}$ for $r > R$.

a) $u = \frac{1}{2}\epsilon_0 E^2 = \frac{1}{2}\epsilon_0\left(\dfrac{1}{4\pi\epsilon_0}\dfrac{Qr}{R^3}\right)^2 = \dfrac{1}{4\pi\epsilon_0}\dfrac{Q^2 r^2}{8\pi R^6}$

b) $u = \frac{1}{2}\epsilon_0 E^2 = \frac{1}{2}\epsilon_0\left(\dfrac{1}{4\pi\epsilon_0}\dfrac{Q}{r^2}\right)^2 = \dfrac{1}{4\pi\epsilon_0}\dfrac{Q^2}{8\pi r^4}$

c) $dU = u\,dV$ where $dV = 4\pi r^2\,dr$ is the volume of a thin spherical shell of radius r and thickness dr.

For $r < R$, $U = \displaystyle\int_0^R u 4\pi r^2\,dr = \dfrac{1}{4\pi\epsilon_0}\dfrac{Q^2}{2R^6}\int_0^R r^4\,dr = \dfrac{1}{4\pi\epsilon_0}\dfrac{Q^2}{2R^6}\left(\dfrac{R^5}{5}\right) = \dfrac{1}{4\pi\epsilon_0}\dfrac{Q^2}{10R}$.

For $r > R$, $U = \displaystyle\int_R^\infty u 4\pi r^2\,dr = \dfrac{1}{4\pi\epsilon_0}\dfrac{Q^2}{2}\int_R^\infty \dfrac{dr}{r^2} = \dfrac{1}{4\pi\epsilon_0}\dfrac{Q^2}{2}\left(-\dfrac{1}{r}\Big|_R^\infty\right) = \dfrac{1}{4\pi\epsilon_0}\dfrac{Q^2}{2R}$.

The total electric field energy is

$$U = \dfrac{1}{4\pi\epsilon_0}\dfrac{Q^2}{10R} + \dfrac{1}{4\pi\epsilon_0}\dfrac{Q^2}{2R} = \dfrac{1}{4\pi\epsilon_0}\dfrac{Q^2}{R}\left(\dfrac{1}{10}+\dfrac{1}{2}\right) = \dfrac{1}{4\pi\epsilon_0}\dfrac{3Q^2}{5R}.$$

25-55 $C = Q/V$, so we need to calculate the effect of the dielectrics on the potential difference between the plates. Let the potential of the positive plate be V_a, the potential of the negative plate be V_c, and the potential midway between the plates where the dielectrics meet be V_b.

$$C = \dfrac{Q}{V_a - V_c} = \dfrac{Q}{V_{ac}}.$$

$$V_{ac} = V_{ab} + V_{bc}.$$

The electric field in the absence of any dielectric is $E_0 = \dfrac{Q}{\epsilon_0 A}$. In the first dielectric the electric field is reduced to

$$E_1 = \dfrac{E_0}{K_1} = \dfrac{Q}{K_1\epsilon_0 A}\quad\text{and}\quad V_{ab} = E_1\left(\dfrac{d}{2}\right) = \dfrac{Qd}{K_1 2\epsilon_0 A}.$$

In the second dielectric the electric field is reduced to

$$E_2 = \frac{E_0}{K_2} = \frac{Q}{K_2 \epsilon_0 A} \text{ and } V_{bc} = E_2 \left(\frac{d}{2}\right) = \frac{Qd}{K_2 2\epsilon_0 A}.$$

Thus $V_{ac} = V_{ab} + V_{bc} = \dfrac{Qd}{K_1 2\epsilon_0 A} + \dfrac{Qd}{K_2 2\epsilon_0 A} = \dfrac{Qd}{2\epsilon_0 A}\left(\dfrac{1}{K_1} + \dfrac{1}{K_2}\right)$

$$V_{ac} = \frac{Qd}{2\epsilon_0 A}\left(\frac{K_1 + K_2}{K_1 K_2}\right).$$

This gives $C = \dfrac{Q}{V_{ac}} = Q\left(\dfrac{2\epsilon_0 A}{Qd}\right)\left(\dfrac{K_1 K_2}{K_1 + K_2}\right) = \dfrac{2\epsilon_0 A}{d}\left(\dfrac{K_1 K_2}{K_1 + K_2}\right).$

CHAPTER 26
CURRENT, RESISTANCE, AND ELECTROMOTIVE FORCE

Exercises 3, 5, 11, 13, 15, 17, 23, 25, 29, 33, 35, 37, 41
Problems 43, 47, 49, 53, 55, 57, 61, 63

Exercises

26-3   **a)** Calculate the drift speed v_d:

$$J = \frac{I}{A} = \frac{I}{\pi r^2} = \frac{4.85 \text{ A}}{\pi (1.025 \text{ m})^2} = 1.469 \times 10^6 \text{ A/m}^2$$

$$v_d = \frac{J}{n|q|} = \frac{1.469 \times 10^6 \text{ A/m}^2}{(8.5 \times 10^{28}/\text{m}^3)(1.602 \times 10^{-19} \text{ C})} = 1.079 \times 10^{-4} \text{ m/s}$$

$$t = \frac{L}{v_d} = \frac{0.710 \text{ m}}{1.079 \times 10^{-4} \text{ m/s}} = 6.58 \times 10^3 \text{ s} = 110 \text{ min.}$$

b) $v_d = \dfrac{I}{\pi r^2 n |q|}$

$$t = \frac{L}{v_d} = \frac{\pi r^2 n |q| L}{I}$$

t is proportional to r^2 and hence to d^2 where $d = 2r$ is the wire diameter.

$$t = (6.58 \times 10^3 \text{ s}) \left(\frac{4.12 \text{ mm}}{2.05 \text{ mm}} \right)^2 = 2.66 \times 10^4 \text{ s} = 440 \text{ min.}$$

c) The drift speed is proportional to the current density and therefore it is inversely proportional to the square of the diameter of the wire. Increasing the diameter by some factor decreases the drift speed by the square of that factor.

26-5   **a)** $dQ = I \, dt$

$$Q = \int_0^{8.0 \text{ s}} (55 \text{ A} - (0.65 \text{ A/s}^2)t^2) \, dt = \left[(55 \text{ A})t - (0.217 \text{ A/s}^2)t^3 \right]_0^{8.0 \text{ s}}$$

$$Q = (55 \text{ A})(8.0 \text{ s}) - (0.217 \text{ A/s}^2)(8.0 \text{ s})^3 = 330 \text{ C}$$

b) $I = \dfrac{Q}{t} = \dfrac{330 \text{ C}}{8.0 \text{ s}} = 41 \text{ A}$

26-11 $E = \dfrac{V}{L} = \rho J$, so $\dfrac{V}{L} = \rho \left(\dfrac{I}{\pi r^2} \right)$

where $\rho = 2.75 \times 10^{-8} \ \Omega \cdot \text{m}$ (Table 26-1).

$$r = \sqrt{\frac{\rho I L}{\pi V}} = \sqrt{\frac{(2.75 \times 10^{-8} \ \Omega \cdot m)(6.00 \ A)(1.20 \ m)}{\pi (1.50 \ V)}} = 0.205 \ mm$$

26-13 a) Eq.(26.5): $\rho = E/J$ so $J = E/\rho$

From Table 26-1 the resistivity for gold is $2.44 \times 10^{-8} \ \Omega \cdot m$.

$$J = \frac{E}{\rho} = \frac{0.49 \ V/m}{2.44 \times 10^{-8} \ \Omega \cdot m} = 2.008 \times 10^{-7} \ A/m^2$$

$$I = JA = J\pi r^2 = (2.008 \times 10^7 \ A/m^2)\pi(0.41 \times 10^{-3} \ m)^2 = 11 \ A$$

b) $V = EL = (0.40 \ V/m)(6.4 \ m) = 3.1 \ V$

c) We can use Ohm's law (Eq.(26-11): $V = IR$.

$$R = \frac{V}{I} = \frac{3.1 \ V}{11 \ A} = 0.28 \ \Omega$$

Or, we can calculate R from the resistivity and the dimensions of the wire (Eq.(26-10):

$$R = \frac{\rho L}{A} = \frac{\rho L}{\pi r^2} = \frac{(2.44 \times 10^{-8} \ \Omega \cdot m)(6.4 \ m)}{\pi (0.42 \times 10^{-3} \ m)^2} = 0.28 \ \Omega, \text{ which checks.}$$

26-15 a) $E = \dfrac{V}{L} = \dfrac{0.938 \ V}{0.750 \ m} = 1.25 \ V/m$

b) $E = \rho J$ so $\rho = \dfrac{E}{J} = \dfrac{1.25 \ V/m}{4.40 \times 10^7 \ A/m^2} = 2.84 \times 10^{-8} \ \Omega \cdot m$

26-17 $R = R_0[1 + \alpha(T - T_0)]$

a) $\alpha = 0.0004 \ (C^\circ)^{-1}$ (Table 26-1)

Let T_0 be $0.0^\circ C$ and T be $11.5^\circ C$.

$$R_0 = \frac{R}{1 + \alpha(T - T_0)} = \frac{100.0 \ \Omega}{1 + (0.0004 \ (C^\circ)^{-1}(11.5^\circ C))} = 99.54 \ \Omega$$

b) $\alpha = -0.0005 \ (C^\circ)^{-1}$ (Table 26-2)

$R = R_0[1 + \alpha(T - T_0)] = 0.0160 \ \Omega[1 + (-0.0005 \ (C^\circ)^{-1})(25.8 \ C^\circ)] = 0.0158 \ \Omega$

26-23 The voltmeter reads the potential difference V_{ab} between the terminals of the battery.

open circuit $I = 0$

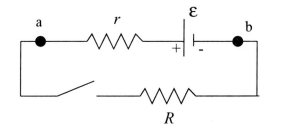

$V_{ab} = \varepsilon = 3.08 \text{ V}$

switch closed

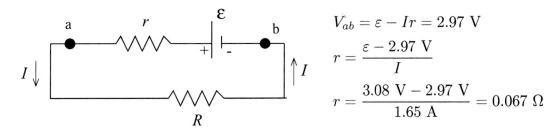

$V_{ab} = \varepsilon - Ir = 2.97 \text{ V}$

$r = \dfrac{\varepsilon - 2.97 \text{ V}}{I}$

$r = \dfrac{3.08 \text{ V} - 2.97 \text{ V}}{1.65 \text{ A}} = 0.067 \ \Omega$

And $V_{ab} = IR$ so $R = \dfrac{V_{ab}}{I} = \dfrac{2.97 \text{ V}}{1.65 \text{ A}} = 1.80 \ \Omega.$

26-25 a) Assume that the current is clockwise.

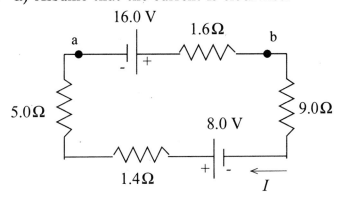

Add up the potential rises and drops as travel clockwise around the circuit:
$16.0 \text{ V} - I(1.6 \ \Omega) - I(9.0 \ \Omega) + 8.0 \text{ V} - I(1.4 \ \Omega) - I(5.0 \ \Omega) = 0$

$I = \dfrac{16.0 \text{ V} + 8.0 \text{ V}}{9.0 \ \Omega + 1.4 \ \Omega + 5.0 \ \Omega + 1.6 \ \Omega} = \dfrac{24.0 \text{ V}}{17.0 \ \Omega} = 1.41 \text{ A, clockwise}$

b) $V_a + 16.0 \text{ V} - I(1.6 \ \Omega) = V_b$

$V_a - V_b = -16.0 \text{ V} + (1.41 \text{ A})(1.6 \ \Omega)$

$V_{ab} = -16.0 \text{ V} + 2.3 \text{ V} = -13.7 \text{ V}$ (point a is at lower potential; it is the negative terminal)

c) $V_a + 16.0 \text{ V} - I(1.6 \text{ }\Omega) - I(9.0 \text{ }\Omega) = V_c$

$V_a - V_c = -16.0 \text{ V} + (1.41 \text{ A})(1.6 \text{ }\Omega + 9.0 \text{ }\Omega)$

$V_{ac} = -16.0 \text{ V} + 15.0 \text{ V} = -1.0 \text{ V}$ (point a is at lower potential than point c)

d) Call the potential zero at point a. Travel clockwise around the circuit.

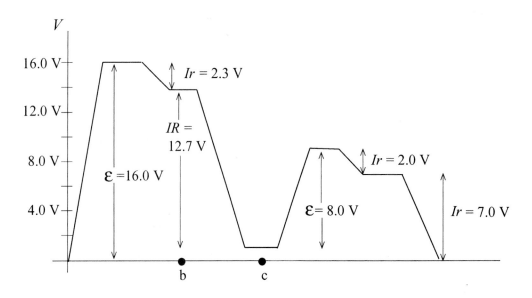

26-29 a)

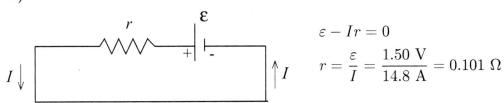

$\varepsilon - Ir = 0$

$r = \dfrac{\varepsilon}{I} = \dfrac{1.50 \text{ V}}{14.8 \text{ A}} = 0.101 \text{ }\Omega$

b) $r = \dfrac{\varepsilon}{I} = \dfrac{1.50 \text{ V}}{6.8 \text{ A}} = 0.22 \text{ }\Omega$

c) $r = \dfrac{\varepsilon}{I} = \dfrac{12.6 \text{ V}}{1000 \text{ A}} = 0.0126 \text{ }\Omega$

26-33 By definition $p = \dfrac{P}{LA}$

a) E is related to V and J is related to I, so use $P = VI$. This gives $p = \dfrac{VI}{LA}$

$\dfrac{V}{L} = E$ and $\dfrac{I}{A} = J$ so $p = EJ$

b) J is related to I and ρ is related to R, so use $P = IR^2$. This gives $p = \dfrac{I^2 R}{LA}$.

$I = JA$ and $R = \dfrac{\rho L}{A}$ so $p = \dfrac{J^2 A^2 \rho L}{LA^2} = \rho J^2$

c) E is related to V and ρ is related to R, so use $P = V^2/R$. This gives $p = \dfrac{V^2}{RLA}$.

$V = EL$ and $R = \dfrac{\rho L}{A}$ so $p = \dfrac{E^2 L^2}{LA}\left(\dfrac{A}{\rho L}\right) = \dfrac{E^2}{\rho}$.

26-35 a) $P = VI = (12\text{ V})(60\text{ A}) = 720\text{ W}$

The battery can provide this for 1.0 h, so the energy the battery has stored is
$U = Pt = (720\text{ W})(3600\text{ s}) = 2.6 \times 10^6\text{ J}$

b) For gasoline the heat of combustion is $L_c = 46 \times 10^6\text{ J/kg}$.

The mass of gasoline that supplies 2.6×10^6 J is $m = \dfrac{2.6 \times 10^6\text{ J}}{46 \times 10^6\text{ J/kg}} = 0.0565\text{ kg}$.

The volume of this mass of gasoline is

$V = \dfrac{m}{\rho} = \dfrac{0.0565\text{ kg}}{900\text{ kg/m}^3} = 6.3 \times 10^{-5}\text{ m}^3 \left(\dfrac{1000\text{ L}}{1\text{ m}^3}\right) = 0.063\text{ L}$

c) $U = Pt, t = \dfrac{U}{P} = \dfrac{2.6 \times 10^6\text{ J}}{450\text{ W}} = 5800\text{ s} = 97\text{ min} = 1.6\text{ h}.$

26-37

Compute I:

$\varepsilon - Ir - IR = 0$

$I = \dfrac{\varepsilon}{r + R} = \dfrac{12.0\text{ V}}{1.0\ \Omega + 5.0\ \Omega} = 2.00\text{ A}$

a) The rate of conversion of chemical energy to electrical energy in the emf of the battery is $P = \varepsilon I = (12.0\text{ V})(2.00\text{ A}) = 24.0\text{ W}$.

b) The rate of dissipation of electrical energy in the internal resistance of the battery is $P = I^2 r = (2.00\text{ A})^2(1.0\ \Omega) = 4.0\text{ W}$.

c) The rate of dissipation of electrical energy in the external resistor R is $P = I^2R = (2.00 \text{ A})^2(5.0 \text{ }\Omega) = 20.0 \text{ W}$.

Note: The rate of production of electrical energy in the circuit is 24.0 W. The total rate of consumption of electrical energy in the circuit is 4.00 W + 20.0 W = 24.0 W. Equal rate of production and consumption of electrical energy are required by energy conservation.

26-41 a) $I = \dfrac{V}{R}$

The total resistance R along the current path is the $R_{\text{ps}} = 2000 \text{ }\Omega$ of the power supply plus the $R_{\text{b}} = 10 \times 10^3 \text{ }\Omega$ body resistance;

$R = R_{\text{ps}} + R_{\text{b}} = 2000 \text{ }\Omega + 10 \times 10^3 \text{ }\Omega$.

$$I = \frac{14 \times 10^3 \text{ V}}{2000 \text{ }\Omega + 10 \times 10^3 \text{ }\Omega} = 1.2 \text{ A}$$

b) $P_{\text{b}} = I^2 R_{\text{b}} = (1.2 \text{ A})^2(10 \times 10^3 \text{ }\Omega) = 1.4 \times 10^4 \text{ W}$

c) $I = 1.00 \text{ mA}$ gives $R = \dfrac{V}{I} = \dfrac{14 \times 10^3 \text{ V}}{1.00 \times 10^{-3} \text{ A}} = 1.4 \times 10^7 \text{ }\Omega$

$R = R_{\text{b}} + R_{\text{ps}}$ so $R_{\text{ps}} = R - R_{\text{b}} = 1.4 \times 10^7 \text{ }\Omega - 10 \times 10^3 \text{ }\Omega = 1.4 \times 10^7 \text{ }\Omega$.

Problems

26-43 a) $R = \dfrac{\rho L}{A}$ so $\rho = \dfrac{RA}{L} = \dfrac{(0.104 \text{ }\Omega)\pi(1.25 \times 10^{-3} \text{ m})^2}{14.0 \text{ m}} = 3.65 \times 10^{-8} \text{ }\Omega \cdot \text{m}$

b) Use $V = IR$ so $I = V/R$.
$V = EL = (1.28 \text{ V/m})(14.0 \text{ m}) = 17.9 \text{ V}$

$$I = \frac{V}{R} = \frac{17.9 \text{ V}}{0.104 \text{ }\Omega} = 172 \text{ A}$$

or

$E = \rho J$ so $J = \dfrac{E}{\rho} = \dfrac{1.28 \text{ V/m}}{3.65 \times 10^{-8} \text{ }\Omega \cdot \text{m}} = 3.51 \times 10^7 \text{ A/m}^2$

$I = JA = (3.51 \times 10^7 \text{ A/m}^2)\pi(1.25 \times 10^{-3} \text{ m})^2 = 172 \text{ A}$, which checks

c) $J = n|q|v_{\text{d}} = nev_{\text{d}}$

$$v_{\rm d} = \frac{J}{ne} = \frac{3.51 \times 10^7 \text{ A/m}^2}{(8.5 \times 10^{28} \text{ m}^{-3})(1.602 \times 10^{-19} \text{ C})} = 2.58 \times 10^{-3} \text{ m/s} = 2.58 \text{ mm/s}$$

26-47 a)

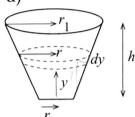

The radius of a truncated cone a distance y above the bottom is given by

$$r = r_2 + (y/h)(r_1 - r_2) = r_2 + y\beta$$
$$\text{with } \beta = (r_1 - r_2)/h$$

Consider a thin slice a distance y above the bottom. The slice has thickness dy and radius r. The resistance of the slice is

$$dR = \frac{\rho \, dy}{A} = \frac{\rho \, dy}{\pi r^2} = \frac{\rho \, dy}{\pi (r_2 + \beta y)^2}$$

The total resistance of the cone is obtained by integrating over these thin slices:

$$R = \int dR = \frac{\rho}{\pi} \int_0^h \frac{dy}{(r_2 + \beta y)^2} = \frac{\rho}{\pi} \left[-\frac{1}{\beta}(r_2 + y\beta)^{-1} \right]_0^h = -\frac{\rho}{\pi \beta} \left[\frac{1}{r_2 + h\beta} - \frac{1}{r_2} \right]$$

But $r_2 + h\beta = r_1$

$$R = \frac{\rho}{\pi \beta} \left[\frac{1}{r_2} - \frac{1}{r_1} \right] = \frac{\rho}{\pi} \left(\frac{h}{r_1 - r_2} \right) \left(\frac{r_1 - r_2}{r_1 r_2} \right) = \frac{\rho h}{\pi r_1 r_2}$$

b) Let $r_1 = r_2 = r$. Then $R = \rho h/\pi r^2 = \rho L/A$ where $A = \pi r^2$ and $L = h$. This agrees with Eq.(26-10).

26-49 Use $E = \rho J$ to calculate the current density between the plates.

Let A be the area of each plate; then $I = JA$.

$$J = \frac{E}{\rho} \text{ and } E = \frac{\sigma}{K\epsilon_0} = \frac{Q}{KA\epsilon_0}$$

Thus $J = \dfrac{Q}{KA\epsilon_0 \rho}$ and $I = JA = \dfrac{Q}{K\epsilon_0 \rho}$, as was to be shown.

26-53

$$V_{ab} = 8.4 \text{ V}$$
$$V_{ab} = \varepsilon - Ir$$
$$\varepsilon - (1.50 \text{ A})r = 8.4 \text{ V}$$

$I = 1.50 \text{ A}$

$V_{ab} = 9.4$ V

$V_{ab} = \varepsilon + Ir$

$\varepsilon + (3.50 \text{ A})r = 9.4$ V

$I = 3.50$ A

a) Solve the first equation for ε and use that result in the second equation:

$\varepsilon = 8.4$ V $+ (1.50 \text{ A})r$

8.4 V $+ (1.50 \text{ A})r + (3.50 \text{ A})r = 9.4$ V

$(5.00 \text{ A})r = 1.0$ V so $r = \dfrac{1.0 \text{ V}}{5.00 \text{ A}} = 0.20 \ \Omega$

b) Then $\varepsilon = 8.4$ V $+ (1.50 \text{ A})r = 8.4$ V $+ (1.50 \text{ A})(0.20 \ \Omega) = 8.7$ V

26-55

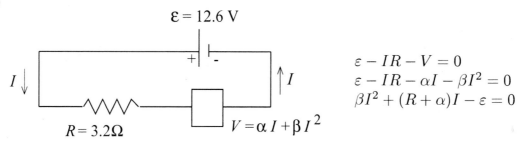

$\varepsilon = 12.6$ V

$\varepsilon - IR - V = 0$

$\varepsilon - IR - \alpha I - \beta I^2 = 0$

$\beta I^2 + (R + \alpha)I - \varepsilon = 0$

$R = 3.2\Omega$ $V = \alpha I + \beta I^2$

The quadratic formula gives $I = (1/2\beta)[-(R + \alpha) \pm \sqrt{(R + \alpha)^2 + 4\beta\varepsilon}]$

I must be positive, so take the $+$ sign

$I = (1/2\beta)[-(R + \alpha) + \sqrt{(R + \alpha)^2 + 4\beta\varepsilon}]$

$I = -2.692$ A $+ 4.116$ A $= 1.42$ A

26-57 a) With the ammeter in the circuit:

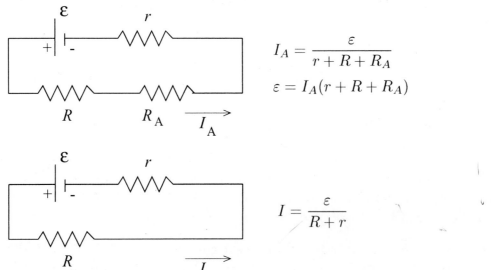

$I_A = \dfrac{\varepsilon}{r + R + R_A}$

$\varepsilon = I_A(r + R + R_A)$

$I = \dfrac{\varepsilon}{R + r}$

Combining the two equations gives

$$I = \left(\frac{1}{R+r}\right) I_A(r + R + R_A) = I_A\left(1 + \frac{R_A}{r+R}\right)$$

b) Want $I_A = 0.990I$. Use this in the result for part (a).

$$I = 0.990I\left(1 + \frac{R_A}{r+R}\right)$$

$$0.010 = 0.990\left(\frac{R_A}{r+R}\right)$$

$$R_A = (r+R)(0.010/0.990) = (0.45\ \Omega + 3.80\ \Omega)(0.010/0.990) = 0.0429\ \Omega$$

c) $I - I_A = \dfrac{\varepsilon}{r+R} - \dfrac{\varepsilon}{r+R+R_A}$

$$I - I_A = \varepsilon\left(\frac{r+R+R_A - r - R}{(r+R)(r+R+R_A)}\right) = \frac{\varepsilon R_A}{(r+R)(r+R+R_A)}.$$

The difference between I and I_A increases as R_A increases. If R_A is larger than the value calculated in part (b) then I_A differs from I by more than 1.0%,

26-61 a)

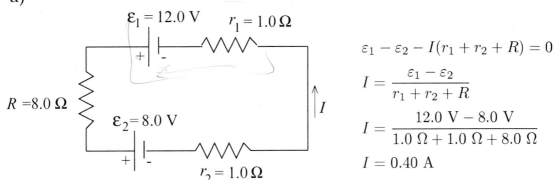

$$\varepsilon_1 - \varepsilon_2 - I(r_1 + r_2 + R) = 0$$

$$I = \frac{\varepsilon_1 - \varepsilon_2}{r_1 + r_2 + R}$$

$$I = \frac{12.0\ \text{V} - 8.0\ \text{V}}{1.0\ \Omega + 1.0\ \Omega + 8.0\ \Omega}$$

$$I = 0.40\ \text{A}$$

b) $P = I^2R + I^2r_1 + I^2r_2 = I^2(R + r_1 + r_2) = (0.40\ \text{A})^2(8.0\ \Omega + 1.0\ \Omega + 1.0\ \Omega)$

$P = 1.6\ \text{W}$

c) Chemical energy is converted to electrical energy in a battery when the current goes through the battery from the negative to the positive terminal, so the electrical energy of the charges increases as the current passes through. This happens in the 12.0 V battery, and the rate of production of electrical energy is $P = \varepsilon_1 I = (12.0\ \text{V})(0.40\ \text{A}) = 4.8\ \text{W}$.

d) Electrical energy is converted to chemical energy in a battery when the current goes through the battery from the positive to the negative terminal, so the electrical energy of the charges decreases as the current passes through. This happens in the

8.0 V battery, and the rate of consumption of electrical energy is $P = \varepsilon_2 I = (8.0 \text{ V})(0.40 \text{ A}) = 3.2 \text{ W}$.

e) Total rate of production of electrical energy $= 4.8 \text{ W}$.

Total rate of consumption of electrical energy $= 1.6 \text{ W} + 3.2 \text{ W} = 4.8 \text{ W}$, which equals the rate of production, as it must.

26-63 **a)**

a

$\sum \vec{F} = m\vec{a}$

$\ominus \xrightarrow{\quad} \atop |q| E$ $|q|E = ma$ and then $|q|/m = a/E$

b) $V_{bc} = EL$ so $E = V_{bc}/L$

Using this in the expression from part (a) gives $|q|/m = aL/V_{bc}$

c)

E

$\ominus \xrightarrow[F_E]{\quad}$

If the acceleration is to the right the force on the free charges must be to the right. For a negative charge the force and electric field are in opposite directions, so $\vec{E}$ is to the left.

$b \bullet\!\!\boxed{\qquad \longleftarrow \quad E \qquad}\!\!\bullet c$

$\vec{E}$ points from high potential so point c is at higher potential.

d) From part (b),

$$a = \frac{V_{bc}}{L}\frac{|q|}{m} = \left(\frac{1.00 \times 10^{-3} \text{ V}}{0.50 \text{ m}}\right)\left(\frac{1.60 \times 10^{-19} \text{ C}}{9.11 \times 10^{-31} \text{ kg}}\right) = 3.5 \times 10^8 \text{ m/s}^2.$$

e) The rotating spool with many turns allows for a very large length L of wire. From $V_{bc} = aL(m/|q|)$ this increases V_{bc} and makes it easier to measure.

CHAPTER 27
DIRECT-CURRENT CIRCUITS

Exercises 3, 5, 7, 9, 11, 13, 15, 17, 23, 31, 35, 37, 39
Problems 41, 45, 47, 49, 51, 53, 55, 61, 63, 65, 69

Exercises

27-3   a)

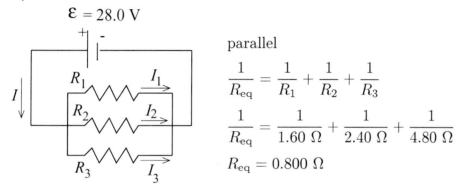

parallel

$$\frac{1}{R_{eq}} = \frac{1}{R_1} + \frac{1}{R_2} + \frac{1}{R_3}$$

$$\frac{1}{R_{eq}} = \frac{1}{1.60 \ \Omega} + \frac{1}{2.40 \ \Omega} + \frac{1}{4.80 \ \Omega}$$

$$R_{eq} = 0.800 \ \Omega$$

b) For resistors in parallel the voltage is the same across each and equal to the applied voltage; $V_1 = V_2 = V_3 = \varepsilon = 28.0$ V

$$V = IR \text{ so } I_1 = \frac{V_1}{R_1} = \frac{28.0 \text{ V}}{1.60 \ \Omega} = 17.5 \text{ A}$$

$$I_2 = \frac{V_2}{R_2} = \frac{28.0 \text{ V}}{2.40 \ \Omega} = 11.7 \text{ A and } I_3 = \frac{V_3}{R_3} = \frac{28.0 \text{ V}}{4.8 \ \Omega} = 5.8 \text{ A}$$

c) The currents through the resistors add to give the current through the battery:
$$I = I_1 + I_2 + I_3 = 17.5 \text{ A} + 11.7 \text{ A} + 5.8 \text{ A} = 35.0 \text{ A}$$

Alternatively, we can use the equivalent resistance R_{eq}:

$$\varepsilon - IR_{eq} = 0$$

$$I = \frac{\varepsilon}{R_{eq}} = \frac{28.0 \text{ V}}{0.800 \ \Omega} = 35.0 \text{ A},$$
which checks

$R_{eq} = 0.800 \Omega$

d) As shown in part (b), the voltage across each resistor is 28.0 V.

e) We can use any of the three expressions for P: $P = VI = I^2 R = V^2/R$. They will all give the same results, if we keep enough significant figures in intermediate calculations. Using $P = V^2/R$,

$$P_1 = V_1^2/R_1 = \frac{(28.0 \text{ V})^2}{1.60 \text{ } \Omega} = 490 \text{ W}, \quad P_2 = V_2^2/R_2 = \frac{(28.0 \text{ V})^2}{2.40 \text{ } \Omega} = 327 \text{ W, and}$$

$$P_3 = V_3^2/R_3 = \frac{(28.0 \text{ V})^2}{4.80 \text{ } \Omega} = 163 \text{ W}$$

Note that the total power dissipated is $P_{\text{out}} = P_1 + P_2 + P_3 = 980$ W. This is the same as the power $P_{\text{in}} = \varepsilon I = (28.0 \text{ V})(35.0 \text{ A}) = 980$ W delivered by the battery.

f) $P = V^2/R$. The resistors in parallel each have the same voltage, so the power P is largest for the one with the least resistance.

27-5   a) $P = V_{ab}^2/R$ (Eq.26-18), so
$$V_{ab} = \sqrt{PR} = \sqrt{((5.0 \text{ W})(15 \times 10^3 \text{ } \Omega)} = 270 \text{ V}.$$

b) $P = \dfrac{V_{ab}^2}{R} = \dfrac{(120 \text{ V})^2}{9.0 \times 10^3 \text{ } \Omega} = 1.6 \text{ W}$

27-7

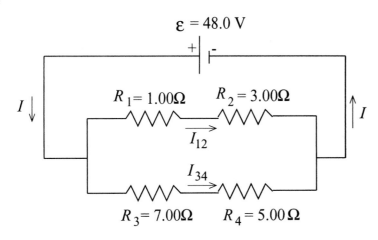

R_1 and R_2 in series have an equivalent resistance of
$$R_{12} = R_1 + R_2 = 4.00 \text{ } \Omega$$

R_3 and R_4 in series have an equivalent resistance of
$$R_{34} = R_3 + R_4 = 12.0 \text{ } \Omega$$

The circuit is equivalent to

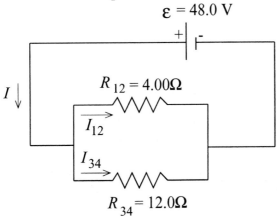

R_{12} and R_{34} in parallel are equivalent to R_{eq} given by

$$\frac{1}{R_{\text{eq}}} = \frac{1}{R_{12}} + \frac{1}{R_{34}} = \frac{R_{12} + R_{34}}{R_{12}R_{34}}$$

$$R_{\text{eq}} = \frac{R_{12}R_{34}}{R_{12} + R_{34}}$$

$$R_{\text{eq}} = \frac{(4.00 \text{ } \Omega)(12.0 \text{ } \Omega)}{4.00 \text{ } \Omega + 12.0 \text{ } \Omega} = 3.00 \text{ } \Omega$$

The voltage across each branch of the parallel combination is ε, so $\varepsilon - I_{12}R_{12} = 0$.

$$I_{12} = \frac{\varepsilon}{R_{12}} = \frac{48.0 \text{ V}}{4.00 \text{ }\Omega} = 12.0 \text{ A}$$

$$\varepsilon - I_{34}R_{34} = 0 \text{ so } I_{34} = \frac{\varepsilon}{R_{34}} = \frac{48.0 \text{ V}}{12.0 \text{ }\Omega} = 4.0 \text{ A}$$

The current is 12.0 A through the 1.00 Ω and 3.00 Ω resistors, and it is 4.0 A through the 7.00 Ω and 5.00 Ω resistors.

Note: The current through the battery is $I = I_{12} + I_{34} = 12.0 \text{ A} + 4.0 \text{ A} = 16.0$ A, and this is equal to $\varepsilon/R_{eq} = 48.0 \text{ V}/3.00 \text{ }\Omega = 16.0$ A.

27-9   a)

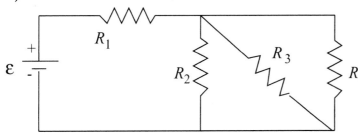

R_2, R_3, and R_4 are in parallel, so their equivalent resistance R_{eq} is given by

$$\frac{1}{R_{eq}} = \frac{1}{R_2} + \frac{1}{R_3} + \frac{1}{R_4}$$

$$\frac{1}{R_{eq}} = \frac{3}{4.50 \text{ }\Omega} \text{ and } R_{eq} = 1.50 \text{ }\Omega.$$

The equivalent circuit is

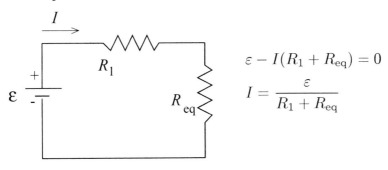

$$\varepsilon - I(R_1 + R_{eq}) = 0$$

$$I = \frac{\varepsilon}{R_1 + R_{eq}}$$

$$I = \frac{9.00 \text{ V}}{4.50 \text{ }\Omega + 1.50 \text{ }\Omega} = 1.50 \text{ A and } I_1 = 1.50 \text{ A}$$

Then $V_1 = I_1 R_1 = (1.50 \text{ A})(4.50 \text{ }\Omega) = 6.75 \text{ V}$

$I_{eq} = 1.50 \text{ A}, V_{eq} = I_{eq} R_{eq} = (1.50 \text{ A})(1.50 \text{ }\Omega) = 2.25 \text{ V}$

For resistors in parallel the voltages are equal and are the same as the voltage across the equivalent resistor, so $V_2 = V_3 = V_4 = 2.25 \text{ V}$.

$$I_2 = \frac{V_2}{R_2} = \frac{2.25 \text{ V}}{4.50 \text{ }\Omega} = 0.500 \text{ A}, I_3 = \frac{V_3}{R_3} = 0.500 \text{ A}, I_4 = \frac{V_4}{R_4} = 0.500 \text{ A}$$

Note that $I_2 + I_3 + I_4 = 1.50$ A, which is I_{eq}. For resistors in parallel the currents add and their sum is the current through the equivalent ressitor.

b) $P = I^2 R$

$P_1 = (1.50 \text{ A})^2 (4.50 \text{ }\Omega) = 10.1$ W

$P_2 = P_3 = P_4 = (0.500 \text{ A})^2 (4.50 \text{ }\Omega) = 1.125$ W, which rounds to 1.12 W.

Note that $P_2 + P_3 + P_4 = 3.37$ W. This equals $P_{eq} = I_{eq}^2 R_{eq} = (1.50 \text{ A})^2 (1.50 \text{ }\Omega) = 3.37$ W, the power dissipated in the equivalent resistor.

c) With R_4 removed the circuit becomes

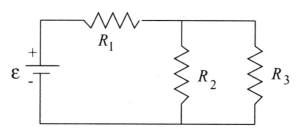

R_2 and R_3 are in parallel and their equivalent resistance R_{eq} is given by

$$\frac{1}{R_{eq}} = \frac{1}{R_2} + \frac{1}{R_3} = \frac{2}{4.50 \text{ }\Omega}$$

and $R_{eq} = 2.25 \text{ }\Omega$

The equivalent circuit is

$$\varepsilon - I(R_1 + R_{eq}) = 0$$

$$I = \frac{\varepsilon}{R_1 + R_{eq}}$$

$$I = \frac{9.00 \text{ V}}{4.50 \text{ }\Omega + 2.25 \text{ }\Omega} = 1.333 \text{ A}$$

$I_1 = 1.33$ A, $V_1 = I_1 R_1 = (1.333 \text{ A})(4.50 \text{ }\Omega) = 6.00$ V

$I_{eq} = 1.33$ A, $V_{eq} = I_{eq} R_{eq} = (1.333 \text{ A})(2.25 \text{ }\Omega) = 3.00$ V and $V_2 = V_3 = 3.00$ V.

$$I_2 = \frac{V_2}{R_2} = \frac{3.00 \text{ V}}{4.50 \text{ }\Omega} = 0.667 \text{ A}, \quad I_3 = \frac{V_3}{R_3} = 0.667 \text{ A}$$

d) $P = I^2 R$

$P_1 = (1.333 \text{ A})^2 (4.50 \text{ }\Omega) = 8.00$ W

$P_2 = P_3 = (0.667 \text{ A})^2 (4.50 \text{ }\Omega) = 2.00$ W.

e) When R_4 is removed, P_1 decreases and P_2 and P_3 increase. Bulb R_1 glows less brightly and bulbs R_2 and R_3 glow more brightly. When R_4 is removed the equivalent resistance of the circuit increases and the current through R_1 decreases. But in the parallel combination this current divides into two equal currents rather than three, so the currents through R_2 and R_3 increase. Can also see this by noting that with R_4 removed and less current through R_1 the voltage drop across R_1 is less so the voltage drop across R_2 and across R_3 must become larger.

27-11 a)

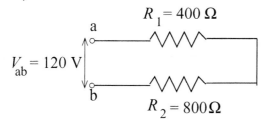

For resistors in series the current is the same through each.
$R_{eq} = R_1 + R_2 = 1200 \ \Omega.$

$$I = \frac{V}{R_{eq}} = \frac{120 \text{ V}}{1200 \ \Omega} = 0.100 \text{ A}.$$ This is the current drawn from the line.

b) $P_1 = I_1^2 R_1 = (0.100 \text{ A})^2 (400 \ \Omega) = 4.0 \text{ W}$

$P_2 = I_2^2 R_2 = (0.100 \text{ A})^2 (800 \ \Omega) = 8.0 \text{ W}$

$P_{out} = P_1 + P_2 = 12.0$ W, the total power dissipated in both bulbs.

Note that $P_{in} = V_{ab} I = (120 \text{ V})(0.100 \text{ A}) = 12.0$ W, the power delivered by the potential source, equals P_{out}.

c)

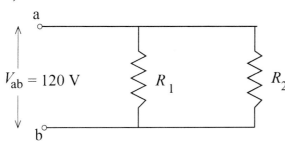

For resistors in parallel the voltage across each resistor is the same.

$$I_1 = \frac{V_1}{R_1} = \frac{120 \text{ V}}{400 \ \Omega} = 0.300 \text{ A}, \quad I_2 = \frac{V_2}{R_2} = \frac{120 \text{ V}}{800 \ \Omega} = 0.150 \text{ A}$$

Note that each current is larger than the current when the resistors are connected in series.

d) $P_1 = I_1^2 R_1 = (0.300 \text{ A})^2 (400 \ \Omega) = 36.0 \text{ W}$

$P_2 = I_2^2 R_2 = (0.150 \text{ A})^2 (800 \ \Omega) = 18.0 \text{ W}$

$P_{out} = P_1 + P_2 = 54.0 \text{ W}$

Note that the total current drawn from the line is $I = I_1 + I_2 = 0.450$ A. The power input from the line is $P_{in} = V_{ab} I = (120 \text{ V})(0.450 \text{ A}) = 54.0$ W, which equals the total power dissipated by the bulbs.

e) The bulb that is dissipating the most power glows most brightly. For the series connection the currents are the same and by $P = I^2 R$ the bulb with the larger R has the larger P; the 800 Ω bulb glows more brightly.

For the parallel combination the voltages are the same and by $P = V^2/R$ the bulb with the smaller R has the larger P; the 400 Ω bulb glows more brightly.

The total power output P_{out} equals $P_{\text{in}} = V_{ab}I$, so P_{out} is larger for the parallel connection where the current drawn from the line is larger (because the equivalent resistance is smaller).

27-13 a)

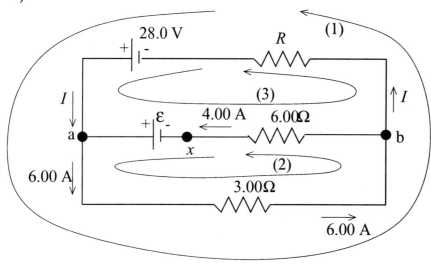

Apply Kirchhoff's point rule to point a: $\sum I = 0$ so $I + 4.00\text{ A} - 6.00\text{ A} = 0$
$I = 2.00$ A (in the direction shown in the diagram)

b) Apply Kirchhoff's loop rule to loop (1):
$-(6.00\text{ A})(3.00\ \Omega) - (2.00\text{ A})R + 28.0\text{ V} = 0$
$-18.0\text{ V} - (2.00\ \Omega)R + 28.0\text{ V} = 0$

$$R = \frac{28.0\text{ V} - 18.0\text{ V}}{2.00\text{ A}} = 5.00\ \Omega$$

c) Apply Kirchhoff's loop rule to loop (2):
$-(6.00\text{ A})(3.00\ \Omega) - (4.00\text{ A})(6.00\ \Omega) + \varepsilon = 0$
$\varepsilon = 18.0\text{ V} + 24.0\text{ V} = 42.0$ V

Note: Can check that the loop rule is satisfied for loop (3), as a check of our work:
$28.0\text{ V} - \varepsilon + (4.00\text{ A})(6.00\ \Omega) - (2.00\text{ A})R = 0$
$28.0\text{ V} - 42.0\text{ V} + 24.0\text{ V} - (2.00\text{ A})(5.00\ \Omega) = 0$
$52.0\text{ V} = 42.0\text{ V} + 10.0\text{ V}$
$52.0\text{ V} = 52.0$ V, so the loop rule is satisfied for this loop.

d) If the circuit is broken at point x there can be no current in the 6.00 Ω resistor. There is now only a single current path:

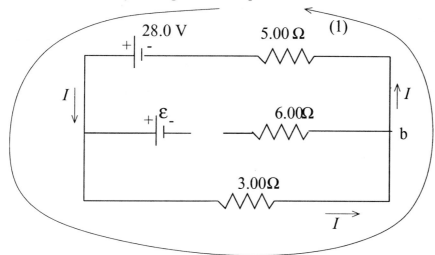

Apply the loop rule: $+28.0 \text{ V} - (3.00 \text{ }\Omega)I - (5.00 \text{ }\Omega)I = 0$

$$I = \frac{28.0 \text{ V}}{8.00 \text{ }\Omega} = 3.50 \text{ A}$$

27-15 a)

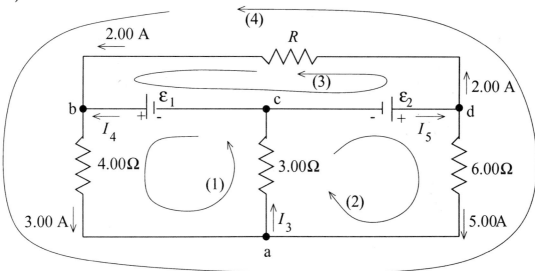

Apply the junction rule to point a:

$3.00 \text{ A} + 5.00 \text{ A} - I_3 = 0$

$I_3 = 8.00 \text{ A}$

Apply the junction rule to point b:

$2.00 \text{ A} + I_4 - 3.00 \text{ A} = 0$

$I_4 = 1.00 \text{ A}$

Apply the junction rule to point c:

$I_3 - I_4 - I_5 = 0$

$I_5 = I_3 - I_4 = 8.00 \text{ A} - 1.00 \text{ A} = 7.00 \text{ A}$

As a check, apply the junction rule to point d:

$I_5 - 2.00 \text{ A} - 5.00 \text{ A} = 0$

$I_5 = 7.00 \text{ A}$

b) Apply the loop rule to loop (1):

$\varepsilon_1 - (3.00 \text{ A})(4.00 \text{ }\Omega) - I_3(3.00 \text{ }\Omega) = 0$

$\varepsilon_1 = 12.0 \text{ V} + (8.00 \text{ A})(3.00 \text{ }\Omega) = 36.0 \text{ V}$

Apply the loop rule to loop (2):

$\varepsilon_2 - (5.00 \text{ A})(6.00 \text{ }\Omega) - I_3(3.00 \text{ }\Omega) = 0$

$\varepsilon_2 = 30.0 \text{ V} + (8.00 \text{ A})(3.00 \text{ }\Omega) = 54.0 \text{ V}$

c) Apply the loop rule to loop (3):

$-(2.00 \text{ A})R - \varepsilon_1 + \varepsilon_2 = 0$

$$R = \frac{\varepsilon_2 - \varepsilon_1}{2.00 \text{ A}} = \frac{54.0 \text{ V} - 36.0 \text{ V}}{2.00 \text{ A}} = 9.00 \text{ }\Omega$$

Note: Apply the loop rule to loop (4) as a check of our calculations:

$-(2.00 \text{ A})R - (3.00 \text{ A})(4.00 \text{ }\Omega) + (5.00 \text{ A})(6.00 \text{ }\Omega) = 0$

$-(2.00 \text{ A})(9.00 \text{ }\Omega) - 12.0 \text{ V} + 30.0 \text{ V} = 0$

$-18.0 \text{ V} + 18.0 \text{ V} = 0$

27-17 a)

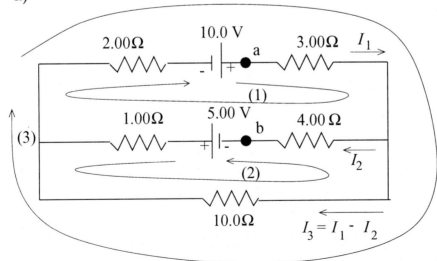

Let I_1 be the current in the 3.00 Ω resistor and I_2 be the current in the 4.00 Ω resistor and assume that these currents are in the directions shown. Then the current in the 10.0 Ω resistor is $I_3 = I_1 - I_2$, in the direction shown, where we have used Kirchhoff's point rule to relate I_3 to I_1 and I_2. If we get a negative answer for any of these currents we know the current is actually in the opposite direction to what we have assumed.

Three loops and directions to travel around the loops are shown in the circuit diagram. Apply Kirchhoff's loop rule to each loop.

loop (1)

$+10.0$ V $- I_1(3.00 \ \Omega) - I_2(4.00 \ \Omega) + 5.00$ V $- I_2(1.00 \ \Omega) - I_1(2.00 \ \Omega) = 0$

15.00 V $- (5.00 \ \Omega)I_1 - (5.00 \ \Omega)I_2 = 0$

3.00 A $- I_1 - I_2 = 0$

loop (2)

$+5.00$ V $- I_2(1.00 \ \Omega) + (I_1 - I_2)10.0 \ \Omega - I_2(4.00 \ \Omega) = 0$

5.00 V $+ (10.0 \ \Omega)I_1 - (15.0 \ \Omega)I_2 = 0$

1.00 A $+ 2.00I_1 - 3.00I_2 = 0$

The first equation says $I_2 = 3.00$ A $- I_1$.

Use this in the second equation: 1.00 A $+ 2.00I_1 - 9.00$ A $+ 3.00I_1 = 0$

$5.00I_1 = 8.00$ A, $\quad I_1 = 1.60$ A

Then $I_2 = 3.00$ A $- I_1 = 3.00$ A $- 1.60$ A $= 1.40$ A.

$I_3 = I_1 - I_2 = 1.60$ A $- 1.40$ A $= 0.20$ A

loop (3) can be used as a check:

$+10.0$ V $- (1.60$ A$)(3.00 \ \Omega) - (0.20$ A$)(10.00 \ \Omega) - (1.60$ A$)(2.00 \ \Omega) = 0$

10.0 V $= 4.8$ V $+ 2.0$ V $+ 3.2$ V

10.0 V $= 10.0$ V

We find that with our calculated currents the loop rule is satisfied for loop (3).

Also, all the currents came out to be positive, so the current directions in the circuit diagram are correct.

b) To find $V_{ab} = V_a - V_b$ start at point b and travel to point a. Many different routes can be taken from b to a and all must yield the same result for V_{ab}.

Travel through the 4.00 Ω resistor and then through the 3.00 Ω resistor:

$V_b + I_2(4.00 \ \Omega) + I_1(3.00 \ \Omega) = V_a$

$V_a - V_b = (1.40$ A$)(4.00 \ \Omega) + (1.60$ A$)(3.00 \ \Omega) = 5.60$ V $+ 4.80$ V $= 10.4$ V (point a is at higher potential than point b)

Or, travel through the 5.00 V emf, the 1.00 Ω resistor, the 2.00 Ω resistor, and the 10.0 V emf:

$V_b + 5.00 \text{ V} - I_2(1.00 \ \Omega) - I_1(2.00 \ \Omega) + 10.0 \text{ V} = V_a$

$V_a - V_b = 15.0 \text{ V} - (1.40 \text{ A})(1.00 \ \Omega) - (1.60 \text{ A})(2.00 \ \Omega) = 15.0 \text{ V} - 1.40 \text{ V} - 3.20 \text{ V} = 10.4 \text{ V}$, the same as before.

27-23

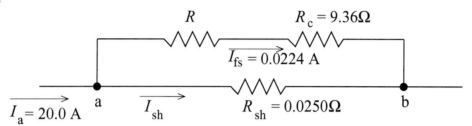

We want that $I_a = 20.0$ A in the external circuit produces $I_{fs} = 0.0224$ A through the galvanometer coil. Applying the junction rule to point a gives

$I_a - I_{fs} - I_{sh} = 0$

$I_{sh} = I_a - I_{fs} = 20.0 \text{ A} - 0.0224 \text{ A} = 19.98 \text{ A}$

The potential difference V_{ab} between points a and b must be the same for both paths between these two points:

$I_{fs}(R + R_c) = I_{sh}R_{sh}$

$$R = \frac{I_{sh}R_{sh}}{I_{fs}} - R_c = \frac{(19.98 \text{ A})(0.0250 \ \Omega)}{0.0224 \text{ A}} - 9.36 \ \Omega = 22.30 \ \Omega - 9.36 \ \Omega = 12.9 \ \Omega$$

27-31 a)

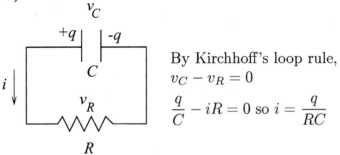

By Kirchhoff's loop rule,

$v_C - v_R = 0$

$\frac{q}{C} - iR = 0$ so $i = \frac{q}{RC}$

Just after the connection is made the charge q on the capacitor hasn't had any time to decrease from its initial value of 6.55×10^{-8} C, so

$$i = \frac{q}{RC} = \frac{6.55 \times 10^{-8} \text{ C}}{(1.28 \times 10^6 \ \Omega)(4.55 \times 10^{-10} \text{ F})} = 1.12 \times 10^{-4} \text{ A}$$

b) Eq.(27-14): $\tau = RC = (1.28 \times 10^6 \ \Omega)(4.55 \times 10^{-10} \text{ F}) = 5.82 \times 10^{-4}$ s

27-35 **a)** With the switch in position 2 the circuit is the charging circuit.

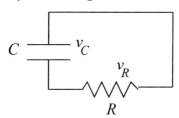

At $t = 0$, $q = 0$.

The charge q on the capacitor is given as a function of time by Eq.(27-12):

$q = C\varepsilon(1 - e^{-t/RC})$

$Q_f = C\varepsilon = (1.50 \times 10^{-5} \text{ F})(18.0 \text{ V}) = 2.70 \times 10^{-4} \text{ C}.$

$RC = (980 \text{ } \Omega)(1.50 \times 10^{-5} \text{ F}) = 0.0147 \text{ s}$

Thus, at $t = 0.0100$ s, $q = (2.70 \times 10^{-4} \text{ C})(1 - e^{-(0.0100 \text{ s})/(0.0147 \text{ s})}) = 133 \text{ } \mu\text{C}.$

b) $v_C = \dfrac{q}{C} = \dfrac{133 \text{ } \mu\text{C}}{1.50 \times 10^{-5} \text{ F}} = 8.87 \text{ V}$

$\varepsilon - v_C - v_R = 0$

$v_R = \varepsilon - v_C = 18.0 \text{ V} - 8.87 \text{ V} = 9.13 \text{ V}$

c) Throwing the switch back to position 1 produces the discharging circuit.

The initial charge Q_0 is the charge calculated in part (b), $Q_0 = 133 \text{ } \mu\text{C}.$

$v_C = \dfrac{q}{C} = \dfrac{133 \text{ } \mu\text{C}}{1.50 \times 10^{-5} \text{ F}} = 8.87 \text{ V}$, the same as just before the switch is thrown.
But now $v_C - v_R = 0$, so $v_R = v_C = 8.87 \text{ V}.$

d) In the discharging circuit the charge on the capacitor as a function of time is given by Eq.(27-16): $q = Q_0 e^{-t/RC}$. $RC = 0.0147$ s, the same as in part (a). Thus at $t = 0.0100$ s,

$q = (133 \text{ } \mu\text{C})e^{-((0.0100 \text{ s})/(0.0147 \text{ s}))} = 67.4 \text{ } \mu\text{C}.$

27-37 Find the current through the heater:

$P = VI$ so $I = P/V = 1500 \text{ W}/120 \text{ V} = 12.5 \text{ A}.$

The maximum total current allowed is 20 A, so the current through the dryer must be less than 20 A $-$ 12.5 A $=$ 7.5 A. The power dissipated by the dryer if the current has this value is $P = VI = (120 \text{ V})(7.5 \text{ A}) = 900 \text{ W}$. For P at this value or larger the cuircuit breaker trips.

27-39 a) When the heater element is first turned on it is at room temperature and has resistance $R = 20$ Ω.

$$I = \frac{V}{R} = \frac{120 \text{ V}}{20 \text{ }\Omega} = 6.0 \text{ A}$$

$$P = \frac{V^2}{R} = \frac{(120 \text{ V})^2}{20 \text{ }\Omega} = 720 \text{ W}$$

b) Find the resistance R_T of the element at the operating temperature of $280°C$: Eq.(26-12) gives $R(T) = R_0(1+\alpha(T-T_0)) = 20 \text{ }\Omega(1+(2.8 \times 10^{-3}(\text{C}°)^{-1})(280°C - 23°C)) = 34.4 \text{ }\Omega$.

$$I = \frac{V}{R} = \frac{120 \text{ V}}{34.4 \text{ }\Omega} = 3.5 \text{ A}$$

$$P = \frac{V^2}{R} = \frac{(120 \text{ V})^2}{34.4 \text{ }\Omega} = 420 \text{ W}$$

Problems

27-41 a) Two of the resistors in series would each dissipate one-half the total, or 1.2 W, which is ok. But the series combination would have an equivalent resistance of 800 Ω, not the 400 Ω that is required. Resistors in parallel have an equivalent resistance that is less than that of the individual resistors, so a solution is two in series in parallel with another two in series.

R_s is the resistance equivalent to two of the 400 Ω resistors in series:

$R_s = R + R = 800$ Ω.

R_{eq} is the resistance equivalent to the two $R_s = 800$ Ω resistors in parallel:

$$\frac{1}{R_{eq}} = \frac{1}{R_s} + \frac{1}{R_s} = \frac{2}{R_s}; \quad R_{eq} = \frac{800 \text{ }\Omega}{2} = 400 \text{ }\Omega.$$

This combination does have the required 400 Ω equivalent resistance. It will be shown in part (b) that a total of 2.4 W can be dissipated without exceeding the power rating of each individual resistor.

Another solution is two resistors in parallel in series with two more in parallel:

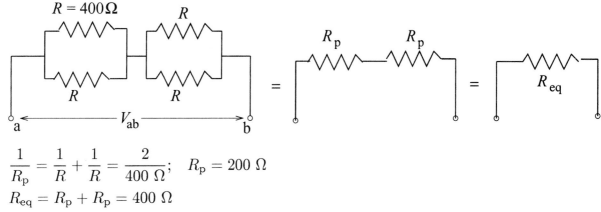

$$\frac{1}{R_p} = \frac{1}{R} + \frac{1}{R} = \frac{2}{400\ \Omega}; \quad R_p = 200\ \Omega$$

$$R_{eq} = R_p + R_p = 400\ \Omega$$

This combination has the required 400 Ω equivalent resistance. It will be shown in part (b) that a total of 2.4 W can be dissipated without exceeding the power rating of each individual resistor.

b) For a combination with equivalent resistance $R_{eq} = 400\ \Omega$ to dissipate 2.4 W the voltage V_{ab} applied to the network must be given by

$P = V_{ab}^2/R_{eq}$ so $V_{ab} = \sqrt{PR_{eq}} = \sqrt{(2.4\ \text{W})(400\ \Omega)} = 31.0$ V and the current through the equivalent resistance is $I = V_{ab}/R = 31.0$ V$/400\ \Omega = 0.0775$ A.

For the first combination this means 31.0 V across each parallel branch and $\frac{1}{2}(31.0\ \text{V}) = 15.5$ V across each 400 Ω resistor. The power dissipated by each individual resistor is then $P = V^2/R = (15.5\ \text{V})^2/400\ \Omega = 0.60$ W, which is less than the maximum allowed value of 1.20 W.

For the second combination this means a voltage of $IR_p = (0.0775\ \text{A})(200\ \Omega) = 15.5$ V across each parallel combination and hence across each separate resistor. The power dissipated by each resistor is again $P = V^2/R = (15.5\ \text{V})^2/400\ \Omega = 0.60$ W, which is less than the maximum allowed value of 1.20 W.

27-45

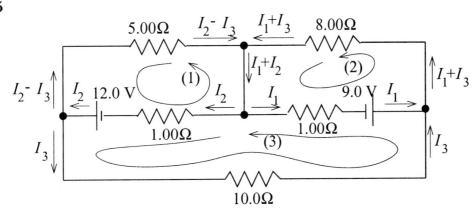

The current in each branch has been written in terms of I_1, I_2 and I_3 such that

the junction rule is satisfied at each junction point.

Apply the loop rule to loop (1):

$-12.0 \text{ V} + I_2(1.00 \ \Omega) + (I_2 - I_3)(5.00 \ \Omega) = 0$

$I_2(6.00 \ \Omega) - I_3(5.00 \ \Omega) = 12.0 \text{ V}$ eq.(1)

Apply the loop rule to loop (2):

$-I_1(1.00 \ \Omega) + 9.00 \text{ V} - (I_1 + I_3)(8.00 \ \Omega) = 0$

$I_1(9.00 \ \Omega) + I_3(8.00 \ \Omega) = 9.00 \text{ V}$ eq.(2)

Apply the loop rule to loop (3):

$-I_3(10.0 \ \Omega) - 9.00 \text{ V} + I_1(1.00 \ \Omega) - I_2(1.00 \ \Omega) + 12.0 \text{ V} = 0$

$-I_1(1.00 \ \Omega) + I_2(1.00 \ \Omega) + I_3(10.0 \ \Omega) = 3.00 \text{ V}$ eq.(3)

Eq.(1) gives $I_2 = 2.00 \text{ A} + \frac{5}{6}I_3$;   eq.(2) gives $I_1 = 1.00 \text{ A} - \frac{8}{9}I_3$

Using these results in eq.(3) gives $-(1.00 \text{ A} - \frac{8}{9}I_3)(1.00 \ \Omega) + (2.00 \text{ A} + \frac{5}{6}I_3)(1.00 \ \Omega) +$
$I_3(10.0 \ \Omega) = 3.00 \text{ V}$

$(\frac{16+15+180}{18})I_3 = 2.00 \text{ A};$ $I_3 = \frac{18}{211}(2.00 \text{ A}) = 0.171 \text{ A}$

Then $I_2 = 2.00 \text{ A} + \frac{5}{6}I_3 = 2.00 \text{ A} + \frac{5}{6}(0.171 \text{ A}) = 2.14 \text{ A}$

and $I_1 = 1.00 \text{ A} - \frac{8}{9}I_3 = 1.00 \text{ A} - \frac{8}{9}(0.171 \text{ A}) = 0.848 \text{ A}.$

27-47

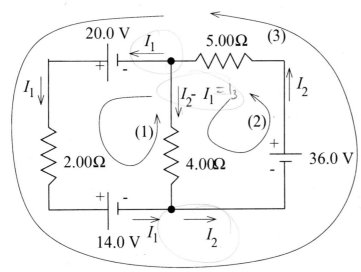

Two unknown currents I_1 (through the 2.00 Ω resistor) and I_2 (through the 5.00 Ω resistor) are labeled on the circuit diagram. The current through the 4.00 Ω resistor has been written as $I_2 - I_1$ using the junction rule.

Apply the loop rule to loops (1) and (2) to get two equations for the unknown currents I_1 and I_2. Loop (3) can then be used to check the results.

loop (1):

$+20.0 \text{ V} - I_1(2.00 \ \Omega) - 14.0 \text{ V} + (I_2 - I_1)(4.00 \ \Omega) = 0$

$6.00I_1 - 4.00I_2 = 6.00 \text{ A}$

$3.00 I_1 - 2.00 I_2 = 3.00$ A   eq.(1)

loop (2):

$+36.0$ V $- I_2 (5.00 \ \Omega) - (I_2 - I_1)(4.00 \ \Omega) = 0$

$-4.00 I_1 + 9.00 I_2 = 36.0$ A   eq.(2)

Solving eq.(1) for I_1 gives $I_1 = 1.00$ A $+ \frac{2}{3} I_2$

Using this in eq.(2) gives $-4.00(1.00$ A $+ \frac{2}{3} I_2) + 9.00 I_2 = 36.0$ A

$(-\frac{8}{3} + 9.00) I_2 = 40.0$ A and $I_2 = 6.32$ A.

Then $I_1 = 1.00$ A $+ \frac{2}{3} I_2 = 1.00$ A $+ \frac{2}{3}(6.32$ A$) = 5.21$ A.

In summary then

Current through the 2.00 Ω resistor: $I_1 = 5.21$ A.

Current through the 5.00 Ω resistor: $I_2 = 6.32$ A.

Current through the 4.00 Ω resistor: $I_2 - I_1 = 6.32$ A $- 5.21$ A $= 1.11$ A.

Loop (3) to check: $+20.0$ V $- I_1 (2.00 \ \Omega) - 14.0$ V $+ 36.0$ V $- I_2 (5.00 \ \Omega) = 0$

$(5.21$ A$)(2.00 \ \Omega) + (6.32$ A$)(5.00 \ \Omega) = 42.0$ V

10.4 V $+ 31.6$ V $= 42.0$ V, so the loop rule is satisfied for this loop.

27-49 a) Break the circuit between points a and b means no current in the middle branch that contains the 3.00 Ω resistor and the 10.0 V battery. The circuit therefore has a single current path. Find the current, so that potential drops across the resistors can be calculated.

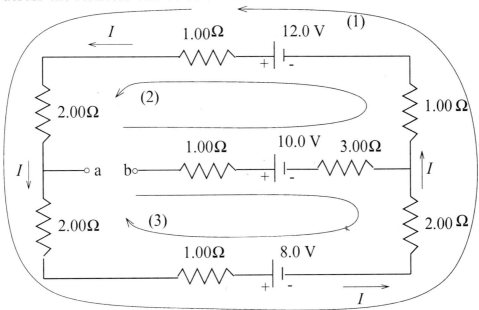

Apply the loop rule to loop (1):

$+12.0 \text{ V} - I(1.00 \ \Omega + 2.00 \ \Omega + 2.00 \ \Omega + 1.00 \ \Omega) - 8.0 \text{ V} - I(2.00 \ \Omega + 1.00 \ \Omega) = 0$

$$I = \frac{12.0 \text{ V} - 8.0 \text{ V}}{9.00 \ \Omega} = 0.4444 \text{ A}.$$

To find V_{ab} start at point b and travel to a, adding up the potential rises and drops. Travel on path (2) shown on the diagram. The 1.00 Ω and 3.00 Ω resistors in the middle branch have no current through them and hence no voltage across them. Therefore,

$V_b - 10.0 \text{ V} + 12.0 \text{ V} - I(1.00 \ \Omega + 1.00 \ \Omega + 2.00 \ \Omega) = V_a;$

thus $V_a - V_b = 2.0 \text{ V} - (0.4444 \text{ A})(4.00 \ \Omega) = +0.22 \text{ V}$

(point a is at higher potential)

As a check on this calculation we also compute V_{ab} by traveling from b to a on path (3):

$V_b - 10.0 \text{ V} + 8.0 \text{ V} + I(2.00 \ \Omega + 1.00 \ \Omega + 2.00 \ \Omega) = V_a$

$V_{ab} = -2.00 \text{ V} + (0.4444 \text{ A})(5.00 \ \Omega) = +0.22 \text{ V}$, which checks.

b) With points a and b connected by a wire there are three current branches:

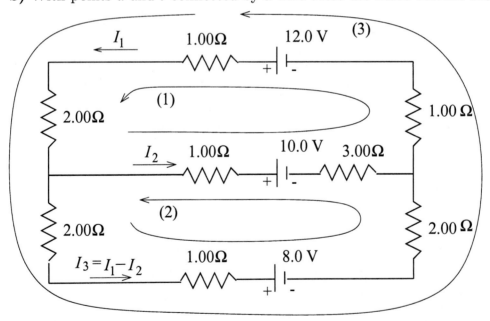

The junction rule has been used to write the third current (in the 8.0 V battery) in terms of the other currents.

Apply the loop rule to loop (1):

$12.0 \text{ V} - I_1(1.00 \ \Omega) - I_1(2.00 \ \Omega) - I_2(1.00 \ \Omega) - 10.0 \text{ V} - I_2(3.00 \ \Omega) - I_1(1.00 \ \Omega) = 0$

$2.0 \text{ V} - I_1(4.00 \ \Omega) - I_2(4.00 \ \Omega) = 0$

$(2.00 \ \Omega)I_1 + (2.00 \ \Omega)I_2 = 1.0 \text{ V} \qquad \text{eq.}(1)$

Apply the loop rule to loop (2):

$-(I_1 - I_2)(2.00 \ \Omega) - (I_1 - I_2)(1.00 \ \Omega) - 8.0 \text{ V} - (I_1 - I_2)(2.00 \ \Omega) + I_2(3.00 \ \Omega) + 10.0 \text{ V} + I_2(1.00 \ \Omega) = 0$

$2.0 \text{ V} - (5.00 \ \Omega)I_1 + (9.00 \ \Omega)I_2 = 0 \qquad \text{eq.}(2)$

Solve eq.(1) for I_2 and use this to replace I_2 in eq.(2):

$I_2 = 0.50 \text{ A} - I_1$

$2.0 \text{ V} - (5.00 \ \Omega)I_1 + (9.00 \ \Omega)(0.50 \text{ A} - I_1) = 0$

$(14.0 \ \Omega)I_1 = 6.50 \text{ V so } I_1 = (6.50 \text{ V})/(14.0 \ \Omega) = 0.464 \text{ A}$

$I_2 = 0.500 \text{ A} - 0.464 \text{ A} = 0.036 \text{ A}.$

The current in the 12.0 V battery is $I_1 = 0.464 \text{ A}$

27-51 a)

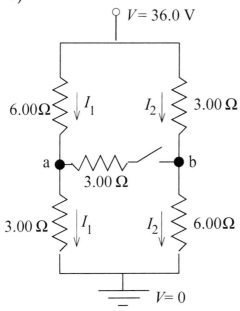

With the switch open there is no current through it and there are only the two currents I_1 and I_2 indicated in the sketch.

The potential drop across each parallel branch is 36.0 V:

$-I_1(6.00 \ \Omega + 3.00 \ \Omega) + 36.0 \text{ V} = 0$

$$I_1 = \frac{36.0 \text{ V}}{6.00 \ \Omega + 3.00 \ \Omega} = 4.00 \text{ A}$$

$-I_2(3.00 \ \Omega + 6.00 \ \Omega) + 36.0 \text{ V} = 0$

$$I_2 = \frac{36.0 \text{ V}}{3.00 \ \Omega + 6.00 \ \Omega} = 4.00 \text{ A}$$

To calculate $V_{ab} = V_a - V_b$ start at point b and travel to point a, adding up all the potential rises and drops along the way. We can do this by going from b up through the 3.00 Ω resistor:

$V_b + I_2(3.00 \ \Omega) - I_1(6.00 \ \Omega) = V_a$

$V_a - V_b = (4.00 \ \text{A})(3.00 \ \Omega) - (4.00 \ \text{A})(6.00 \ \Omega) = 12.0 \ \text{V} - 24.0 \ \text{V} = -12.0 \ \text{V}$

$V_{ab} = -12.0 \ \text{V}$ (point a is 12.0 V lower in potential than point b)

Alternatively, we can go from point b down through the 6.00 Ω resistor:

$V_b - I_2(6.00 \ \Omega) + I_1(3.00 \ \Omega) = V_a$

$V_a - V_b = -(4.00 \ \text{A})(6.00 \ \Omega) + (4.00 \ \text{A})(3.00 \ \Omega) = -24.0 \ \text{V} + 12.0 \ \text{V} = -12.0 \ \text{V}$

b)

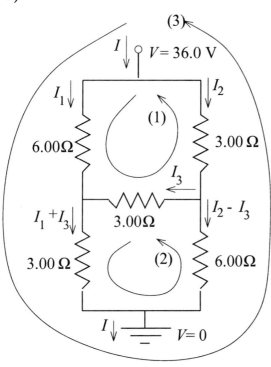

The three unknown currents I_1, I_2, and I_3 are labeled on the sketch.

Apply the loop rule to loops (1), (2), and (3).

loop (1):
$-I_1(6.00 \ \Omega) + I_3(3.00 \ \Omega) + I_2(3.00 \ \Omega) = 0$

$I_2 = 2I_1 - I_3$ eq.(1)

loop (2):
$-(I_1 + I_3)(3.00 \ \Omega) + (I_2 - I_3)(6.00 \ \Omega) - I_3(3.00 \ \Omega) = 0$

$6I_2 - 12I_3 - 3I_1 = 0$ so $2I_2 - 4I_3 - I_1 = 0$

Use eq.(1) to replace I_2:

$4I_1 - 2I_3 - 4I_3 - I_1 = 0$

$3I_1 = 6I_3$ and $I_1 = 2I_3$ eq.(2)

loop (3) (This loop is completed through the battery [not shown], in the direction from the − to the + terminal.):

$-I_1(6.00 \ \Omega) - (I_1 + I_3)(3.00 \ \Omega) + 36.0 \ \text{V} = 0$

$9I_1 + 3I_3 = 36.0 \ \text{A}$ and $3I_1 + I_3 = 12.0 \ \text{A}$ eq.(3)

Use eq.(2) in eq.(3) to replace I_1:

$3(2I_3) + I_3 = 12.0 \ \text{A}$

$I_3 = 12.0 \ \text{A}/7 = 1.71 \ \text{A}$

$I_1 = 2I_3 = 3.42 \ \text{A}$

$I_2 = 2I_1 - I_3 = 2(3.42 \ \text{A}) - 1.71 \ \text{A} = 5.13 \ \text{A}$

The current through the switch is $I_3 = 1.71 \ \text{A}$.

c) From the results in part (a) the current through the battery is $I = I_1 + I_2 = 3.42 \ \text{A} + 5.13 \ \text{A} = 8.55 \ \text{A}$. The equivalent circuit is a single resitor that produces the same current through the 36.0 V battery.

36.0 V

$I = 8.55 \ \text{A}$

R

$-IR + 36.0 \ \text{V} = 0$

$R = \dfrac{36.0 \ \text{V}}{I} = \dfrac{36.0 \ \text{V}}{8.55 \ \text{A}} = 4.21 \ \Omega$

27-53 a)

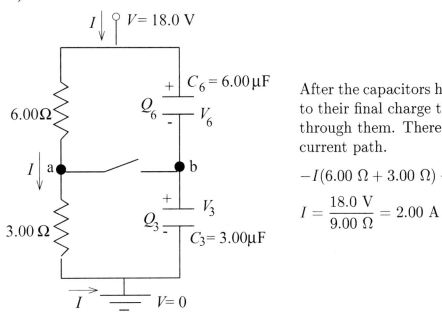

After the capacitors have been charged to their final charge there is no current through them. There is only one current path.

$-I(6.00 \ \Omega + 3.00 \ \Omega) + 18.0 \ \text{V} = 0$

$I = \dfrac{18.0 \ \text{V}}{9.00 \ \Omega} = 2.00 \ \text{A}$

There is also a potential difference of 18.0 V applied across the two capacitors in series. The capacitors have the same charge: $Q_3 = Q_6 = Q$.

$V_6 + V_3 = 18.0$ V and $V = Q/C$ so

$$Q\left(\frac{1}{C_6} + \frac{1}{C_3}\right) = 18.0 \text{ V} \text{ so } Q\left(\frac{C_3 + C_6}{C_3 C_6}\right) = 18.0 \text{ V}$$

$$Q = \left(\frac{C_3 C_6}{C_3 + C_6}\right)(18.0 \text{ V}) = \left[\frac{(3.00 \times 10^{-6} \text{ F})(6.00 \times 10^{-6} \text{ F})}{3.00 \times 10^{-6} \text{ F} + 6.00 \times 10^{-6} \text{ F}}\right]18.0 \text{ V} = 36.0 \text{ } \mu\text{C}$$

And then $V_6 = \dfrac{Q}{C_6} = \dfrac{36.0 \text{ } \mu\text{C}}{6.00 \text{ } \mu\text{C}} = 6.00$ V, $V_3 = \dfrac{Q}{C_3} = \dfrac{36.0 \text{ } \mu\text{C}}{3.00 \text{ } \mu\text{C}} = 12.0$ V

$V_b + V_6 - I(6.00 \text{ } \Omega) = V_a$

$V_a - V_b = 6.00 \text{ V} - (2.00 \text{ A})(6.00 \text{ } \Omega) = -6.00$ V

Or, $V_b - V_3 + I(3.00 \text{ } \Omega) = V_a$

$V_a - V_b = (2.00 \text{ A})(3.00 \text{ } \Omega) - 12.0 \text{ V} = -6.00$ V, which checks

b) $V_{ab} < 0$; point b is at higher potential

c)

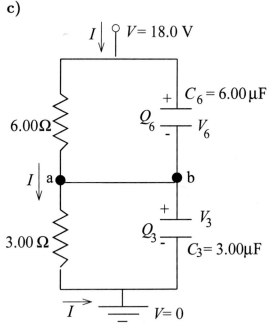

The only current path is still the one through both resistors, so $I = 2.00$ A as calculated in part (a).

The effect of closing the switch is to put points a and b at the same potential, so that $I(6.00 \text{ } \Omega) = 12.0$ V is the potential V_6 across the 6.00 μF capacitor and $I(3.00 \text{ } \Omega) = 6.00$ V is the potential V_3 across the 3.00 μF capacitor.

The potential of point b above ground is $V_b = I(3.00 \text{ } \Omega) = V_3 = 6.00$ V.

d) The charges on the capacitors after the switch is closed are

$Q_6 = C_6 V_6 = (6.00 \text{ } \mu\text{F})(12.0 \text{ V}) = 72.0 \text{ } \mu\text{C}$

$Q_3 = C_3 V_3 = (3.00 \text{ } \mu\text{F})(6.00 \text{ V}) = 18.0 \text{ } \mu\text{C}$

From part (a), with the switch open the charge on each capacitor was 36.0 μC.

<u>switch open</u>

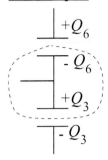

The net charge on the conductor enclosed by the dashed line is zero.

<u>switch closed</u>

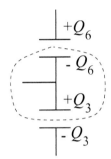

The net charge on the conductor enclosed by the dashed line is now $-Q_6 + Q_3 =$
$-72.0 \ \mu C + 18.0 \ \mu C = -54.0 \ \mu C$

The charge $-54.0 \ \mu$C must have flowed through the switch when it was closed.

27-55 For each range setting the circuit has the form

$I_{fs} = 1.00 \times 10^{-3} A$

$R_G = 40.0\Omega$ R

V

<u>3.00 V</u>
For $V = 3.00$ V, $R = R_1$ and the total meter resistance R_m is $R_m = R_G + R_1$.

$V = I_{fs} R_m$ so $R_m = \dfrac{V}{I_{fs}} = \dfrac{3.00 \text{ V}}{1.00 \times 10^{-3} \text{ A}} = 3.00 \times 10^3 \ \Omega$.

$R_m = R_G + R_1$ so $R_1 = R_m - R_G = 3.00 \times 10^3 \ \Omega - 40.0 \ \Omega = 2960 \ \Omega$

<u>15.0 V</u>
For $V = 15.0$ V, $R = R_1 + R_2$ and the total meter resistance is $R_m = R_G + R_1 + R_2$.

$V = I_{fs} R_m$ so $R_m = \dfrac{V}{I_{fs}} = \dfrac{15.0 \text{ V}}{1.00 \times 10^{-3} \text{ A}} = 1.50 \times 10^4 \ \Omega$.

$R_2 = R_m - R_G - R_1 = 1.50 \times 10^4 \ \Omega - 40.0 \ \Omega - 2960 \ \Omega = 1.20 \times 10^4 \ \Omega$

<u>150 V</u>

For $V = 150$ V, $R = R_1 + R_2 + R_3$ and the total meter resistance is $R_m = R_G + R_1 + R_2 + R_3$.

$V = I_{fs} R_m$ so $R_m = \dfrac{V}{I_{fs}} = \dfrac{150 \text{ V}}{1.00 \times 10^{-3} \text{ A}} = 1.50 \times 10^5 \ \Omega.$

$R_3 = R_m - R_G - R_1 - R_2 = 1.50 \times 10^5 \ \Omega - 40.0 \ \Omega - 2960 \ \Omega - 1.20 \times 10^4 \ \Omega = 1.35 \times 10^5$

27-61 a) $I = I_1 = I_2$

$I = \dfrac{90.0 \text{ V}}{R_1 + R_2} = \dfrac{90.0 \text{ V}}{224 \ \Omega + 589 \ \Omega} = 0.1107$ A

$V_1 = I_1 R_1 = (0.1107 \text{ A})(224 \ \Omega) = 24.8$ V; $V_2 = I_2 R_2 = (0.1107 \text{ A})(589 \ \Omega) = 65.2$ V

b)

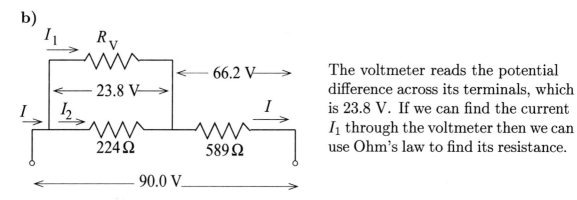

The voltmeter reads the potential difference across its terminals, which is 23.8 V. If we can find the current I_1 through the voltmeter then we can use Ohm's law to find its resistance.

The voltage drop across the 589 Ω resistor is 90.0 V $-$ 23.8 V $= 66.2$ V, so

$I = \dfrac{V}{R} = \dfrac{66.2 \text{ V}}{589 \ \Omega} = 0.1124$ A.

The voltage drop across the 224 Ω resistor is 23.8 V, so $I_2 = \dfrac{V}{R} = \dfrac{23.8 \text{ V}}{224 \ \Omega} = 0.1062$ A.

Then $I = I_1 + I_2$ gives $I_1 = I - I_2 = 0.1124$ A $-$ 0.1062 A $= 0.0062$ A.

$R_V = \dfrac{V}{I_1} = \dfrac{23.8 \text{ V}}{0.0062 \text{ A}} = 3840 \ \Omega$

c)

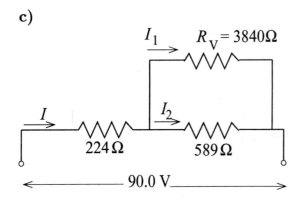

Replace the two resistors in parallel by their equivalent:

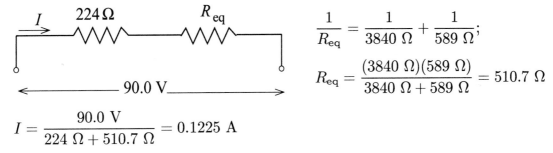

$$\frac{1}{R_{eq}} = \frac{1}{3840\ \Omega} + \frac{1}{589\ \Omega};$$

$$R_{eq} = \frac{(3840\ \Omega)(589\ \Omega)}{3840\ \Omega + 589\ \Omega} = 510.7\ \Omega$$

$$I = \frac{90.0\ \text{V}}{224\ \Omega + 510.7\ \Omega} = 0.1225\ \text{A}$$

The potential drop across the 224 Ω resistor then is $IR = (0.1125\ \text{A})(224\ \Omega) = 27.4$ V, so the potential drop across the 589 Ω resistor and across the voltmeter (what the voltmeter reads) is $90.0\ \text{V} - 27.4\ \text{V} = 62.6\ \text{V}$.

d) No, any real voltmeter will draw some current and thereby reduce the current through the resistance whose voltage is being measured. Thus the presence of the voltmeter connected in parallel with the resistance lowers the voltage drop across that resistance.

27-63 a)

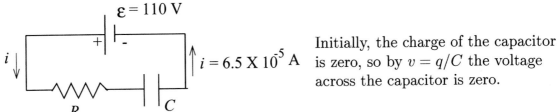

$$v_C = \frac{q}{C}$$

$$v_C = \frac{815 \times 10^{-6}\ \text{C}}{3.40 \times 10^{-6}\ \text{F}} = 239.7\ \text{V}$$

$$+v_C - iR - \varepsilon = 0$$

$$iR = v_C - \varepsilon = 239.7\ \text{V} - 180\ \text{V} = 59.7\ \text{V}$$

$$i = \frac{59.7\ \text{V}}{7.25 \times 10^3\ \Omega} = 8.24 \times 10^{-3}\ \text{A} = 8.24\ \text{mA, toward the negative plate.}$$

b) After a long time $i = 0$ so $v_C = \varepsilon$.

$q = \varepsilon C = (180\ \text{V})(3.40 \times 10^{-6}\ \text{F}) = 6.12 \times 10^{-4}\ \text{C} = 612\ \mu\text{C}$.

27-65

$\varepsilon = 110$ V

$i = 6.5 \times 10^5$ A

Initially, the charge of the capacitor is zero, so by $v = q/C$ the voltage across the capacitor is zero.

The loop rule therefore gives $\varepsilon - iR = 0$ and

$$R = \frac{\varepsilon}{i} = \frac{110\ \text{V}}{6.5 \times 10^{-5}\ \text{A}} = 1.7 \times 10^6\ \Omega$$

The time constant is given by $\tau = RC$ (Eq.27-14), so

$$C = \frac{\tau}{R} = \frac{6.2\ \text{s}}{1.7 \times 10^6\ \Omega} = 3.6\ \mu\text{F}.$$

27-69 Eq.(27-13) $i = \dfrac{\varepsilon}{R} e^{-t/RC}$

a) $P = \varepsilon i$

The total energy supplied by the battery is

$$\int_0^\infty P\,dt = \int_0^\infty \varepsilon i\,dt = (\varepsilon^2/R) \int_0^\infty e^{-t/RC}\,dt = (\varepsilon^2/R) \left[-RCe^{-t/RC} \right]_0^\infty = C\varepsilon^2.$$

b) $P = i^2 R$

The total energy dissipated in the resistor is

$$\int_0^\infty P\,dt = \int_0^\infty i^2 R\,dt = (\varepsilon^2/R) \int_0^\infty e^{-2t/RC}\,dt = (\varepsilon^2/R) \left[-(RC/2)e^{-2t/RC} \right]_0^\infty = \tfrac{1}{2}C\varepsilon^2.$$

c) The final charge on the capacitor is $Q = C\varepsilon$. The energy stored is
$U = Q^2/(2C) = \tfrac{1}{2}C\varepsilon^2.$

The final energy stored in the capacitor $\left(\tfrac{1}{2}C\varepsilon^2 \right)$ = total energy supplied by the battery $\left(C\varepsilon^2 \right)$ - energy dissipated in the resistor $\left(\tfrac{1}{2}C\varepsilon^2 \right)$

d) $\tfrac{1}{2}$ of the energy supplied by the batery is stored in the capacitor. This fraction is independent of R.

CHAPTER 28

MAGNETIC FIELD AND MAGNETIC FORCES

Exercises 1, 3, 9, 11, 13, 15, 19, 23, 25, 29, 33, 35, 37
Problems 39, 41, 43, 47, 51, 53, 57, 59, 61, 65, 67

Exercises

28-1 $\vec{v} = (+4.19 \times 10^4 \text{ m/s})\hat{i} + (-3.85 \times 10^4 \text{ m/s})\hat{j}$

a) $\vec{B} = (1.40 \text{ T})\hat{i}$

$\vec{F} = q\vec{v}\mathbf{X}\vec{B} = (-1.24\times10^{-8} \text{ C})(1.40 \text{ T})[(+4.19\times10^4 \text{ m/s})\hat{i}\mathbf{X}\hat{i} - (3.85\times10^4 \text{ m/s})\hat{j}\mathbf{X}\hat{i}]$

$\hat{i}\mathbf{X}\hat{i} = 0, \quad \hat{j}\mathbf{X}\hat{i} = -\hat{k}$

$\vec{F} = (-1.24 \times 10^{-8} \text{ C})(1.40 \text{ T})(-3.85 \times 10^4 \text{ m/s})(-\hat{k}) = (-6.68 \times 10^{-4} \text{ N})\hat{k}$

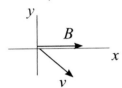

The right-hand rule gives that $\vec{v}\mathbf{X}\vec{B}$ is directed out of the paper ($+z$-direction). The charge is negative so $\vec{F}$ is opposite to $\vec{v}\mathbf{X}\vec{B}$;

$\vec{F}$ is in the $-z$-direction. This agrees with the direction calculated with unit vectors.

b) $\vec{B} = (1.40 \text{ T})\hat{k}$

$\vec{F} = q\vec{v}\mathbf{X}\vec{B} = (-1.24\times10^{-8} \text{ C})(1.40 \text{ T})[(+4.19\times10^4 \text{ m/s})\hat{i}\mathbf{X}\hat{k} - (3.85\times10^4 \text{ m/s})\hat{j}\mathbf{X}\hat{k}]$

$\hat{i}\mathbf{X}\hat{k} = -\hat{j}, \quad \hat{j}\mathbf{X}\hat{k} = \hat{i}$

$\vec{F} = (-7.27 \times 10^{-4} \text{ N})(-\hat{j}) + (6.68 \times 10^{-4} \text{ N})\hat{i} = (6.68 \times 10^{-4} \text{ N})\hat{i} + (7.27 \times 10^{-4} \text{ N})\hat{j}]$

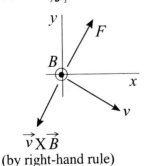

The direction of $\vec{F}$ is opposite to $\vec{v}\mathbf{X}\vec{B}$ since q is negative. The direction of $\vec{F}$ computed from the right-hand rule agrees qualitatively with the direction calculated with unit vectors.

28-3 The force must be in the direction the particle is deflected.

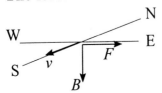

Or, if view from above,

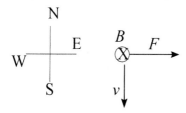

$\vec{F}$ is in the direction of $\vec{v} \times \vec{B}$ as given by the right-hand rule, so q is positive.

28-9 $\Phi_B = \int \vec{B} \cdot d\vec{A}$

Circular area in the xy-plane, so $A = \pi r^2 = \pi (0.0650 \text{ m})^2 = 0.01327 \text{ m}^2$ and $d\vec{A}$ is in the z-direction.

a) $\vec{B} = (0.230 \text{ T})\hat{k}$; $\vec{B}$ and $d\vec{A}$ are parallel ($\phi = 0°$) so $\vec{B} \cdot d\vec{A} = B \, dA$.
B is constant over the circular area so
$\Phi_B = \int \vec{B} \cdot d\vec{A} = \int B \, dA = B \int dA = BA = (0.230 \text{ T})(0.01327 \text{ m}^2) = $
$3.05 \times 10^{-3} \text{ Wb}$

b)

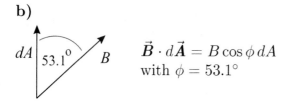

$\vec{B} \cdot d\vec{A} = B \cos \phi \, dA$
with $\phi = 53.1°$

B and ϕ are constant over the circular area so
$\Phi_B = \int \vec{B} \cdot d\vec{A} = \int B \cos \phi \, dA = B \cos \phi \int dA = B \cos \phi A$
$\Phi_B = (0.230 \text{ T}) \cos 53.1° (0.01327 \text{ m}^2) = 1.83 \times 10^{-3} \text{ Wb}$

c)

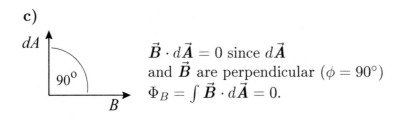

$\vec{B} \cdot d\vec{A} = 0$ since $d\vec{A}$
and $\vec{B}$ are perpendicular ($\phi = 90°$)
$\Phi_B = \int \vec{B} \cdot d\vec{A} = 0.$

28-11 **a)** $\vec{B} = (\beta - \gamma y^2)\hat{j}$ with $\beta = 0.300 \text{ T}$ and $\gamma = 2.00 \text{ T/m}^2$

surface *abcd*
$d\vec{A} = dA(-\hat{i})$ $\vec{B} \cdot d\vec{A} = -(\beta - \gamma y^2)dA(\hat{j} \cdot \hat{i}) = 0$, so $\Phi_B = 0.$

surface *befc*

$$d\vec{A} = dA(-\hat{k}) \qquad \vec{B} \cdot d\vec{A} = -(\beta - \gamma y^2)dA(\hat{j} \cdot \hat{k}) = 0, \text{ so } \Phi_B = 0.$$

surface $aefd$

$$d\vec{A} = dA(\hat{i} + \hat{k})/\sqrt{2} \qquad \vec{B} \cdot d\vec{A} = (\beta - \gamma y^2)dA(1/\sqrt{2})(\hat{i} + \hat{k}) \cdot \hat{j} = 0, \text{ so } \Phi_B = 0.$$

surface abe

$$d\vec{A} = dA\hat{j} \qquad \vec{B} \cdot d\vec{A} = (\beta - \gamma y^2)dA\hat{j} \cdot \hat{j} = (\beta - \gamma y^2)dA$$

$\Phi_B = \int \vec{B} \cdot d\vec{A} = \int (\beta - \gamma y^2)\, dA = (\beta - \gamma y^2) \int dA = (\beta - \gamma y^2)A$ (since y is constant on the surface)

$y = 0.300$ m, $\quad A = \frac{1}{2}(0.400 \text{ m})(0.300 \text{ m}) = 0.0600 \text{ m}^2$

$\Phi_B = (0.300 \text{ T} - (2.00 \text{ T/m}^2)(0.300 \text{ m})^2)(0.0600 \text{ m}^2) = 0.00720$ Wb

surface cfd

On this surface $y = 0$ so $\vec{B} = \beta\hat{j}$. $d\vec{A} = -dA\hat{j}$, so $\vec{B} \cdot d\vec{A} = -\beta\, dA$. Then $\Phi_B = -\beta A = -(0.300 \text{ T})(0.0600 \text{ m}^2) = -0.0180$ Wb.

The net flux is the sum of the fluxes through each surface. Only for surfaces abe and cfd is the flux nonzero, so the net flux is $\Phi_B = +0.00720 \text{ Wb} - 0.0180 \text{ Wb} = -0.0108$ Wb.

b) Gauss's law for magnetism (Eq.(28-8)) says $\oint \vec{B} \cdot d\vec{A} = 0$. But in part (a) we found $\oint \vec{B} \cdot d\vec{A} = $ net $\Phi_B = -0.0108$ Wb. The magnetic field produces a flux that violates Gauss's law for magnetism so is not possible.

28-13 a)

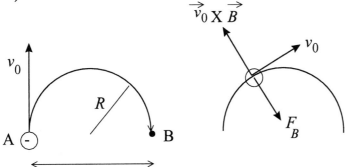

As the electron moves in the semicircle, its velocity is tangent to the circular path.

For circular motion the acceleration of the electron $\vec{a}_{\text{rad}}$ is directed in toward the center of the circle. Thus the force $\vec{F}_B$ exerted by the magnetic field, since it is the only force on the electron, must be radially inward. Since q is negative, $\vec{F}_B$ is opposite to the direction given by the right-hand rule for $\vec{v}_0 \times \vec{B}$. Thus $\vec{B}$ is directed into the page.

Apply Newton's 2nd law to calculate the magnitude of $\vec{B}$:

$\sum \vec{F} = m\vec{a}$ gives $\sum F_{\text{rad}} = ma$

$$F_B = m(v^2/R)$$

$$F_B = |q|vB\sin\phi = |q|vB, \text{ so } |q|vB = m(v^2/R)$$

$$B = \frac{mv}{|q|R} = \frac{(9.109 \times 10^{-31}\text{ kg})(1.41 \times 10^6\text{ m/s})}{(1.602 \times 10^{-19}\text{ C})(0.050\text{ m})} = 1.60 \times 10^{-4}\text{ T}$$

b) The speed of the electron as it moves along the path is constant. ($\vec{F}_B$ changes the direction of $\vec{v}$ but not its magnitude.) The time is given by the distance divided by v_0. The distance along the semicircular path is πR, so

$$t = \frac{\pi R}{v_0} = \frac{\pi(0.050\text{ m})}{1.41 \times 10^6\text{ m/s}} = 1.11 \times 10^{-7}\text{ s}$$

28-15   a) $\sum \vec{F} = m\vec{a}$ says $|q|vB = m(v^2/R)$

$$v = \frac{|q|BR}{m} = \frac{(1.602 \times 10^{-19}\text{ C})(2.50\text{ T})(6.96 \times 10^{-3}\text{ m})}{3.34 \times 10^{-27}\text{ kg}} = 8.35 \times 10^5\text{ m/s}$$

b) $t = \dfrac{\pi R}{v} = \dfrac{\pi(6.96 \times 10^{-3}\text{ m})}{8.35 \times 10^5\text{ m/s}} = 2.62 \times 10^{-8}\text{ s}$

c) kinetic energy gained = electric potential energy lost

$$\tfrac{1}{2}mv^2 = |q|V$$

$$V = \frac{mv^2}{2|q|} = \frac{(3.34 \times 10^{-27}\text{ kg})(8.35 \times 10^5\text{ m/s})^2}{2(1.602 \times 10^{-19}\text{ C})} = 7.27 \times 10^3\text{ V} = 7.27\text{ kV}$$

28-19   a) In the plane perpendicular to $\vec{B}$ (the yz-plane) the motion is circular. But there is a velocity component in the direction of $\vec{B}$, so the motion is a helix. The electric field in the $+\hat{i}$ direction exerts a force in the $+\hat{i}$ direction. This force produces an acceleration in the $+\hat{i}$ direction and this causes the pitch of the helix to vary. The force does not affect the circular motion in the yz-plane, so the electric field does not affect the radius of the helix.

b) Calculate the period T:

$$\omega = |q|B/m$$

$$T = \frac{2\pi}{\omega} = \frac{2\pi m}{|q|B} = \frac{2\pi(1.67 \times 10^{-27}\text{ kg})}{(1.60 \times 10^{-19}\text{ C})(0.500\text{ T})} = 1.312 \times 10^{-7}\text{ s}.$$

Then $t = T/2 = 6.56 \times 10^{-8}$ s.

$$v_{0x} = 1.50 \times 10^5\text{ m/s}$$

$$a_x = \frac{F_x}{m} = \frac{(1.60 \times 10^{-19} \text{ C})(2.00 \times 10^4 \text{ V/m})}{1.67 \times 10^{-27} \text{ kg}} = +1.916 \times 10^{12} \text{ m/s}^2$$

$$x - x_0 = v_{0x}t + \tfrac{1}{2}a_x t^2$$

$x - x_0 = (1.50 \times 10^5 \text{ m/s})(6.56 \times 10^{-8} \text{ s}) + \tfrac{1}{2}(1.916 \times 10^{12} \text{ m/s}^2)(6.56 \times 10^{-8} \text{ s})^2 = 1.40$
cm

28-23 In the velocity selector $|q|E = |q|vB$.

$$v = \frac{E}{B} = \frac{1.12 \times 10^5 \text{ V/m}}{0.540 \text{ T}} = 2.074 \times 10^5 \text{ m/s}$$

In the region of the circular path $\sum \vec{F} = m\vec{a}$ gives

$|q|vB = m(v^2/R)$ so $m = |q|RB/v$

Singly charged ion, so $|q| = +e = 1.602 \times 10^{-19}$ C

$$m = \frac{(1.602 \times 10^{-19} \text{ C})(0.310 \text{ m})(0.540 \text{ T})}{2.074 \times 10^5 \text{ m/s}} = 1.29 \times 10^{-25} \text{ kg}$$

Mass number = mass in atomic mass units, so is $\dfrac{1.29 \times 10^{-25} \text{ kg}}{1.66 \times 10^{-27} \text{ kg}} = 78.$

28-25 The force $\vec{F}_B$ on the current due to the magnetic field must be upward and and equal in magnitude to the weight mg of the rod. This force has its maximum magnitude $F_B = IlB = ILB$ when the magnetic field is pependicular to the current direction. Then

$ILB = mg$ and $B = mg/IL$

b)

To produce a force that opposes gravity the magnetic field must be directed to the west.

28-29 $\vec{F} = I\vec{l}\times\vec{B}$

$I = 3.50$ A, $\vec{l} = -(0.0100 \text{ m})\hat{i}$ (since the current is in the $-x$-direction)

a) $\vec{B} = -(0.65 \text{ T})\hat{j}$

$\vec{F} = (3.50 \text{ A})(-0.0100 \text{ m})(-0.65 \text{ T})\hat{i}\times\hat{j}$

$\hat{i} \times \hat{j} = \hat{k}$, so $\vec{F} = +(0.023 \text{ N})\hat{k}$

b) $\vec{B} = +(0.56 \text{ T})\hat{k}$

$\vec{F} = (3.50 \text{ A})(-0.0100 \text{ m})(0.56 \text{ T})\hat{i} \times \hat{k}$

$\hat{i} \times \hat{k} = -\hat{j}$, so $\vec{F} = +(0.020 \text{ N})\hat{j}$

c) $\vec{B} = -(0.31 \text{ T})\hat{i}$

$\vec{F} = (3.50 \text{ A})(-0.0100 \text{ m})(-0.31 \text{ T})\hat{i} \times \hat{i}$

$\hat{i} \times \hat{i} = 0$, so $F = 0$

d) $\vec{B} = (0.33 \text{ T})\hat{i} - (0.28 \text{ T})\hat{k}$

$\vec{F} = (3.50 \text{ A})(-0.0100 \text{ m})[(0.33 \text{ T})\hat{i} \times \hat{i} - (0.28 \text{ T})\hat{i} \times \hat{k}]$

$\hat{i} \times \hat{i} = 0, \hat{i} \times \hat{k} = -\hat{j}$ so $\vec{F} = (3.50 \text{ A})(-0.0100 \text{ m})(-0.28 \text{ T})(-\hat{j}) = -(0.0098 \text{ N})\hat{j}$

e) $\vec{B} = (0.74 \text{ T})\hat{j} - (0.36 \text{ T})\hat{k}$

$\vec{F} = (3.50 \text{ A})(-0.0100 \text{ m})[(0.74 \text{ T})\hat{i} \times \hat{j} - (0.36 \text{ T})\hat{i} \times \hat{k}]$

$\hat{i} \times \hat{j} = \hat{k}, \hat{i} \times \hat{k} = -\hat{j}$ so

$\vec{F} = -(0.026 \text{ N})\hat{k} + (0.013 \text{ N})(-\hat{j}) = -(0.013 \text{ N})\hat{j} - (0.026 \text{ N})\hat{k}$

28-33 $U = -\vec{\mu} \cdot \vec{B}$

For $\vec{\mu}$ and $\vec{B}$ parallel, $\phi = 0°$ and $\vec{\mu} \cdot \vec{B} = \mu B \cos \phi = \mu B$.

For $\vec{\mu}$ and $\vec{B}$ antiparallel, $\phi = 180°$ and $\vec{\mu} \cdot \vec{B} = \mu B \cos \phi = -\mu B$.

$U_1 = +\mu B, U_2 = -\mu B$

$\Delta U = U_2 - U_1 = -2\mu B = -2(1.45 \text{ A} \cdot \text{m}^2)(0.835 \text{ T}) = -2.42 \text{ J}$

28-35

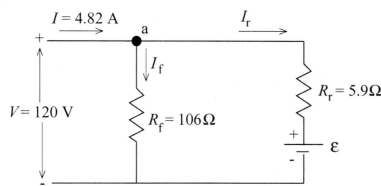

ε is the induced emf developed by the motor. It is directed so as to oppose the current through the rotor.

a) The field coils and the rotor are in parallel with the applied potential difference V, so $V = I_f R_f$.

$I_f = \dfrac{V}{R_f} = \dfrac{120 \text{ V}}{106 \text{ }\Omega} = 1.13 \text{ A}.$

b) Applying the junction rule to point a in the circuit diagram gives $I - I_f - I_r = 0$.
$I_r = I - I_f = 4.82 \text{ A} - 1.13 \text{ A} = 3.69 \text{ A}$.

c) The potential drop across the rotor, $I_r R_r + \varepsilon$, must equal the applied potential difference V: $V = I_r R_r + \varepsilon$

$\varepsilon = V - I_r R_r = 120 \text{ V} - (3.69 \text{ A})(5.9 \ \Omega) = 98.2 \text{ V}$

d) The mechanical power output is the electrical power input minus the rate of dissipation of electrical energy in the resistance of the motor:

electrical power input to the motor
$P_{\text{in}} = IV = (4.82 \text{ A})(120 \text{ V}) = 578 \text{ W}$

electrical power loss in the two resistances
$P_{\text{loss}} = I_f^2 R_f + I_r^2 R = (1.13 \text{ A})^2 (106 \ \Omega) + (3.69 \text{ A})^2 (5.9 \ \Omega) = 216 \text{ W}$

mechanical power output
$P_{\text{out}} = P_{\text{in}} - P_{\text{loss}} = 578 \text{ W} - 216 \text{ W} = 362 \text{ W}$

The mechanical power output is the power associated with the induced emf ε
$P_{\text{out}} = P_\varepsilon = \varepsilon I_r = (98.2 \text{ V})(3.69 \text{ A}) = 362 \text{ W}$, which agrees with the above calculation.

28-37 a)

$J_x = n|q|v_d$

so $v_d = \dfrac{J_x}{n|q|}$

$J_x = \dfrac{I}{A} = \dfrac{I}{y_1 z_1} = \dfrac{120 \text{ A}}{(0.23 \times 10^{-3} \text{ m})(0.0118 \text{ m})} = 4.42 \times 10^7 \text{ A/m}^2$

$v_d = \dfrac{J_x}{n|q|} = \dfrac{4.42 \times 10^7 \text{ A/m}^2}{(5.85 \times 10^{28}/\text{ m}^3)(1.602 \times 10^{-19} \text{ C})} = 4.7 \times 10^{-3} \text{ m/s} = 4.7 \text{ mm/s}$

b) <u>magnitude of $\vec{E}$</u>

$|q|E_z = |q|v_d B_y$

$E_z = v_d B_y = (4.7 \times 10^{-3} \text{ m/s})(0.95 \text{ T}) = 4.5 \times 10^{-3} \text{ V/m}$

<u>direction of $\vec{E}$</u>
The drift velocity of the electrons is in the opposite direction to the current.

$$\vec{v}\mathbf{X}\vec{B}\ \uparrow$$

$$\vec{F}_B = q\vec{v}\mathbf{X}\vec{B} = -e\vec{v}\mathbf{X}\vec{B}\ \downarrow$$

$\vec{F}_E$ must oppose $\vec{F}_B$
so $\vec{F}_E$ is in the $-z$-direction

$\vec{F}_E = q\vec{E} = -e\vec{E}$ so $\vec{E}$ is opposite to the direction of $\vec{F}_E$ and thus $\vec{E}$ is in the $+z$-direction.

c) The Hall emf is the potential difference between the two edges of the strip (at $z = 0$ and $z = z_1$) that results from the electric field calculated in part (b).

$\varepsilon_{\text{Hall}} = Ez_1 = (4.5 \times 10^{-3}\ \text{V/m})(0.0118\ \text{m}) = 53\ \mu\text{V}$

Problems

28-39 a)

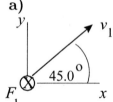

$\vec{F} = q\vec{v}\mathbf{X}\vec{B}$ says
that $\vec{F}$ is perpendicular to $\vec{v}$ and $\vec{B}$.

The information given here means that
$\vec{B}$ can have no z-component.

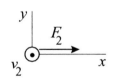

$\vec{F}$ is perpendicular to $\vec{v}$ and $\vec{B}$, so
$\vec{B}$ can have no x-component.

Both pieces of information taken together say that $\vec{B}$ is in the y-direction; $\vec{B} = B_y\hat{j}$.

Use the information given about $\vec{F}_2$ to calculate F_y: $\vec{F}_2 = F_2\hat{i}$, $\vec{v}_2 = v_2\hat{k}$, $\vec{B} = B_y\hat{j}$.

$\vec{F}_2 = q\vec{v}_2\mathbf{X}\vec{B}$ says $F_2\hat{i} = qv_2B_y\hat{k}\mathbf{X}\hat{j} = qv_2B_y(-\hat{i})$ and $F_2 = -qv_2B_y$

$B_y = -F_2/(qv_2) = -F_2/(qv_1)$

$\vec{B}$ has magnitude $F_2/(qv_1)$ and is in the $-y$-direction.

b) $F_1 = qvB\sin\phi = qv_1|B_y|/\sqrt{2} = F_2/\sqrt{2}$

28-41 $\vec{v}_0 = (5.85 \times 10^3 \text{ m/s})\hat{j}, \quad \vec{B} = -(1.35 \text{ T})\hat{k}$

a) $q = +0.640 \times 10^{-9}$ C

$\vec{F}_B = q\vec{v}_0 \times \vec{B} = -(qv_0B)(\hat{j} \times \hat{k}) = -qv_0B\hat{i}$

$\vec{F}_E = q\vec{E}$

Undeflected requires $\sum \vec{F} = 0$, so $\vec{F}_E = -\vec{F}_B$

$q\vec{E} = qv_0B\hat{i}$ so $\vec{E} = v_0B\hat{i}$

$\vec{E}$ is in the $+x$-direction and has magnitude $E = v_0B = (5.85 \times 10^3 \text{ m/s})(1.35 \text{ T}) = 7.90 \times 10^3$ V/m.

b) $q = -0.320 \times 10^{-9}$ C

$\vec{E} = v_0B\hat{i}$, independent of q

$\vec{E}$ is the same as calculated in part (a): $E = 7.90 \times 10^3$ V/m and is in the $+x$-direction.

28-43 a) The maximum radius of the orbit determines the maximum speed v of the protons.

$\sum \vec{F} = m\vec{a}$ gives $|q|vB = m(v^2/R)$

$v = \dfrac{|q|BR}{m} = \dfrac{(1.60 \times 10^{-19} \text{ C})(0.85 \text{ T})(0.40 \text{ m})}{1.67 \times 10^{-27} \text{ kg}} = 3.257 \times 10^7$ m/s.

The kinetic energy of a proton moving with this speed is

$K = \frac{1}{2}mv^2 = \frac{1}{2}(1.67 \times 10^{-27} \text{ kg})(3.257 \times 10^7 \text{ m/s})^2 = 8.9 \times 10^{-13}$ J $= 5.6$ MeV

b) The time for one revolution is the period

$T = \dfrac{2\pi R}{v} = \dfrac{2\pi(0.40 \text{ m})}{3.257 \times 10^7 \text{ m/s}} = 7.7 \times 10^{-8}$ s

c) $K = \frac{1}{2}mv^2 = \frac{1}{2}m\left(\dfrac{|q|BR}{m}\right)^2 = \frac{1}{2}\dfrac{|q|^2B^2R^2}{m}$. Or, $B = \dfrac{\sqrt{2Km}}{|q|R}$.

B is proportional to $\sqrt{K}$, so if K is increased by a factor of 2 then B must be increased by a factor of $\sqrt{2}$.

$B = \sqrt{2}(0.85 \text{ T}) = 1.2$ T.

d) $v = \dfrac{|q|BR}{m} = \dfrac{(3.20 \times 10^{-19} \text{ C})(0.85 \text{ T})(0.40 \text{ m})}{6.65 \times 10^{-27} \text{ kg}} = 1.636 \times 10^7$ m/s

$K = \frac{1}{2}mv^2 = \frac{1}{2}(6.65 \times 10^{-27} \text{ kg})(1.636 \times 10^7 \text{ m/s})^2 = 8.9 \times 10^{-13}$ J $= 5.5$ MeV, the same as the maximum energy for protons. We can see that the maximum energy must be approximately the same as follows:

From part (c), $K = \frac{1}{2}m\left(\frac{|q|BR}{m}\right)^2$. For alpha particles $|q|$ is larger by a factor of 2 and m is larger by a factor of 4 (approximately). Thus $|q|^2/m$ is unchanged and K is the same.

28-47 $\vec{B} = -(0.120 \text{ T})\hat{k}$, $\vec{v} = (1.05 \times 10^6 \text{ m/s})(-3\hat{i} + 4\hat{j} + 12\hat{k})$, $F = 1.25$ N

a) $\vec{F} = q\vec{v}\mathbf{X}\vec{B}$

$\vec{F} = q(-0.120 \text{ T})(1.05 \times 10^6 \text{ m/s})(-3\hat{i}\mathbf{X}\hat{k} + 4\hat{j}\mathbf{X}\hat{k} + 12\hat{k}\mathbf{X}\hat{k})$

$\hat{i}\mathbf{X}\hat{k} = -\hat{j}, \hat{j}\mathbf{X}\hat{k} = \hat{i}, \hat{k}\mathbf{X}\hat{k} = 0$

$\vec{F} = -q(1.26 \times 10^5 \text{ N/C})(+3\hat{j} + 4\hat{i}) = -q(1.26 \times 10^5 \text{ N/C})(+4\hat{i} + 3\hat{j})$

The magnitude of the vector $+4\hat{i} + 3\hat{j}$ is $\sqrt{3^2 + 4^2} = 5$.

Thus $F = -q(1.26 \times 10^5 \text{ N/C})(5)$.

$q = -\frac{F}{5(1.26 \times 10^5 \text{ N/C})} = -\frac{1.25 \text{ N}}{5(1.26 \times 10^5 \text{ N/C})} = -1.98 \times 10^{-6}$ C

b) $\sum \vec{F} = m\vec{a}$ so $\vec{a} = \vec{F}/m$

$\vec{F} = -q(1.26\times 10^5 \text{ N/C})(+4\hat{i} + 3\hat{j}) = -(-1.98\times 10^{-6} \text{ C})(1.26\times 10^5 \text{ N/C})(+4\hat{i} + 3\hat{j}) = +0.250 \text{ N}(+4\hat{i} + 3\hat{j})$

Then $\vec{a} = \vec{F}/m = \left(\frac{0.250 \text{ N}}{2.58 \times 10^{-15} \text{ kg}}\right)(+4\hat{i} + 3\hat{j}) =$

$(9.69 \times 10^{13} \text{ m/s}^2)(+4\hat{i} + 3\hat{j})$

c) $\vec{F}$ is in the xy-plane, so in the z-direction the particle moves with constant speed 12.6×10^6 m/s.

In the xy-plane the force $\vec{F}$ causes the particle to move in a circle, with $\vec{F}$ directed in towards the center of the circle.

$\sum \vec{F} = m\vec{a}$ gives $F = m(v^2/R)$ and $R = mv^2/F$

$v^2 = v_x^2 + v_y^2 = (-3.15 \times 10^6 \text{ m/s})^2 + (+4.20 \times 10^6 \text{ m/s})^2 = 2.756 \times 10^{13} \text{ m}^2/\text{s}^2$

$F = \sqrt{F_x^2 + F_y^2} = (0.250 \text{ N})\sqrt{4^2 + 3^2} = 1.25$ N

$R = \frac{mv^2}{F} = \frac{(2.58 \times 10^{-15} \text{ kg})(2.756 \times 10^{13} \text{ m}^2/\text{s}^2)}{1.25 \text{ N}} = 0.0569 \text{ m} = 5.69$ cm

d) Eq.(28-12): cyclotron frequency $f = \omega/2\pi = v/2\pi R$.

The circular motion is in the xy-plane, so $v = \sqrt{v_x^2 + v_y^2} = 5.25 \times 10^6$ m/s.

$$f = \frac{v}{2\pi R} = \frac{5.25 \times 10^6 \text{ m/s}}{2\pi (0.0569 \text{ m})} = 1.47 \times 10^7 \text{ Hz, and } \omega = 2\pi f = 9.23 \times 10^7 \text{ rad/s}$$

e) The period of the motion in the xy-plane is given by

$$T = \frac{1}{f} = \frac{1}{1.47 \times 10^7 \text{ Hz}} = 6.80 \times 10^{-8} \text{ s}$$

In $t = 2T$ the particle has returned to the same x and y coordinates. The z-component of the motion is motion with a constant velocity of $v_z = +12.6 \times 10^6$ m/s. Thus $z = z_0 + v_z t = 0 + (12.6 \times 10^6 \text{ m/s})(2)(6.80 \times 10^{-8} \text{ s}) = +1.71$ m.

The coordinates at $t = 2T$ are $x = R$, $y = 0$, $z = +1.71$ m.

28-51 $\vec{F} = I\vec{l} \mathbf{X} \vec{B}$, $\vec{B} = (0.860 \text{ T})\hat{i}$, $I = 6.58$ A

a) <u>segment ab</u>

$\vec{l} = (0.750 \text{ m})\hat{j}$

$\vec{F}_{ab} = I\vec{l} \mathbf{X} \vec{B} = (6.58 \text{ A})(0.750 \text{ m})(0.860 \text{ T})\hat{j}\mathbf{X}\hat{i} = (4.24 \text{ N})(-\hat{k}) = -(4.24 \text{ N})\hat{k}.$

The force has magnitude 4.24 N and is in the $-z$-direction.

<u>segment bc</u>

$\vec{l} = (0.750 \text{ m})(\hat{i} - \hat{k})$

$\vec{F}_{bc} = I\vec{l} \mathbf{X} \vec{B} = (6.58 \text{ A})(0.750 \text{ m})(0.860 \text{ T})(\hat{i}\mathbf{X}\hat{i} - \hat{k}\mathbf{X}\hat{i})$

$\hat{i}\mathbf{X}\hat{i} = 0$, $\hat{k}\mathbf{X}\hat{i} = \hat{j}$

$\vec{F}_{bc} = -(4.24 \text{ N})\hat{j}.$

The force has magnitdue 4.24 N and is in the $-y$-direction.

<u>segment cd</u>

$\vec{l} = (0.750 \text{ m})(-\hat{j} + \hat{k})$

$\vec{F}_{cd} = I\vec{l} \mathbf{X} \vec{B} = (6.58 \text{ A})(0.750 \text{ m})(0.860 \text{ T})(-\hat{j}\mathbf{X}\hat{i} + \hat{k}\mathbf{X}\hat{i})$

$\hat{j}\mathbf{X}\hat{i} = -\hat{k}$, $\hat{k}\mathbf{X}\hat{i} = \hat{j}$

$\vec{F}_{cd} = (4.24 \text{ N})(\hat{j} + \hat{k})$, so $F_{cd} = 4.24 \text{ N}\sqrt{1^2 + 1^2} = 6.00$ N.

The force has magnitude 6.00 N and is directed midway between the $+y$-axis and the $+z$-axis.

<u>segment de</u>

$\vec{l} = -(0.750 \text{ m})\hat{k}$

$\vec{F}_{de} = I\vec{l} \mathbf{X} \vec{B} = -(6.58 \text{ A})(0.750 \text{ m})(0.860 \text{ T})\hat{k}\mathbf{X}\hat{i} = -(4.24 \text{ N})\hat{j}$

The force has magnitude 4.24 N and is in the $-y$-direction.

<u>segment ef</u>

$$\vec{l} = -(0.750 \text{ m})\hat{i}$$

$$\vec{F}_{ef} = I\vec{l}\mathbf{X}\vec{B} = -(6.58 \text{ A})(0.750 \text{ m})(0.860 \text{ T})\hat{i}\mathbf{X}\hat{i} = 0.$$

b) $\vec{F}_{\text{tot}} = \vec{F}_{ab} + \vec{F}_{bc} + \vec{F}_{cd} + \vec{F}_{de} + \vec{F}_{ef}$

$$\vec{F}_{\text{tot}} = (4.24 \text{ N})(-\hat{k} - \hat{j} + \hat{j} + \hat{k} - \hat{j}) = -(4.24 \text{ N})\hat{j}$$

The total force has magnitude 4.24 N and is in the $-y$-direction.

28-53 For the wire to remain at rest the force exerted on it by the magnetic field must have a component directed up the incline. To produce a force in this direction, the current in the wire must be directed from right to left in Fig.28-47.

Or, viewing the wire from its left-hand end:

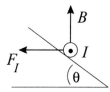

Free-body diagram for the wire:

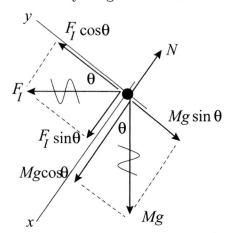

$\sum F_y = 0$
$F_I \cos\theta - Mg\sin\theta = 0$

$F_I = ILB\sin\phi$
$\phi = 90°$ since $\vec{B}$ is
perpendicular to the current direction.

Thus $(ILB)\cos\theta - Mg\sin\theta = 0$ and

$$I = \frac{Mg\tan\theta}{LB}$$

28-57

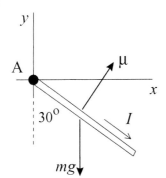

Use $\sum \tau_A = 0$, where
point A is at the origin.

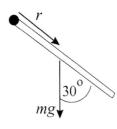

$\tau_{mg} = mgr \sin \phi = mg(0.0400 \text{ m}) \sin 30.0°$
The torque is clockwise; $\vec{\tau}_{mg}$ is
directed into the paper.

For the loop to be in equilibrium the torque due to $\vec{B}$ must be counterclockwise
(opposite to $\vec{\tau}_{mg}$) and it must be that $\tau_B = \tau_{mg}$.

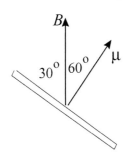

$\vec{\tau}_B = \vec{\mu} \mathbf{X} \vec{B}.$

For this torque to be counterclockwise
($\vec{\tau}_B$ directed out of the paper),
$\vec{B}$ must be in the $+y$-direction.

$\tau_B = \mu B \sin \phi = IAB \sin 60.0°$

$\tau_B = \tau_{mg}$ gives $IAB \sin 60.0° = mg(0.0400 \text{ m}) \sin 30.0°$

$m = (0.15 \text{ g/cm})2(8.00 \text{ cm} + 6.00 \text{ cm}) = 4.2 \text{ g} = 4.2 \times 10^{-3} \text{ kg}$

$A = (0.0800 \text{ m})(0.0600 \text{ m}) = 4.80 \times 10^{-3} \text{ m}^2$

$B = \dfrac{mg(0.0400 \text{ m})(\sin 30.0°)}{IA \sin 60.0°}$

$B = \dfrac{(4.2 \times 10^{-3} \text{ kg})(9.80 \text{ m/s}^2)(0.0400 \text{ m}) \sin 30.0°}{(8.2 \text{ A})(4.80 \times 10^{-3} \text{ m}^2) \sin 60.0°} = 0.024 \text{ T}$

28-59 The restoring torque applied by the magnetic field is given by
$\tau = -NIAB \sin \phi.$

For small angles, $\sin \phi \approx \phi$, when ϕ is measured in radians.

Thus $\tau = -NIAB\phi.$

Use this in $\tau = I\alpha$, and write $\alpha = \dfrac{d^2\phi}{dt^2}$:

$$-(NIAB)\phi = I_{\mathrm{S}}\frac{d^2\phi}{dt^2}$$

$$\frac{d^2\phi}{dt^2} = -\left(\frac{NIAB}{I_{\mathrm{S}}}\right)\phi = -\omega^2\phi, \text{ where } \omega \text{ is the angular frequency of the oscillations}$$

and we have used the results of Section 13-7.

Thus $\omega = \sqrt{\dfrac{NIAB}{I_{\mathrm{S}}}}$ and $T = \dfrac{2\pi}{\omega} = 2\pi\sqrt{\dfrac{I_{\mathrm{S}}}{NIAB}}$

28-61

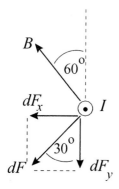

Consider the force $d\vec{F}$ on a short segment dl at the left-hand side of the coil, as viewed in Fig.28-51. The current at this point is directed out of the page. $d\vec{F}$ is perpendicular both to $\vec{B}$ and to the direction of I.

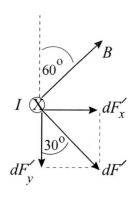

Consider also the force $d\vec{F}'$ on a short segment on the opposite side of the coil, at the right-hand side of the coil in Fig.28-51. The current at this point is directed into the page.

The two sketches show that the x-components cancel and that the y-components add. This is true for all pairs of short segments on opposite sides of the coil. The net magnetic force on the coil is in the y-direction and its magnitude is given by $F = \int dF_y$.

$dF = I\,dl\,B\sin\phi$. But $\vec{B}$ is perpendicular to the current direction so $\phi = 90°$.

$dF_y = dF\cos 30.0° = IB\cos 30.0°\,dl$

$F = \int dF_y = IB\cos 30.0° \int dl$

But $\int dl = N(2\pi r)$, the total length of wire in the coil.

$F = IB\cos 30.0°\,N(2\pi r) = (0.950\text{ A})(0.220\text{ T})(\cos 30.0°)(50)2\pi(0.0078\text{ m}) = 0.444\text{ N}$

$$\vec{F} = -(0.444 \text{ N})\hat{j}$$

28-65 a) After the wire leaves the mercury its acceleration is g, downward.

The wire travels upward a total distance of 0.350 m from its initial position. Its ends lose contact with the mercury after the wire has traveled 0.025 m, so the wire travels upward 0.325 m after it leaves the mercury. Consider the motion of the wire after it leaves the mercury. Take $+y$ to be upward and take the origin at the position of the wire as it leaves the mercury.

$a_y = -9.80 \text{ m/s}^2, \quad y - y_0 = +0.325 \text{ m}, \quad v_y = 0 \text{ (at maximum height)}, \quad v_{0y} = ?$

$v_y^2 = v_{0y}^2 + 2a_y(y - y_0)$

$v_{0y} = \sqrt{-2a_y(y - y_0)} = \sqrt{-2(-9.80 \text{ m/s}^2)(0.325 \text{ m})} = 2.52 \text{ m/s}$

b) Now consider the motion of the wire while it is in contact with the mercury. Take $+y$ to be upward and the origin at the initial position of the wire.

Calculate the acceleration:

$y - y_0 = +0.025 \text{ m}, \quad v_{0y} = 0 \text{ (starts from rest)}, \quad v_y = +2.52 \text{ m/s (from part (a))},$
$a_y = ?$

$v_y^2 = v_{0y}^2 + 2a_y(y - y_0)$

$a_y = \dfrac{v_y^2}{2(y - y_0)} = \dfrac{(2.52 \text{ m/s})^2}{2(0.025 \text{ m})} = 127 \text{ m/s}^2$

Free-body diagram for the wire:

$\sum F_y = ma_y$

$F_B - mg = ma_y$

$IlB = m(g + a_y)$

$I = \dfrac{m(g + a_y)}{lB}$

l is the length of the horizontal section of the wire; $l = 0.150$ m

$I = \dfrac{(5.40 \times 10^{-5} \text{ kg})(9.80 \text{ m/s}^2 + 127 \text{ m/s}^2)}{(0.150 \text{ m})(0.00650 \text{ T})} = 7.58 \text{ A}$

c) $V = IR$ so $R = \dfrac{V}{I} = \dfrac{1.50 \text{ V}}{7.58 \text{ A}} = 0.198 \ \Omega$

28-67 a)

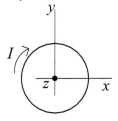

$\vec{\mu}$ is into the plane of the paper, in the $-z$-direction

$$\vec{\mu} = -\mu\hat{k} = -IA\hat{k}$$

b) $\vec{\tau} = D(+4\hat{i} - 3\hat{j})$, where $D > 0$.

$\vec{\mu} = -IA\hat{k}$, $\vec{B} = B_x\hat{i} + B_y\hat{j} + B_z\hat{k}$

$\vec{\tau} = \vec{\mu}\mathbf{X}\vec{B} = (-IA)(B_x\hat{k}\mathbf{X}\hat{i} + B_y\hat{k}\mathbf{X}\hat{j} + B_z\hat{k}\mathbf{X}\hat{k}) = IAB_y\hat{i} - IAB_x\hat{j}$

Compare this to the expression given for $\vec{\tau}$:

$IAB_y = 4D$ so $B_y = 4D/IA$ and $-IAB_x = -3D$ so $B_x = 3D/IA$

B_z doesn't contribute to the torque since $\vec{\mu}$ is along the z-direction.

But $B = B_0$ and $B_x^2 + B_y^2 + B_z^2 = B_0^2$, with $B_0 = 13D/IA$.

Thus $B_z = \pm\sqrt{B_0^2 - B_x^2 - B_y^2} = \pm(D/IA)\sqrt{169 - 9 - 16} = \pm12(D/IA)$

That $U = -\vec{\mu}\cdot\vec{B}$ is negative determines the sign of B_z:

$U = -\vec{\mu}\cdot\vec{B} = -(-IA\hat{k})\cdot(B_x\hat{i} + B_y\hat{j} + B_z\hat{k}) = +IAB_z$

So U negative says that B_z is negative, and thus $B_z = -12D/IA$.

CHAPTER 29
SOURCES OF MAGNETIC FIELD

Exercises 1, 3, 7, 9, 13, 15, 19, 21, 25, 27, 31, 37, 39
Problems 41, 43, 45, 49, 57, 61, 65, 67, 69

Exercises

29-1 $\vec{B} = \dfrac{\mu_0}{4\pi} \dfrac{q\vec{v} \times \hat{r}}{r^2} = \dfrac{\mu_0}{4\pi} \dfrac{q\vec{v} \times \vec{r}}{r^3}$, since $\hat{r} = \dfrac{\vec{r}}{r}$.

$\vec{v} = (8.00 \times 10^6 \text{ m/s})\hat{j}$ and $\vec{r}$ is the vector from the charge to the point where the field is calculated.

a) $\vec{r} = (0.500 \text{ m})\hat{i}$, $r = 0.500$ m

$\vec{v} \times \vec{r} = vr\hat{j}\times\hat{i} = -vr\hat{k}$

$\vec{B} = -\dfrac{\mu_0}{4\pi} \dfrac{qv}{r^2} \hat{k} = -(1 \times 10^{-7} \text{ T} \cdot \text{m/A})\dfrac{(6.00 \times 10^{-6} \text{ C})(8.00 \times 10^6 \text{ m/s})}{(0.500 \text{ m})^2} \hat{k}$

$\vec{B} = -(1.92 \times 10^{-5} \text{ T})\hat{k}$

b) $\vec{r} = -(0.500 \text{ m})\hat{j}$, $r = 0.500$ m

$\vec{v} \times \vec{r} = -vr\hat{j}\times\hat{j} = 0$ and $\vec{B} = 0$.

c) $\vec{r} = (0.500 \text{ m})\hat{k}$, $r = 0.500$ m

$\vec{v} \times \vec{r} = vr\hat{j}\times\hat{k} = vr\hat{i}$

$\vec{B} = (1 \times 10^{-7} \text{ T} \cdot \text{m/A})\dfrac{(6.00 \times 10^{-6} \text{ C})(8.00 \times 10^6 \text{ m/s})}{(0.500 \text{ m})^2}\hat{i} = +(1.92 \times 10^{-5} \text{ T})\hat{i}$

d) $\vec{r} = -(0.500 \text{ m})\hat{j} + (0.500 \text{ m})\hat{k}$, $r = \sqrt{(0.500 \text{ m})^2 + (0.500 \text{ m})^2} = 0.7071$ m

$\vec{v} \times \vec{r} = v(0.500 \text{ m})(-\hat{j}\times\hat{j} + \hat{j}\times\hat{k}) = (4.00 \times 10^6 \text{ m}^2/\text{s})\hat{i}$

$\vec{B} = (1 \times 10^{-7} \text{ T} \cdot \text{m/A})\dfrac{(6.00 \times 10^{-6} \text{ C})(4.00 \times 10^6 \text{ m}^2/\text{s})}{(0.7071 \text{ m})^3}\hat{i} = +(6.79 \times 10^{-6} \text{ T})\hat{i}$

29-3

$\vec{B}_{\text{tot}} = \vec{B}_q + \vec{B}'_q$

$v = 2.00 \times 10^5$ m/s, $v' = 8.00 \times 10^5$ m/s

$q = +4.00 \times 10^{-6}$ C, $q' = -1.50 \times 10^{-6}$ C

$$\vec{B} = \frac{\mu_0}{4\pi} \frac{q\vec{v} \times \vec{r}}{r^3}$$

field $\vec{B}_q$ due to q:

$\vec{v} = (2.00 \times 10^5 \text{ m/s})\hat{\imath}, \quad \vec{r} = -(0.300 \text{ m})\hat{\jmath}$

$\vec{v} \times \vec{r} = -(2.00 \times 10^5 \text{ m/s})(0.300 \text{ m})(\hat{\imath}\mathbf{x}\hat{\jmath}) = -(6.00 \times 10^4 \text{ m}^2/\text{s})\hat{k}$

$$\vec{B}_q = -(1 \times 10^{-7} \text{ T·m/A})\frac{(4.00 \times 10^{-6} \text{ C})(6.00 \times 10^4 \text{m}^2/\text{s})}{(0.300 \text{ m})^3}\hat{k} = -(8.889 \times 10^{-7} \text{ T})\hat{k}$$

field $\vec{B}'_q$ due to q':

$\vec{v} = (8.00 \times 10^5 \text{ m/s})\hat{\jmath}, \quad \vec{r} = -(0.400 \text{ m})\hat{\imath}$

$\vec{v} \times \vec{r} = -(8.00 \times 10^5 \text{ m/s})(0.400 \text{ m})(\hat{\jmath}\mathbf{x}\hat{\imath}) = +(3.20 \times 10^5 \text{ m}^2/\text{s})\hat{k}$

$$\vec{B}'_q = (1 \times 10^{-7} \text{ T·m/A})\frac{(-1.50 \times 10^{-6} \text{ C})(3.20 \times 10^5 \text{ m}^2/\text{s})}{(0.400 \text{ m})^3}\hat{k} = -(7.50 \times 10^{-7} \text{ T})\hat{k}$$

$$\vec{B}_{\text{tot}} = \vec{B}_q + \vec{B}'_q = (-8.889 \times 10^{-7} \text{ T} - 7.50 \times 10^{-7} \text{ T})\hat{k} = -(1.64 \times 10^{-6})\hat{k}$$

The resultant field at the origin has magnitude 1.64×10^{-6} T and is in the $-z$-direction.

29-7 $d\vec{B} = \frac{\mu_0}{4\pi} \frac{I\vec{l} \times \hat{r}}{r^2}$

As in Example 29-2 use this equation for the finite 0.500-mm segment of wire since the $\Delta l = 0.500$ mm length is much smaller than the distances to the field points.

$$\vec{B} = \frac{\mu_0}{4\pi} \frac{I\Delta\vec{l} \times \hat{r}}{r^2} = \frac{\mu_0}{4\pi} \frac{I\Delta\vec{l} \times \vec{r}}{r^3}$$

I is in the $+z$-direction, so $\Delta\vec{l} = (0.500 \times 10^{-3} \text{ m})\hat{k}$

a) Field point is at $x = 2.00$ m, $y = 0$, $z = 0$ so the vector $\vec{r}$ from the source point (at the origin) to the field point is $\vec{r} = (2.00 \text{ m})\hat{\imath}$.

$\Delta\vec{l}\mathbf{x}\vec{r} = (0.500 \times 10^{-3} \text{ m})(2.00 \text{ m})\hat{k}\mathbf{x}\hat{\imath} = +(1.00 \times 10^{-3} \text{ m}^2)\hat{\jmath}$

$$\vec{B} = \frac{(1 \times 10^{-7} \text{ T·m/A})(4.00 \text{ A})(1.00 \times 10^{-3} \text{ m}^2)}{(2.00 \text{ m})^3}\hat{\jmath} = (5.00 \times 10^{-11} \text{ T})\hat{\jmath}$$

b) $\vec{r} = (2.00 \text{ m})\hat{\jmath}, \ r = 2.00$ m.

$\Delta\vec{l}\mathbf{x}\vec{r} = (0.500 \times 10^{-3} \text{ m})(2.00 \text{ m})\hat{k}\mathbf{x}\hat{\jmath} = -(1.00 \times 10^{-3} \text{ m}^2)\hat{\imath}$

$$\vec{B} = -\frac{(1 \times 10^{-7} \text{ T·m/A})(4.00 \text{ A})(1.00 \times 10^{-3} \text{ m}^2)}{(2.00 \text{ m})^3}\hat{\imath} = -(5.00 \times 10^{-11} \text{ T})\hat{\imath}$$

c) $\vec{r} = (2.00 \text{ m})(\hat{i} + \hat{j})$, $r = \sqrt{2}(2.00 \text{ m})$.

$\Delta \vec{l} \times \vec{r} = (0.500 \times 10^{-3} \text{ m})(2.00 \text{ m})\hat{k} \times (\hat{i} + \hat{j}) = (1.00 \times 10^{-3} \text{ m}^2)(\hat{j} - \hat{i})$

$$\vec{B} = \frac{(1 \times 10^{-7} \text{ T} \cdot \text{m/A})(4.00 \text{ A})(1.00 \times 10^{-3} \text{ m}^2)}{[\sqrt{2}(2.00 \text{ m})]^3}(\hat{j} - \hat{i}) =$$

$(-1.77 \times 10^{-11} \text{ T})(\hat{i} - \hat{j})$

d) $\vec{r} = (2.00 \text{ m})\hat{k}$, $r = 2.00 \text{ m}$.

$\Delta \vec{l} \times \vec{r} = (0.500 \times 10^{-3} \text{ m})(2.00 \text{ m})\hat{k} \times \hat{k} = 0;$ $\quad \vec{B} = 0.$

29-9 For each wire $B = \dfrac{\mu_0 I}{2\pi r}$ (Eq.29-9), and the direction

of $\vec{B}$ is given by the right-hand rule (Fig.29-6).

a)

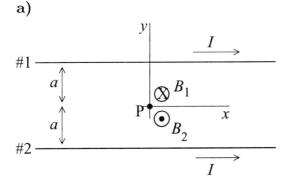

At point P midway between the two wires the fields $\vec{B}_1$ and $\vec{B}_2$ due to the two currents are in opposite directions, so $B = B_2 - B_1$.

But $B_1 = B_2 = \dfrac{\mu_0 I}{2\pi a}$, so $B = 0$.

b)

At point Q above the upper wire $\vec{B}_1$ and $\vec{B}_2$ are both directed out of the page ($+z$-direction), so $B = B_1 + B_2$.

$B_1 = \dfrac{\mu_0 I}{2\pi a}$, $B_2 = \dfrac{\mu_0 I}{2\pi(3a)}$

$$B = \frac{\mu_0 I}{2\pi a}(1 + \frac{1}{3}) = \frac{2\mu_0 I}{3\pi a}; \quad \vec{B} = \frac{2\mu_0 I}{3\pi a}\hat{k}$$

c)

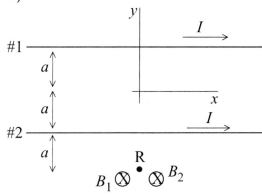

At point R below the lower wire $\vec{B}_1$ and $\vec{B}_2$ are both directed into the page ($-z$-direction), so $B = B_1 + B_2$.

$$B_1 = \frac{\mu_0 I}{2\pi(3a)}, \quad B_2 = \frac{\mu_0 I}{2\pi a}$$

$$B_1 = \frac{\mu_0 I}{2\pi a}(1 + \frac{1}{3}) = \frac{2\mu_0 I}{3\pi a}; \quad \vec{B} = -\frac{2\mu_0 I}{3\pi a}\hat{k}$$

29-13 a) Viewed from above:

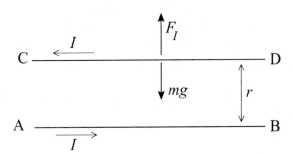

Directly below the wire the direction of the magnetic field due to the current in the wire is east.

$$B = \frac{\mu_0 I}{2\pi r} = (2 \times 10^{-7} \text{ T} \cdot \text{m/A})\left(\frac{800 \text{ A}}{5.50 \text{ m}}\right) = 2.91 \times 10^{-5} \text{ T}$$

b) B from the current is nearly equal in magnitude to the earth's field, so, yes, the current really is a problem.

29-15 The wire CD rises until the upward force F_I due to the currents balances the downward force of gravity.

Currents in opposite directions so the force is repulsive and F_I is upward, as shown.

Eq.(29-11) says $F_I = \dfrac{\mu_0 I^2 L}{2\pi h}$ where L is the length of wire CD and h is the distance between the wires.

$$mg = \lambda L g$$

Thus $F_I - mg = 0$ says $\dfrac{\mu_0 I^2 L}{2\pi h} = \lambda L g$

$$h = \dfrac{\mu_0 I^2}{2\pi g \lambda}$$

29-19 First consider the field at P produced by the current I_1

in the upper semicircle of wire.

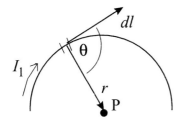

Consider the three parts of this wire
a: long straight section,
b: semicircle
c: long, straight section

Apply the Biot-Savart law $d\vec{B} = \dfrac{\mu_0}{4\pi} \dfrac{I\, d\vec{l} \times \hat{r}}{r^2} = \dfrac{\mu_0}{4\pi} \dfrac{I\, d\vec{l} \times \vec{r}}{r^3}$ to each piece.

<u>part a</u>

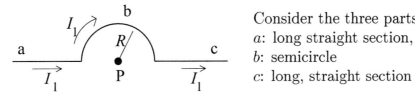

$d\vec{l} \times \vec{r} = 0$,
so $dB = 0$

The same is true for all the infinitesimal segments that make up this piece of the wire, so $B = 0$ for this piece.

<u>part c</u>

$d\vec{l} \times \vec{r} = 0$,
so $dB = 0$ and $B = 0$ for this piece.

<u>part b</u>

$d\vec{l} \times \vec{r}$ is directed into the paper for all infinitesimal segments that make up this semicircular piece, so $\vec{B}$ is directed into the paper and $B = \int dB$ (the vector sum of the $d\vec{B}$ is obtained by adding their magnitudes since they are in the same direction).

$|d\vec{l} \times \vec{r}| = r\, dl \sin\theta$. The angle θ between $d\vec{l}$ and $\vec{r}$ is $90°$ and $r = R$, the radius of the semicircle. Thus $|d\vec{l} \times \vec{r}| = R\, dl$

$$dB = \frac{\mu_0}{4\pi} \frac{I \left| d\vec{l} \times \vec{r} \right|}{r^3} = \frac{\mu_0 I_1}{4\pi} \frac{R}{R^3} dl = \left(\frac{\mu_0 I_1}{4\pi R^2} \right) dl$$

$$B = \int dB = \left(\frac{\mu_0 I_1}{4\pi R^2} \right) \int dl = \left(\frac{\mu_0 I_1}{4\pi R^2} \right) (\pi R) = \frac{\mu_0 I_1}{4R}$$

(We used that $\int dl$ is equal to πR, the length of wire in the semicircle.)

We have shown that the two straight sections make zero contribution to $\vec{B}$, so $B_1 = \mu_0 I_1 / 4R$ and is directed into the page.

For 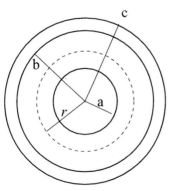 a similar analysis gives

$B_2 = \mu_0 I_2 / 4R$,
out of the paper

$\vec{B}_1$ and $\vec{B}_2$ are in opposite directions, so the magnitude of the net field at P is
$$B = |B_1 - B_2| = \frac{\mu_0 |I_1 - I_2|}{4R}.$$

When $I_1 = I_2$, $B = 0$.

29-21 The magnetic field at a point on the axis of N circular loops is given by
$$B = \frac{\mu_0 N I a^2}{2(x^2 + a^2)^{3/2}}$$

Thus $N = \dfrac{2B(x^2 + a^2)^{3/2}}{\mu_0 I a^2} =$

$$\frac{2(6.39 \times 10^{-4} \text{ T})[(0.0600 \text{ m})^2 + (0.0600 \text{ m})^2]^{3/2}}{(4\pi \times 10^{-7} \text{ T} \cdot \text{m/A})(2.50 \text{ A})(0.0600 \text{ m})^2} = 69.$$

29-25 **a)** $\underline{a < r < b}$
end view

Apply Ampere's law to a circle of radius r, where $a < r < b$. Take currents I_1 and I_2 to be directed into the page. Take this direction to be positive, so go around the integration path in the clockwise direction.

$$\oint \vec{B} \cdot d\vec{l} = \mu_0 I_{encl}$$

$$\oint \vec{B} \cdot d\vec{l} = B(2\pi r), \quad I_{\text{encl}} = I_1$$

Thus $B(2\pi r) = \mu_0 I_1$ and $B = \dfrac{\mu_0 I_1}{2\pi r}$

b) $\underline{r > c}$

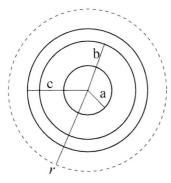

Apply Ampere's law to a circle of radius r, where $r > c$. Both currents are in the positive direction.

$$\oint \vec{B} \cdot d\vec{l} = \mu_0 I_{\text{encl}}$$
$$\oint \vec{B} \cdot d\vec{l} = B(2\pi r), \quad I_{\text{encl}} = I_1 + I_2$$

Thus $B(2\pi r) = \mu_0 (I_1 + I_2)$ and $B = \dfrac{\mu_0 (I_1 + I_2)}{2\pi r}$

29-27 a) The magnetic field near the center of a long solenoid is given by Eq.(29-23): $B = \mu_0 n I$.

turns per unit length $n = \dfrac{B}{\mu_0 I} = \dfrac{0.0270 \text{ T}}{(4\pi \times 10^{-7} \text{ T} \cdot \text{m/A})(12.0 \text{ A})} = 1790$ turns/m

b) $N = nL = (1790 \text{ turns/m})(0.400 \text{ m}) = 716$ turns

Each turn of radius R has a length $2\pi R$ of wire. The total length of wire required is

$N(2\pi R) = (716)(2\pi)(1.40 \times 10^{-2} \text{ m}) = 63.0$ m.

29-31 $B = \dfrac{K_m \mu_0 N I}{2\pi r}$ (Eq.29-24, with μ_0 replaced by $K_m \mu_0$)

a) $K_m = 1400$

$$I = \dfrac{2\pi r B}{\mu_0 K_m N} = \dfrac{(2.90 \times 10^{-2} \text{ m})(0.350 \text{ T})}{(2 \times 10^{-7} \text{ T} \cdot \text{m/A})(1400)(500)} = 0.0725 \text{ A}$$

b) $K_m = 5200$

$$I = \frac{2\pi r B}{\mu_0 K_m N} = \frac{(2.90 \times 10^{-2} \text{ m})(0.350 \text{ T})}{(2 \times 10^{-7} \text{ T} \cdot \text{m/A})(5200)(500)} = 0.0195 \text{ A}$$

29-37 a) $i_C = i_D$, so $j_D = \dfrac{i_D}{A} = \dfrac{i_C}{A} = \dfrac{0.280 \text{ A}}{\pi r^2} = \dfrac{0.280 \text{ A}}{\pi (0.0400 \text{ m})^2} = 55.7 \text{ A/m}^2$

b) $j_D = \epsilon_0 \dfrac{dE}{dt}$ so $\dfrac{dE}{dt} = \dfrac{j_D}{\epsilon_0} = \dfrac{55.7 \text{ A/m}^2}{8.854 \times 10^{-12} \text{ C}^2/\text{N} \cdot \text{m}^2} = 6.29 \times 10^{12} \text{ V/m·s}$

c) Apply Ampere's law $\oint \vec{B} \cdot d\vec{l} = \mu_0 (i_C + i_D)_{\text{encl}}$ (Eq.(29-36)) to a circular path with radius $r = 0.0200$ m.

end view

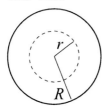

By symmetry the magnetic field is tangent to the path and constant around it.

Thus $\oint \vec{B} \cdot d\vec{l} = \oint B \, dl = B \int dl = B(2\pi r)$.

$i_C = 0$ (no conduction current flows through the air space between the plates)

The displacement current enclosed by the path is $j_D \pi r^2$.

Thus $B(2\pi r) = \mu_0 (j_D \pi r^2)$ and

$B = \frac{1}{2}\mu_0 j_D r = \frac{1}{2}(4\pi \times 10^{-7} \text{ T} \cdot \text{m/A})(55.7 \text{ A/m}^2)(0.0200 \text{ m}) = 7.00 \times 10^{-7} \text{ T}$

d) $B = \frac{1}{2}\mu_0 j_D r$. Now r is $\frac{1}{2}$ the value in (c), so B is $\frac{1}{2}$ also:

$B = \frac{1}{2}(7.00 \times 10^{-7} \text{ T}) = 3.50 \times 10^{-7} \text{ T}$

29-39 a) $i_C = 1.80 \times 10^{-3}$ A

$q = 0$ at $t = 0$

The amount of charge brought to the plates by the charging current in time t is

$q = i_C t = (1.80 \times 10^{-3} \text{ A})(0.500 \times 10^{-6} \text{ s}) = 9.00 \times 10^{-10} \text{ C}$

$$E = \frac{\sigma}{\epsilon_0} = \frac{q}{\epsilon_0 A} = \frac{9.00 \times 10^{-10} \text{ C}}{(8.854 \times 10^{-12} \text{ C}^2/\text{N} \cdot \text{m}^2)(5.00 \times 10^{-4} \text{ m}^2)} = 2.03 \times 10^5 \text{ V/m}$$

$V = Ed = (2.03 \times 10^5 \text{ V/m})(2.00 \times 10^{-3} \text{ m}) = 406 \text{ V}$

b) $E = q/\epsilon_0 A$

$$\frac{dE}{dt} = \frac{dq/dt}{\epsilon_0 A} = \frac{i_C}{\epsilon_0 A} =$$

$$\frac{1.80 \times 10^{-3} \text{ A}}{(8.854 \times 10^{-12} \text{ C}^2/\text{N} \cdot \text{m}^2)(5.00 \times 10^{-4} \text{ m}^2)} = 4.07 \times 10^{11} \text{ V/m·s}$$

Since i_C is constant dE/dt does not vary in time.

c) $j_D = \epsilon_0 \dfrac{dE}{dt}$ (Eq.29-37, with ϵ replaced by ϵ_0 since there is vacuum between the plates.)

$j_D = (8.854 \times 10^{-12} \text{ C}^2/\text{N} \cdot \text{m}^2)(4.07 \times 10^{11} \text{ V/m} \cdot \text{s}) = 3.60 \text{ A/m}^2$

$i_D = j_D A = (3.60 \text{ A/m}^2)(5.00 \times 10^{-4} \text{ m}^2) = 1.80 \times 10^{-3} \text{ A}; \; i_D = i_C$

Problems

29-41

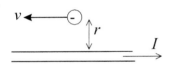

At the electron's position the magnetic field $\vec{B}$ due to the current in the wire has magnitude

$$B = \frac{\mu_0 I}{2\pi r} = \frac{(2 \times 10^{-7} \text{ T} \cdot \text{m/A})(2.50 \text{ A})}{0.0450 \text{ m}} = 1.111 \times 10^{-5} \text{ T}$$

By the right hand rule, $\vec{B}$ is out of the page at this point.

But, since the electron has negative charge, $\vec{F} = q\vec{v} \times \vec{B} = -e\vec{v} \times \vec{B}$ has direction $\downarrow$ (toward the wire)

$F = |q|vB \sin\phi = evB = (1.602 \times 10^{-19} \text{ C})(6.00 \times 10^4 \text{ m/s})(1.111 \times 10^{-5} \text{ T}) = 1.07 \times 10^{-19} \text{ N}$

The force has magnitude 1.07×10^{-19} N and is directed toward the wire.

29-43 $\vec{B} = B_0(x/a)\hat{i}$

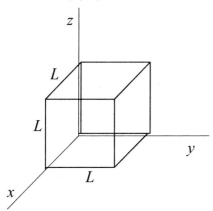

Apply Gauss's law for magnetic fields to a cube with side length L, one corner at the origin, and sides parallel to the x, y and z axes.

Since $\vec{B}$ is parallel to the x-axis the only sides that have nonzero flux are the front side (parallel to the yz-plane at $x = L$) and the back side (parallel to the yz-plane at $x = 0$.)

<u>front</u>

$\Phi_B = \int \vec{B} \cdot d\vec{A} = B_0(x/a) \int dA(\hat{i} \cdot \hat{i}) = B_0(x/a) \int dA$

$x = L$ on this face so $\vec{B} \cdot d\vec{A} = B_0(L/a) \, dA$

$\Phi_B = B_0(L/a) \int dA = B_0(L/a)L^2 = B_0(L^3/a)$

<u>back</u>

On the back face $x = 0$ so $B = 0$ and $\Phi_B = 0$.

The total flux through the cubical Gaussian surface is $\Phi_B = B_0(L^3/a)$. But this violates Eq.(28-8), which says that $\Phi_B = 0$ for any closed surface. The claimed $\vec{B}$ is impossible because it has been shown to violate Gauss's law for magnetism.

29-45 a)

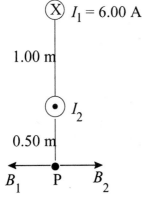

$\vec{B}_1$ and $\vec{B}_2$ must be equal and opposite for the resultant field at P to be zero. $\vec{B}_2$ is to the right so I_2 is out of the page.

$B_1 = \dfrac{\mu_0 I_1}{2\pi r_1} = \dfrac{\mu_0}{2\pi}\left(\dfrac{6.00 \text{ A}}{1.50 \text{ m}}\right) \qquad B_2 = \dfrac{\mu_0 I_2}{2\pi r_2} = \dfrac{\mu_0}{2\pi}\left(\dfrac{I_2}{0.50 \text{ m}}\right)$

$B_1 = B_2$ says $\dfrac{\mu_0}{2\pi}\left(\dfrac{6.00 \text{ A}}{1.50 \text{ m}}\right) = \dfrac{\mu_o}{2\pi}\left(\dfrac{I_2}{0.50 \text{ m}}\right)$

$I_2 = \left(\dfrac{0.50 \text{ m}}{1.50 \text{ m}}\right)(6.00 \text{ A}) = 2.00 \text{ A}$

b)

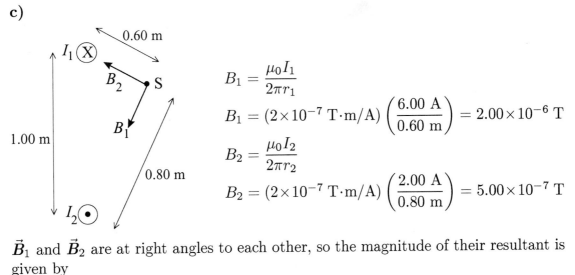

$B_1 = \dfrac{\mu_0 I_1}{2\pi r_1}$

$B_1 = (2 \times 10^{-7} \text{ T} \cdot \text{m/A})\left(\dfrac{6.00 \text{ A}}{0.50 \text{ m}}\right) = 2.40 \times 10^{-6} \text{ T}$

$B_2 = \dfrac{\mu_0 I_2}{2\pi r_2}$

$B_2 = (2 \times 10^{-7} \text{ T} \cdot \text{m/A})\left(\dfrac{2.00 \text{ A}}{1.50 \text{ m}}\right) = 2.67 \times 10^{-7} \text{ T}$

$\vec{B}_1$ and $\vec{B}_2$ are in opposite directions and $B_1 > B_2$ so
$B = B_1 - B_2 = 2.40 \times 10^{-6} \text{ T} - 2.67 \times 10^{-7} \text{ T} = 2.13 \times 10^{-6} \text{ T}$, and $\vec{B}$ is to the right.

c)

$B_1 = \dfrac{\mu_0 I_1}{2\pi r_1}$

$B_1 = (2 \times 10^{-7} \text{ T} \cdot \text{m/A})\left(\dfrac{6.00 \text{ A}}{0.60 \text{ m}}\right) = 2.00 \times 10^{-6} \text{ T}$

$B_2 = \dfrac{\mu_0 I_2}{2\pi r_2}$

$B_2 = (2 \times 10^{-7} \text{ T} \cdot \text{m/A})\left(\dfrac{2.00 \text{ A}}{0.80 \text{ m}}\right) = 5.00 \times 10^{-7} \text{ T}$

$\vec{B}_1$ and $\vec{B}_2$ are at right angles to each other, so the magnitude of their resultant is given by
$$B = \sqrt{B_1^2 + B_2^2} = \sqrt{(2.00 \times 10^{-6} \text{ T})^2 + (5.00 \times 10^{-7} \text{ T})^2} = 2.06 \times 10^{-6} \text{ T}$$

29-49 The force diagram for one of the wires is

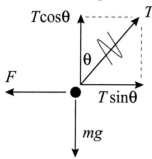

The force one wire exerts on the other is

$$F = \left(\frac{\mu_0 I^2}{2\pi r}\right) L,$$

where $r = 2(0.040 \text{ m}) \sin \theta = 8.362 \times 10^{-3}$ m is the distance between the two wires.

$\sum F_y = 0$ gives $T \cos \theta = mg$ and $T = mg/\cos \theta$

$\sum F_x = 0$ gives $F = T \sin \theta = (mg/\cos \theta) \sin \theta = mg \tan \theta$

And $m = \lambda L$, so $F = \lambda L g \tan \theta$

$$\left(\frac{\mu_0 I^2}{2\pi r}\right) L = \lambda L g \tan \theta$$

$$I = \sqrt{\frac{\lambda g r \tan \theta}{(\mu_0/2\pi)}}$$

$$I = \sqrt{\frac{(0.0125 \text{ kg/m})(9.80 \text{ m/s}^2)(\tan 6.00°)(8.362 \times 10^{-3} \text{ m})}{2 \times 10^{-7} \text{ T} \cdot \text{m/A}}} = 23.2 \text{ A}$$

29-57 a) We can't say $I = JA = J\pi R^2$, since J varies across the cross section.

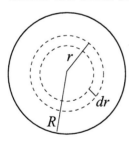

To integrate J over the cross section of the wire divide the wire cross section up into thin concentric rings of radius r and width dr.

The area of such a ring is dA, and the current through it is $dI = J \, dA$;

$dA = 2\pi r \, dr$ and $dI = J \, dA = \alpha r (2\pi r \, dr) = 2\pi \alpha r^2 \, dr$

$$I = \int dI = 2\pi \alpha \int_0^R r^2 \, dr = 2\pi \alpha (R^3/3) \text{ so } \alpha = \frac{3I}{2\pi R^3}$$

b) (i)$r \leq R$

Apply Ampere's law to a circle of radius $r < R$:

$\oint \vec{B} \cdot d\vec{l} = \oint B \, dl = B \oint dl = B(2\pi r)$, by the symmetry and direction of $\vec{B}$.

The current pasing through the path is $I_{\mathrm{encl}} = \int dI$, where the integration is from 0 to r.

$$I_{\mathrm{encl}} = 2\pi\alpha \int_0^r r^2 \, dr = \frac{2\pi\alpha r^3}{3} = \frac{2\pi}{3}\left(\frac{3I}{2\pi R^3}\right) r^3 = \frac{Ir^3}{R^3}.$$

Thus $\oint \vec{B} \cdot d\vec{l} = \mu_0 I_{\mathrm{encl}}$ gives

$$B(2\pi r) = \mu_0 \left(\frac{Ir^3}{R^3}\right) \text{ and } B = \frac{\mu_0 I r^2}{2\pi R^3}$$

(ii) $r \geq R$

Apply Ampere's law to a circle of radius $r > R$:

$\oint \vec{B} \cdot d\vec{l} = \oint B \, dl = B \oint dl = B(2\pi r)$

$I_{\mathrm{encl}} = I$; all the current in the wire passes through this path.

Thus $\oint \vec{B} \cdot d\vec{l} = \mu_0 I_{\mathrm{encl}}$ gives

$$B(2\pi r) = \mu_0 I \text{ and } B = \frac{\mu_0 I}{2\pi r}$$

Note that at $r = R$ the expression in (i) (for $r \leq R$) gives $B = \dfrac{\mu_0 I}{2\pi R}$.

At $r = R$ the expression in (ii) (for $r \geq R$) gives $B = \dfrac{\mu_0 I}{2\pi R}$, which is the same.

29-61 a)

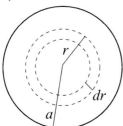

Divide the cross section of the cylinder into thin concentric rings of radius r and width dr. The current through each ring is
$dI = J \, dA = J 2\pi r \, dr$.

$$dI = \frac{2I_0}{\pi a^2}\left[1 - (r/a)^2\right] 2\pi r \, dr = \frac{4I_0}{a^2}\left[1 - (r/a)^2\right] r \, dr.$$

The total current I is obtained by integrating dI over the cross section

$$I = \int_0^a dI = \left(\frac{4I_0}{a^2}\right) \int_0^a (1 - r^2/a^2) r \, dr = \left(\frac{4I_0}{a^2}\right) \left[\frac{1}{2}r^2 - \frac{1}{4}r^4/a^2\right]_0^a = I_0, \text{ as was}$$

to be shown.

b) Apply Ampere's law to a path that is a circle of radius $r > a$.

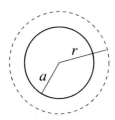

$$\oint \vec{B} \cdot d\vec{l} = B(2\pi r)$$

$I_{encl} = I_0$ (the path encloses the entire cylinder)

Thus $\oint \vec{B} \cdot d\vec{l} = \mu_0 I_{encl}$ says $B(2\pi r) = \mu_0 I_0$ and $B = \dfrac{\mu_0 I_0}{2\pi r}$.

c)

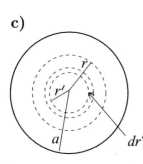

Divide the cross section of the cylinder into concentric rings of radius r' and width dr', as was done in part (a).
The current dI through each ring is

$$dI = \frac{4I_0}{a^2}\left[1 - \left(\frac{r'}{a}\right)^2\right] r'\, dr'$$

The current I is obtained by integrating dI from $r' = 0$ from $r' = r$:

$$I = \int dI = \frac{4I_0}{a^2}\int_0^r \left[1 - \left(\frac{r'}{a}\right)^2\right] r'\, dr' = \frac{4I_0}{a^2}\left[\frac{1}{2}(r')^2 - \frac{1}{4}(r')^4/a^2\right]_0^r$$

$$I = \frac{4I_0}{a^2}(r^2/2 - r^4/4a^2) = \frac{I_0 r^2}{a^2}\left(2 - \frac{r^2}{a^2}\right)$$

d) Apply Ampere's law to a path that is a circle of radius $r < a$.

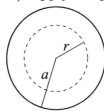

$$\oint \vec{B} \cdot d\vec{l} = B(2\pi r)$$

$$I_{encl} = \frac{I_0 r^2}{a^2}\left(2 - \frac{r^2}{a^2}\right) \text{ (from part (c))}$$

Thus $\oint \vec{B} \cdot d\vec{l} = \mu_0 I_{encl}$ says $B(2\pi r) = \mu_0 \dfrac{I_0 r^2}{a^2}(2 - r^2/a^2)$

and $B = \dfrac{\mu_0 I_0}{2\pi}\dfrac{r}{a^2}(2 - r^2/a^2)$

Note: Result in part (b) evaluated at $r = a$: $B = \dfrac{\mu_0 I_0}{2\pi a}$.

Result in part (d) evaluated at $r = a$: $B = \dfrac{\mu_0 I_0}{2\pi}\dfrac{a}{a^2}(2 - a^2/a^2) = \dfrac{\mu_0 I_0}{2\pi a}$.

The two results, one for $r > a$ and the other for $r < a$, agree at $r = a$.

29-65 Do parts (a) and (b) together.

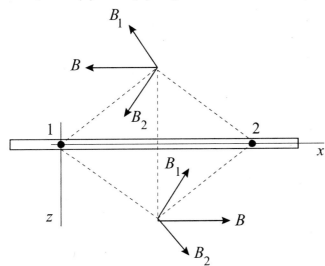

Consider the individual currents in pairs, where the currents in each pair are equidistant on either side of the point where $\vec{B}$ is being calculated.

The sketch shows that for each pair the z-components cancel, and that above the sheet the field is in the $-x$-direction and that below the sheet it is in the $+x$-direction.

Also, by symmetry the magnitude of $\vec{B}$ a distance a above the sheet must equal the magnitude of $\vec{B}$ a distance a below the sheet.

Now that we have deduced the symmetry of $\vec{B}$, apply Ampere's law. Use a path that is a rectangle:

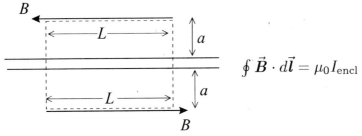

$$\oint \vec{B} \cdot d\vec{l} = \mu_0 I_{\text{encl}}$$

I is directed out of the page, so for I to be positive the integral around the path is taken in the counterclockwise direction.

Since $\vec{B}$ is parallel to the sheet, on the sides of the rectangle that have length $2a$, $\oint \vec{B} \cdot d\vec{l} = 0$. On the long sides of length L, $\vec{B}$ is parallel to the side, in the direction we are integrating around the path, and has the same magnitude, B, on each side. Thus $\oint \vec{B} \cdot d\vec{l} = 2BL$.

n conductors per unit length and current I out of the page in each conductor gives $I_{\text{encl}} = InL$.

Ampere's law then gives $2BL = \mu_0 InL$ and $B = \frac{1}{2}\mu_0 In$.

Note that B is independent of the distance a from the sheet. Compare this result to the electric field due to an infinite sheet of charge (Example 23-7).

29-67 Eq.(29-28): $M = \dfrac{\mu_{\text{total}}}{V}$, so $\mu_{\text{total}} = MV$

The average magnetic moment per atom is $\mu_{\text{atom}} = \mu_{\text{total}}/N = MV/N$, where N is the number of atoms in volume V.

The mass of volume V is $m = \rho V$, where ρ is the density. ($\rho_{\text{iron}} = 7.8 \times 10^3$ kg/m^3).

The number of moles of iron in volume V is

$$n = \frac{m}{55.847 \times 10^{-3} \text{ kg/mol}} = \frac{\rho V}{55.847 \times 10^{-3} \text{ kg/mol}}, \text{ where } 55.847 \times 10^{-3} \text{ kg/mol}$$
is the atomic mass of iron from appendix D.

$N = nN_{\text{A}}$, where $N_{\text{A}} = 6.022 \times 10^{23}$ atoms/mol is Avogadro's number.

Thus $N = nN_{\text{A}} = \dfrac{\rho V N_{\text{A}}}{55.847 \times 10^{-3} \text{ kg/mol}}$.

Thus $\mu_{\text{atom}} = \dfrac{MV}{N} = MV \left(\dfrac{55.847 \times 10^{-3} \text{ kg/mol}}{\rho V N_{\text{A}}} \right) = \dfrac{M(55.847 \times 10^{-3} \text{ kg/mol})}{\rho N_{\text{A}}}$.

$$\mu_{\text{atom}} = \frac{(6.50 \times 10^4 \text{ A/m})(55.847 \times 10^{-3} \text{ kg/mol})}{(7.8 \times 10^3 \text{ kg/m}^3)(6.022 \times 10^{23} \text{ atoms/mol})}$$

$\mu_{\text{atom}} = 7.73 \times 10^{-25}$ A·m^2 = 7.73×10^{-25} J/T

$\mu_{\text{B}} = 9.274 \times 10^{-24}$ A·m^2, so $\mu_{\text{atom}} = 0.0834\mu_{\text{B}}$.

29-69 a) Apply Ohm's law to the dielectric:

$$i(t) = \frac{v(t)}{R}$$

$$v(t) = \frac{q(t)}{C} \text{ and } C = K\frac{\epsilon_0 A}{d}$$

$$v(t) = \left(\frac{d}{K\epsilon_0 A} \right) q(t)$$

The resistance R of the dielectric slab is $R = \rho d/A$.

Thus $i(t) = \dfrac{v(t)}{R} = \left(\dfrac{q(t)d}{K\epsilon_0 A} \right) \left(\dfrac{A}{\rho d} \right) = \dfrac{q(t)}{K\epsilon_0 \rho}$.

But the current $i(t)$ in the dielectric is related to the rate of change dq/dt of the charge $q(t)$ on the plates by $i(t) = -dq/dt$ (a positive i in the direction from the $+$ to the $-$ plate of the capacitor corresponds to a decrease in the charge).

Using this in the above gives $-\dfrac{dq}{dt} = \left(\dfrac{1}{K\rho\epsilon_0} \right) q(t)$.

$$\frac{dq}{q} = -\frac{dt}{K\rho\epsilon_0}$$

Integrate both sides of this equation from $t = 0$, where $q = Q_0$, to a later time t when the charge is $q(t)$

$$\int_{Q_0}^{q} \frac{dq}{q} = -\left(\frac{1}{K\rho\epsilon_0}\right) \int_0^t dt$$

$$\ln\left(\frac{q}{Q_0}\right) = -\frac{t}{K\rho\epsilon_0} \quad \text{and} \quad q(t) = Q_0 e^{-t/K\rho\epsilon_0}.$$

Then $i(t) = -\dfrac{dq}{dt} = \left(\dfrac{Q_0}{K\rho\epsilon_0}\right) e^{-t/K\rho\epsilon_0}$

and $j_C = \dfrac{i(t)}{A} = \left(\dfrac{Q_0}{AK\rho\epsilon_0}\right) e^{-t/K\rho\epsilon_0}$

The conduction current flows from the positive to the negative plate of the capacitor.

b) $E(t) = \dfrac{q(t)}{\epsilon A} = \dfrac{q(t)}{K\epsilon_0 A}$

$$j_D(t) = \epsilon\frac{dE}{dt} = K\epsilon_0\frac{dE}{dt} = K\epsilon_0\frac{dq(t)/dt}{K\epsilon_0 A} = -\frac{i_C(t)}{A} = -j_C(t)$$

The minus sign means that $j_D(t)$ is directed from the negative towards the positive plate. $\vec{E}$ is from $+$ to $-$ but dE/dt is negative (E decreases) so $j_D(t)$ is from $-$ to $+$.

CHAPTER 30
ELECTROMAGNETIC INDUCTION

Exercises 1, 3, 5, 13, 17, 19, 21, 23, 27
Problems 31, 37, 39, 41, 43, 45

Exercises

30-1 The magnitude of the average induced emf is $|\varepsilon_{av}| = N\left|\dfrac{\Delta \Phi_B}{\Delta t}\right|$.

$\Delta \Phi_B = \Phi_{Bf} - \Phi_{Bi}$

<u>initial</u>

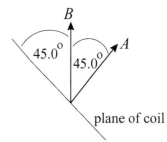

$\Phi_{Bi} = BA \cos \phi$
$\Phi_{Bi} = (1.10 \text{ T})(0.250 \text{ m})(0.400 \text{ m}) \cos 45.0°$
$\Phi_{Bi} = 0.07778 \text{ Wb}$

<u>final</u>

$\Phi_{Bf} = BA \cos \phi$
$\Phi_{Bf} = (1.10 \text{ T})(0.250 \text{ m})(0.400 \text{ m}) \cos 0° = 0.1100 \text{ Wb}$

Then $|\varepsilon_{av}| = N\left|\dfrac{\Delta \Phi_B}{\Delta t}\right| = 80 \left(\dfrac{0.1100 \text{ Wb} - 0.07778 \text{ Wb}}{0.0600 \text{ s}}\right) = 43.0 \text{ V}.$

30-3 **a)** The magnitude of the average emf induced in the coil is

$|\varepsilon_{av}| = N\left|\dfrac{\Delta \Phi_B}{\Delta t}\right|.$

Initially, $\Phi_{Bi} = BA \cos \phi = BA.$

The final flux is zero, so $|\varepsilon_{av}| = N\dfrac{|\Phi_{Bf} - \Phi_{Bi}|}{\Delta t} = \dfrac{NBA}{\Delta t}.$

The average induced current is $I = \dfrac{|\varepsilon_{av}|}{R} = \dfrac{NBA}{R \Delta t}.$

The total charge that flows through the coil is $Q = I \Delta t = \left(\dfrac{NBA}{R \Delta t}\right) \Delta t = \dfrac{NBA}{R}.$

b) The magnetic stripe consists of a pattern of magnetic fields. The pattern of charges that flow in the reader coil tell the card reader the magnetic field pattern and hence the digital information coded onto the card.

c) According to the result in part (a) the charge that flows depends only on the change in the magnetic flux and it does not depend on the rate at which this flux changes.

30-5 Exercise 30-3 derived the result $Q = NBA/R$.

In the present exercise the flux changes from its maximum value of $\Phi_B = BA$ to zero, so this equation applies. R is the total resistance so here $R = 60.0\ \Omega + 45.0\ \Omega = 105.0\ \Omega$.

$$Q = \frac{NBA}{R} \text{ says } B = \frac{QR}{NA} = \frac{(3.56 \times 10^{-5}\ \text{C})(105.0\ \Omega)}{120(3.20 \times 10^{-4}\ \text{m}^2)} = 0.0973\ \text{T}.$$

30-13 **a)** With the switch closed the magnetic field of coil A is to the right at the location of coil B. When the switch is opened the magnetic field of coil A goes away. Hence by Lenz's law the field of the current induced in coil B is to the right, to oppose the decrease in the flux in this direction. To produce magnetic field that is to the right the current in the circuit with coil B must flow through the resistor in the direction a to b.

b) With the switch closed the magnetic field of coil A is to the right at the location of coil B. This field is stronger at points closer to coil A so when coil B is brought closer the flux through coil B increases. By Lenz's law the field of the induced current in coil B is to the left, to oppose the increase in flux to the right. To produce magnetic field that is to the left the current in the circuit with coil B must flow through the resistor in the direction b to a.

c) With the switch closed the magnetic field of coil A is to the right at the location of coil B. The current in the circuit that included coil A increases when R is decreased and the magnetic field of coil A increases when the current through the coil increases. By Lenz's law the field of the induced current in coil B is to the left, to oppose the increase in flux to the right. To produce magnetic field that is to the left the current in the circuit with coil B must flow through the resistor in the direction b to a.

30-17 a)

The magnitude of the induced emf is given by $|\varepsilon| = vLB$.

Thus $v = \dfrac{|\varepsilon|}{LB} = \dfrac{0.620 \text{ V}}{(0.850 \text{ m})(0.850 \text{ T})} = 0.858 \text{ m/s}$.

b) $I = \dfrac{|\varepsilon|}{R} = \dfrac{0.620 \text{ V}}{0.750 \text{ }\Omega} = 0.827 \text{ A}$

c) The direction of the induced emf and induced current can be determined as discussed in Sect.30-3:

Let positive $\vec{A}$ be into the page (in the same direction as $\vec{B}$).

$\vec{B}$ is in the positive direction (the direction of $\vec{A}$) so Φ_B is positive. The magnitude of the flux through the circuit is increasing as the area of the circuit is getting larger, so $d\Phi_B/dt$ is positive. Then by $\varepsilon = -d\Phi_B/dt$, ε is negative. With our chosen direction of $\vec{A}$, clockwise is positive so a negative ε is counterclockwise and I is counterclockwise.

Force on the current in the rod: $\vec{F} = I\vec{l}\textbf{X}\vec{B}$.

The force on the rod due to the induced current is to the left.

$F = IlB \sin\phi = IlB = (0.827 \text{ A})(0.850 \text{ m})(0.850 \text{ T}) = 0.598 \text{ N}$.

30-19

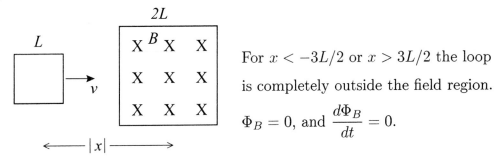

For $x < -3L/2$ or $x > 3L/2$ the loop is completely outside the field region.

$\Phi_B = 0$, and $\dfrac{d\Phi_B}{dt} = 0$.

Thus $\varepsilon = 0$ and $I = 0$, so there is no force from the magnetic field and the external force F necessary to maintain constant velocity is zero.

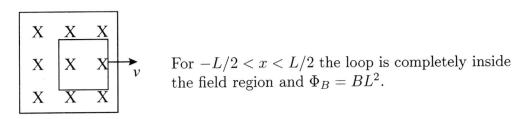

For $-L/2 < x < L/2$ the loop is completely inside the field region and $\Phi_B = BL^2$.

But $\dfrac{d\Phi_B}{dt} = 0$ so $\varepsilon = 0$ and $I = 0$. There is no force $\vec{F} = I\vec{l}\mathbf{X}\vec{B}$ from the magnetic field and the external force F necessary to maintian constant velocity is zero.

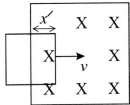

For $-3L/2 < x < -L/2$ the loop is entering the field region. Let x' be the length of the loop that is within the field.

Then $|\Phi_B| = BLx'$ and $|\dfrac{d\Phi_B}{dt}| = Blv$.

The magnitude of the induced emf is $|\varepsilon| = |\dfrac{d\Phi_B}{dt}| = BLv$ and the induced current is

$$I = \frac{|\varepsilon|}{R} = \frac{BLv}{R}.$$

Direction of I: Let $\vec{A}$ be directed into the plane of the figure. Then Φ_B is positive. The flux is positive and increasing in magnitude, so $\dfrac{d\Phi_B}{dt}$ is positive. Then by Faraday's law ε is negative, and with our choice for direction of $\vec{A}$ a negative ε is counterclockwise. The current induced in the loop is counterclockwise.

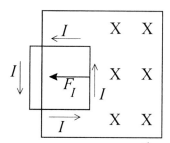

Then $\vec{F}_I = I\vec{l}\mathbf{X}\vec{B}$ gives that the force $\vec{F}_I$ exerted on the loop by the magnetic field is to the left and has magnitude

$$F_I = ILB = \left(\frac{BLv}{R}\right)LB = \frac{B^2L^2v}{R}.$$

The external force $\vec{F}$ needed to move the loop at constant speed is equal in magnitude and opposite in direction to $\vec{F}_I$ so is to the right and has this same magnitude.

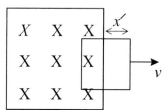

For $L/2 < x < 3L/2$ the loop is leaving the field region. Let x' be the length of the loop that is outside the field.

Then $|\Phi_B| = BL(L - x')$ and $|\frac{d\Phi_B}{dt}| = BLv$.

The magnitude of the induced emf is $|\varepsilon| = |\frac{d\Phi_B}{dt}| = BLv$ and the induced current is

$$I = \frac{|\varepsilon|}{R} = \frac{BLv}{R}.$$

Direction of I: Again let $\vec{A}$ be directed into the plane of the figure. Then Φ_B is positive and decreasing in magnitude, so $\frac{d\Phi_B}{dt}$ is negative. Then by Faraday's law ε is positive, and with our choice for direction of $\vec{A}$ a positive ε is clockwise. The current induced in the loop is clockwise.

Then $\vec{F}_I = I\vec{l}\mathbf{X}\vec{B}$ gives that the force $\vec{F}_I$ exerted on the loop by the magnetic field is to the left and has magnitude

$$F_I = ILB = \left(\frac{BLv}{R}\right)LB = \frac{B^2L^2v}{R}.$$

The external force $\vec{F}$ needed to move the loop at constant speed is equal in magnitude and opposite in direction to $\vec{F}_I$ so is to the right and has this same magnitude.

a)

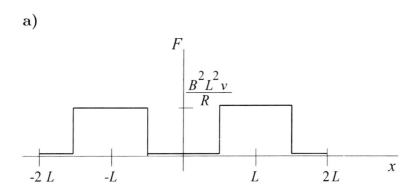

b)

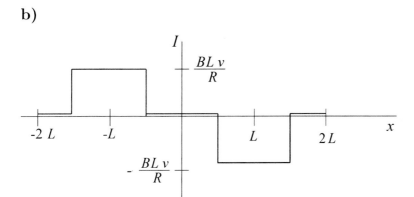

30-21 Away from the ends of the solenoid, $B = \mu_0 n I$ inside and $B = 0$ outside.

a) end view

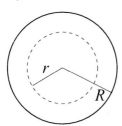

Let R be the radius of the solenoid.

Apply $\oint \vec{E} \cdot d\vec{l} = -\dfrac{d\Phi_B}{dt}$ to an integration path that is a circle of radius r, where $r < R$. We need to calculate just the magnitude of E so we can take absolute values.

$$|\oint \vec{E} \cdot d\vec{l}| = E(2\pi r)$$

$$\Phi_B = B\pi r^2, \quad |-\frac{d\Phi_B}{dt}| = \pi r^2 |\frac{dB}{dt}|$$

$$|\oint \vec{E} \cdot d\vec{l}| = |-\frac{d\Phi_B}{dt}| \text{ implies } E(2\pi r) = \pi r^2 |\frac{dB}{dt}|$$

$$E = \tfrac{1}{2}r\left|\frac{dB}{dt}\right|$$

$$B = \mu_0 n I, \text{ so } \frac{dB}{dt} = \mu_0 n \frac{dI}{dt}$$

Thus $E = \tfrac{1}{2}r\mu_0 n \dfrac{dI}{dt} = \dfrac{1}{2}(0.00500 \text{ m})(4\pi \times 10^{-7} \text{ T} \cdot \text{m/A})(900 \text{ m}^{-1})(60.0 \text{ A/s}) = 1.70 \times 10^{-4} \text{ V/m}$

b) $r = 0.0100$ cm is still inside the solenoid so the expression in part (a) applies.

$$E = \tfrac{1}{2}r\mu_0 n \frac{dI}{dt} = \frac{1}{2}(0.0100 \text{ m})(4\pi \times 10^{-7} \text{ T} \cdot \text{m/A})(900 \text{ m}^{-1})(60.0 \text{ A/s}) = 3.39 \times 10^{-4} \text{ V/m}$$

Inside the solenoid E is proportional to r, so E doubles when r doubles.

30-23 **a)** Because of the axial symmetry and the absence of any electric charge, the field lines are concentric circles.

b)

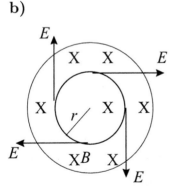

$\vec{E}$ is tangent to the ring. The direction of $\vec{E}$ (clockwise or counterclockwise) is the direction in which current will be induced in the ring.

Use the sign convention for Faraday's law to deduce this direction. Let $\vec{A}$ be into the paper. Then Φ_B is positive. B decreasing then means $\dfrac{d\Phi_B}{dt}$ is negative, so by $\varepsilon = -\dfrac{d\Phi_B}{dt}$, ε is positive and therefore clockwise. Thus $\vec{E}$ is clockwise around the ring.

To calculate E apply $\oint \vec{E} \cdot d\vec{l} = -\dfrac{d\Phi_B}{dt}$ to a circular path that coincides with the ring.

$$\oint \vec{E} \cdot d\vec{l} = E(2\pi r)$$

$$\Phi_B = B\pi r^2; \quad \left|\frac{d\Phi_B}{dt}\right| = \pi r^2 \left|\frac{dB}{dt}\right|$$

$$E(2\pi r) = \pi r^2 \left|\frac{dB}{dt}\right| \text{ and } E = \tfrac{1}{2}r\left|\frac{dB}{dt}\right| = \frac{1}{2}(0.100 \text{ m})(0.0350 \text{ T/s}) = 1.75 \times 10^{-3} \text{ V/m}$$

c) The induced emf has magnitude

$$\varepsilon = \oint \vec{E} \cdot d\vec{l} = E(2\pi r) = (1.75 \times 10^{-3} \text{ V/m})(2\pi)(0.100 \text{ m}) = 1.100 \times 10^{-3} \text{ V}.$$

Then $I = \dfrac{\varepsilon}{R} = \dfrac{1.100 \times 10^{-3} \text{ V}}{4.00 \text{ }\Omega} = 2.75 \times 10^{-4} \text{ A}.$

d) Points a and b are separated by a distance around the ring of πr so

$$\varepsilon = E(\pi r) = (1.75 \times 10^{-3} \text{ V/m})(\pi)(0.100 \text{ m}) = 5.50 \times 10^{-4} \text{ V}$$

e) The ends are separated by a distance around the ring of $2\pi r$ so

$\varepsilon = 1.10 \times 10^{-3}$ V as calculated in part (c).

30-27 Apply Ampere's law to a circular path of radius $r < R$,
where R is the radius of the wire.

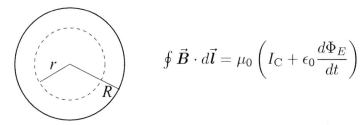

$$\oint \vec{B} \cdot d\vec{l} = \mu_0 \left(I_C + \epsilon_0 \frac{d\Phi_E}{dt} \right)$$

There is no displacement current, so $\oint \vec{B} \cdot d\vec{l} = \mu_0 I_C$

The magnetic field inside the superconducting material is zero, so $\oint \vec{B} \cdot d\vec{l} = 0$. But then Ampere's law says that $I_C = 0$; there can be no conduction current through the path. This same argument applies to any circular path with $r < R$, so all the current must be at the surface of the wire.

Problems

30-31 a)

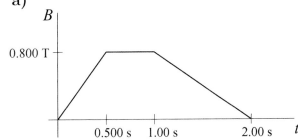

b)

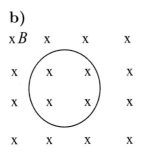

When $\vec{B}$ is increasing in magnitude ($t = 0$ to $t = 0.500$ s), by Lenz's law $\vec{B}_{\text{induced}}$ is directed out of the page inside the coil and the induced current I is counterclockwise.

When $\vec{B}$ is decreasing in magnitude ($t = 1.00$ s to $t = 2.00$ s), by Lenz's law $\vec{B}_{\text{induced}}$ is directed into the page inside the coil and the induced current I is clockwise.

When $\vec{B}$ is constant ($t = 0.500$ s to $t = 1.00$ s), the induced current is zero.

<u>Magnitude of I</u>

$$|\varepsilon| = N\left|\frac{d\Phi_B}{dt}\right| = N\pi r^2 \left|\frac{dB}{dt}\right|, \ I = \frac{|\varepsilon|}{R}$$

$t = 0$ to $t = 0.500$ s: $\left|\dfrac{dB}{dt}\right| = \dfrac{0.800 \text{ T}}{0.500 \text{ s}} = 1.60$ T/s

$|\varepsilon| = 20\pi(0.500 \text{ m})^2(1.60 \text{ T/s}) = 25.13$ V

$I = \dfrac{|\varepsilon|}{R} = \dfrac{25.13 \text{ V}}{1.57 \ \Omega} = 16.0$ A

$t = 1.00$ to $t = 2.00$ s: $\left|\dfrac{dB}{dt}\right| = \dfrac{0.800 \text{ T}}{1.00 \text{ s}} = 0.800$ T/s

$|\varepsilon| = 20\pi(0.500 \text{ m})^2(0.800 \text{ T/s}) = 12.57$ V

$I = \dfrac{|\varepsilon|}{R} = \dfrac{12.57 \text{ V}}{1.57 \ \Omega} = 8.0$ A

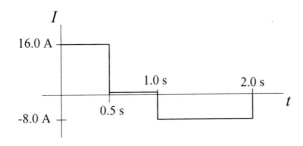

c) $N \oint \vec{E} \cdot d\vec{l} = \varepsilon$ ($\oint \vec{E} \cdot d\vec{l} = \varepsilon/N$ is the emf in one turn of the coil)

$NE(2\pi r) = \varepsilon$ and $E = \dfrac{\varepsilon}{N 2\pi r}$

E is maximum when ε is maximum. From part (b) we have that $\varepsilon_{\max} = 25.13$ V.

$$E_{\max} = \frac{\varepsilon_{\max}}{N2\pi r} = \frac{25.13 \text{ V}}{20(2\pi)(0.500 \text{ m})} = 0.400 \text{ V/m}$$

30-37 a)

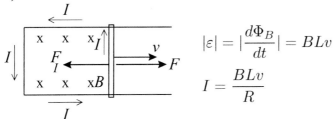

$$|\varepsilon| = |\frac{d\Phi_B}{dt}| = BLv$$

$$I = \frac{BLv}{R}$$

Use $\varepsilon = -\dfrac{d\Phi_B}{dt}$ to find the direction of I:

Let $\vec{A}$ be into the page. Then $\Phi_B > 0$. The area of the circuit is increasing, so $\dfrac{d\Phi_B}{dt} > 0$. Then $\varepsilon < 0$ and with our direction for $\vec{A}$ this means that ε and I are counterclockwise, as shown in the sketch. The force F_I on the rod due to the induced current is given by $\vec{F}_I = I\vec{l}\mathbf{X}\vec{B}$. This gives $\vec{F}_I$ to the left with magnitude $F_I = ILB = (BLv/R)LB = B^2L^2v/R$. Note that $\vec{F}_I$ is directed to oppose the motion of the rod, as required by Lenz's law.

The net force on the rod is $F - F_I$, so its acceleration is $a = (F - F_I)/m = (F - B^2L^2v/R)/m$. The rod starts with $v = 0$ and $a = F/m$. As the speed v increases the acceleration a decreases. When $a = 0$ the rod has reached its terminal speed v_t.

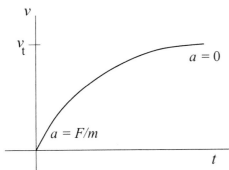

(Recall that a is the slope of the tangent to the v versus t curve.)

b) $v = v_t$ when $a = 0$ so $\dfrac{F - B^2L^2v_t/R}{m} = 0$ and $v_t = \dfrac{RF}{B^2L^2}$.

30-39 a) and b)

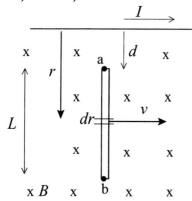

The magnetic field of the wire is given by $B = \dfrac{\mu_0 I}{2\pi r}$ and varies along the length of the bar. At every point along the bar $\vec{B}$ has direction into the page. Divide the bar up into thin slices.

The emf $d\varepsilon$ induced in each slice is given by $d\varepsilon = \vec{v}\mathbf{X}\vec{B}\cdot d\vec{l}$. $\vec{v}\mathbf{X}\vec{B}$ is directed toward the wire, so

$$d\varepsilon = -vB\,dr = -v\left(\frac{\mu_0 I}{2\pi r}\right)dr.$$

The total emf induced in the bar is

$$V_{ba} = \int_a^b d\varepsilon = -\int_d^{d+L}\left(\frac{\mu_0 I v}{2\pi r}\right)dr = -\frac{\mu_0 I v}{2\pi}\int_d^{d+L}\frac{dr}{r} = -\frac{\mu_0 I v}{2\pi}\left[\ln(r)\right]_d^{d+L}$$

$$V_{ba} = -\frac{\mu_0 I v}{2\pi}\left(\ln(d+L) - \ln(d)\right) = -\frac{\mu_0 I v}{2\pi}\ln(1 + L/d)$$

The minus sign means that V_{ba} is negative; point a is at higher potential than point b. (The force $\vec{F} = q\vec{v}\mathbf{X}\vec{B}$ on positive charge carriers in the bar is towards a, so a is at higher potential.)

c)

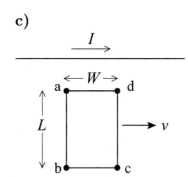

As the loop moves to the right the magnetic flux through it doesn't change.

Thus $\varepsilon = -\dfrac{d\Phi_B}{dt} = 0$

and $I = 0$.

This result can also be understood as follows. The induced emf in section ab puts point a at higher potential; the induced emf in section dc puts point d at higher potnetial. If you travel around the loop then these two induced emf's sum to zero. There is no emf in the loop and hence no current.

30-41 a)

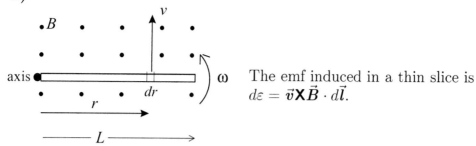

The emf induced in a thin slice is
$d\varepsilon = \vec{v}\mathbf{X}\vec{B} \cdot d\vec{l}$.

Assume that $\vec{B}$ is directed out of the page. Then $\vec{v}\mathbf{X}\vec{B}$ is directed radially outward and $dl = dr$, so $\vec{v}\mathbf{X}\vec{B} \cdot d\vec{l} = vB\,dr$

$v = r\omega$ so $d\varepsilon = \omega Br\,dr$.

The $d\varepsilon$ for all the thin slices that make up the rod are in series so they add:

$$\varepsilon = \int d\varepsilon = \int_0^L \omega Br\,dr = \frac{1}{2}\omega BL^2 = \frac{1}{2}(8.80 \text{ rad/s})(0.650 \text{ T})(0.240 \text{ m})^2 = 0.165 \text{ V}$$

b) No current flows so there is no IR drop in potential. Thus the potential difference between the ends equals the emf of 0.165 V calculated in part (a).

c)

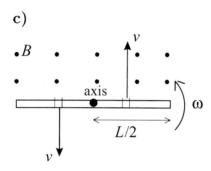

The emf between the center of the rod and each end is $\varepsilon = \frac{1}{2}\omega B(L/2)^2$
$= \frac{1}{4}(0.165 \text{ V}) = 0.0412 \text{ V}$, with the direction of the emf from the center of the rod toward each end. The emfs in each half of the rod thus oppose each other and there is no net emf between the ends of the rod.

30-43 a)

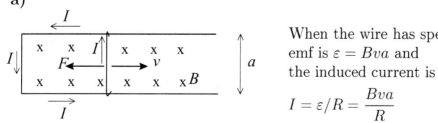

When the wire has speed v the induced emf is $\varepsilon = Bva$ and the induced current is

$$I = \varepsilon/R = \frac{Bva}{R}$$

The induced current flows upward in the wire as shown, so the force $\vec{F} = I\vec{l}\mathbf{X}\vec{B}$ exerted by the magnetic field on the induced current is to the left. $\vec{F}$ opposes the motion of the wire, as it must by Lenz's law. The magnitude of the force is $F = IaB = B^2a^2v/R$.

b) Apply $\sum \vec{F} = m\vec{a}$ to the wire. Take $+x$ to be toward the right and let the origin be at the location of the wire at $t = 0$, so $x_0 = 0$.

$\sum F_x = ma_x$ says $-F = ma_x$

$$a_x = -\frac{F}{m} = -\frac{B^2a^2v}{mR}$$

Use this expression to solve for $v(t)$:

$$a_x = \frac{dv}{dt} = -\frac{B^2a^2v}{mR} \text{ and } \frac{dv}{v} = -\frac{B^2a^2}{mR}dt$$

$$\int_{v_0}^{v} \frac{dv}{v} = -\frac{B^2a^2}{mR} \int_0^t dt$$

$$\ln(v) - \ln(v_0) = -\frac{B^2a^2t}{mR}$$

$$\ln\left(\frac{v}{v_0}\right) = -\frac{B^2a^2t}{mR} \text{ and } v = v_0e^{-B^2a^2t/mR}$$

Note: At $t = 0$, $v = v_0$ and $v \rightarrow 0$ when $t \rightarrow \infty$

Now solve for $x(t)$:

$$v = \frac{dx}{dt} = v_0e^{-B^2a^2t/mR} \text{ so } dx = v_0e^{-B^2a^2t/mR}\,dt$$

$$\int_0^x dx = \int_0^t v_0e^{-B^2a^2t/mR}dt$$

$$x = v_0\left(-\frac{mR}{B^2a^2}\right)\left[e^{-B^2a^2t/mR}\right]_0^t = \frac{mRv_0}{B^2a^2}\left(1 - e^{-B^2a^2t/mR}\right)$$

Comes to rest implies $v = 0$. This happens when $t \rightarrow \infty$.

$t \rightarrow \infty$ gives $x = \dfrac{mRv_0}{B^2a^2}$. Thus this is the distance the wire travels before coming to rest.

30-45 Calculate the induced electric field at each point and then use $\vec{F} = q\vec{E}$.

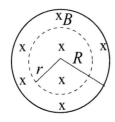

Apply $\oint \vec{E} \cdot d\vec{l} = -\dfrac{d\Phi_B}{dt}$

to a concentric circle of radius r.
Take $\vec{A}$ to be into the
page, in the direction of $\vec{B}$.

B increasing then gives $\dfrac{d\Phi_B}{dt} > 0$, so $\oint \vec{E} \cdot d\vec{l}$ is negative. This means that E is tangent to the circle in the counterclockwise direction.

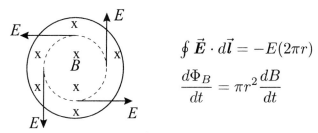

$$\oint \vec{E} \cdot d\vec{l} = -E(2\pi r)$$

$$\frac{d\Phi_B}{dt} = \pi r^2 \frac{dB}{dt}$$

$$-E(2\pi r) = -\pi r^2 \frac{dB}{dt} \quad \text{so} \quad E = \tfrac{1}{2} r \frac{dB}{dt}$$

point a

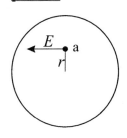

$$F = qE = \tfrac{1}{2} q r \frac{dB}{dt}$$

$\vec{F}$ is to the left
($\vec{F}$ is in the same direction as $\vec{E}$ since q is positive.)

point b

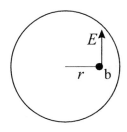

$$F = qE = \tfrac{1}{2} q r \frac{dB}{dt}$$

$\vec{F}$ is to the toward the top of the page.

point c

$r = 0$ here, so $E = 0$ and $F = 0$.

CHAPTER 31

INDUCTANCE

Exercises 1, 3, 9, 11, 13, 17, 19, 21, 27, 29, 37
Problems 43, 47, 49, 51, 53, 55, 57, 59, 61

Exercises

31-1 **a)** $|\varepsilon_2| = M|\frac{di_1}{dt}| = (3.25 \times 10^{-4} \text{ H})(830 \text{ A/s}) = 0.270 \text{ V}$;
yes, it is constant.

b) $|\varepsilon_1| = M|\frac{di_2}{dt}|$; M is a property of the pair of coils so is the same as in part (a). Thus $|\varepsilon_1| = 0.270 \text{ V}$.

31-3 **a)** $M = \frac{N_2\Phi_{B2}}{i_1} = \frac{400(0.0320 \text{ Wb})}{6.52 \text{ A}} = 1.96 \text{ H}$

b) $M = \frac{N_1\Phi_{B1}}{i_2}$ so $\Phi_{B1} = \frac{Mi_2}{N_1} = \frac{(1.96 \text{ H})(2.54 \text{ A})}{700} = 7.11 \times 10^{-3} \text{ Wb}$

31-9 $L = \frac{N\Phi_B}{i}$

If the magnetic field is uniform inside the solenoid $\Phi_B = BA$.

From Eq.(29-23), $B = \mu_0 ni = \mu_0 \left(\frac{N}{l}\right) i$ so $\Phi_B = \frac{\mu_0 NiA}{l}$.

Then $L = \frac{N}{i}\left(\frac{\mu_0 NiA}{l}\right) = \frac{\mu_0 N^2 A}{l}$.

31-11 **a)** $|\varepsilon| = L|\frac{di}{dt}|$ so $L = \frac{|\varepsilon|}{|di/dt|} = \frac{0.0160 \text{ V}}{0.0640 \text{ A/s}} = 0.250 \text{ H}$

b) $N\Phi_B = Li$ so $\Phi_B = \frac{Li}{N} = \frac{(0.250 \text{ H})(0.720 \text{ A})}{400} = 4.50 \times 10^{-4} \text{ Wb}$

31-13 $U = \frac{1}{2}LI^2$ so $L = \frac{2U}{I^2} = \frac{2(0.390 \text{ J})}{(12.0 \text{ A})^2} = 5.417 \times 10^{-3} \text{ H}$

Example 31-3 gives the inductance of a toroidal solenoid to be $L = \frac{\mu_0 N^2 A}{2\pi r}$.

Thus $N = \sqrt{\dfrac{2\pi r L}{\mu_0 A}} = \sqrt{\dfrac{2\pi (0.150 \text{ m})(5.417 \times 10^{-3} \text{ H})}{(4\pi \times 10^{-7} \text{ T} \cdot \text{m/A})(5.00 \times 10^{-4} \text{ m}^2)}} = 2850$

31-17 a) The energy density (energy per unit volume) in a

magnetic field (in vacuum) is given by $u = \dfrac{U}{V} = \dfrac{B^2}{2\mu_0}$ (Eq.31-10).

Thus $V = \dfrac{2\mu_0 U}{B^2} = \dfrac{2(4\pi \times 10^{-7} \text{ T} \cdot \text{m/A})(3.60 \times 10^6 \text{ J})}{(0.600 \text{ T})^2} = 25.1 \text{ m}^3$.

b) $u = \dfrac{U}{V} = \dfrac{B^2}{2\mu_0}$

$B = \sqrt{\dfrac{2\mu_0 U}{V}} = \sqrt{\dfrac{2(4\pi \times 10^{-7} \text{ T} \cdot \text{m/A})(3.60 \times 10^6 \text{ J})}{(0.400 \text{ m})^3}} = 11.9 \text{ T}$

31-19

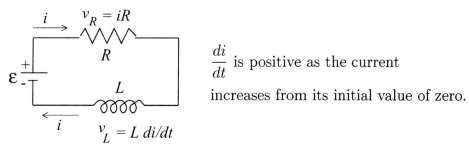

$\dfrac{di}{dt}$ is positive as the current

increases from its initial value of zero.

$\varepsilon - v_R - v_L = 0$

$\varepsilon - iR - L\dfrac{di}{dt} = 0$ so $i = \dfrac{\varepsilon}{R}\left(1 - e^{-(R/L)t}\right)$

a) Initially $(t = 0)$, $i = 0$ so $\varepsilon - L\dfrac{di}{dt} = 0$

$\dfrac{di}{dt} = \dfrac{\varepsilon}{L} = \dfrac{6.00 \text{ V}}{2.50 \text{ H}} = 2.40 \text{ A/s}$

b) $\varepsilon - iR - L\dfrac{di}{dt} = 0$ (Use this equation rather than Eq.(31-15) since i rather than t is given.)

Thus $\dfrac{di}{dt} = \dfrac{\varepsilon - iR}{L} = \dfrac{6.00 \text{ V} - (0.500 \text{ A})(8.00 \text{ }\Omega)}{2.50 \text{ H}} = 0.800 \text{ A/s}$

c) $i = \dfrac{\varepsilon}{R}\left(1 - e^{-(R/L)t}\right) = \left(\dfrac{6.00 \text{ V}}{8.00 \text{ }\Omega}\right)\left(1 - e^{-(8.00 \text{ }\Omega/2.50 \text{ H})(0.250 \text{ s})}\right) =$

$0.750 \text{ A} \left(1 - e^{-0.800}\right) = 0.413 \text{ A}$

d) Final steady state means $t \to \infty$ and $\frac{di}{dt} \to 0$, so $\varepsilon - iR = 0$.

$$i = \frac{\varepsilon}{R} = \frac{6.00 \text{ V}}{8.00 \text{ }\Omega} = 0.750 \text{ A}$$

31-21

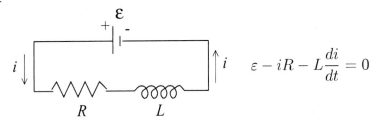

$$\varepsilon - iR - L\frac{di}{dt} = 0$$

Constant current established means $\frac{di}{dt} = 0$.

$$i = \frac{\varepsilon}{R} = \frac{60.0 \text{ V}}{240 \text{ }\Omega} = 0.250 \text{ A}$$

a)

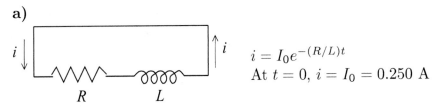

$$i = I_0 e^{-(R/L)t}$$
At $t = 0$, $i = I_0 = 0.250 \text{ A}$

The inductor prevents an instantaneous change in the current; the current in the inductor just after S_2 is closed and S_1 is opened equals the current in the inductor just before this is done.

b) $i = I_0 e^{-(R/L)t} = (0.250 \text{ A})e^{-(240 \text{ }\Omega/0.160 \text{ H})(4.00\times10^{-4} \text{ s})} = (0.250 \text{ A})e^{-0.600} = 0.137 \text{ A}$

c)

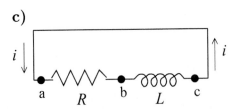

If we trace around the loop in the direction of the current the potential falls as we travel through the resistor so it must rise as we pass through the inductor: $v_{ab} > 0$ and $v_{bc} < 0$. So point c is at higher potential than point b.

$v_{ab} + v_{bc} = 0$ and $v_{bc} = -v_{ab}$

Or, $v_{cb} = v_{ab} = iR = (0.137 \text{ A})(240 \text{ }\Omega) = 32.9 \text{ V}$

d) $i = I_0 e^{-(R/L)t}$

$i = \frac{1}{2}I_0$ says $\frac{1}{2}I_0 = I_0 e^{-(R/L)t}$ and $\frac{1}{2} = e^{-(R/L)t}$

Taking natural logs of both sides of this equation gives $\ln(\frac{1}{2}) = -Rt/L$

$$t = \left(\frac{0.160 \text{ H}}{240 \text{ }\Omega}\right) \ln 2 = 4.62 \times 10^{-4} \text{ s}$$

31-27 a) Eq.(31-22) $\omega = \dfrac{1}{\sqrt{LC}} = \dfrac{1}{\sqrt{(1.50 \text{ H})(6.00 \times 10^{-5} \text{ F})}} = 105.4 \text{ rad/s}$,

which rounds to 105 rad/s.

The period is given by $T = \dfrac{2\pi}{\omega} = \dfrac{2\pi}{105.4 \text{ rad/s}} = 0.0596 \text{ s}$

b)

$\varepsilon = 12.0 \text{ V}$

$\varepsilon - \dfrac{Q}{C} = 0$

$Q = \varepsilon C = (12.0 \text{ V})(6.00 \times 10^{-5} \text{ F}) = 7.20 \times 10^{-4} \text{ C}$

c) $U = \frac{1}{2}CV^2 = \frac{1}{2}(6.00 \times 10^{-5} \text{ F})(12.0 \text{ V})^2 = 4.32 \times 10^{-3} \text{ J}$

d) $q = Q\cos(\omega t + \phi)$ (Eq.31-21).

$q = Q$ at $t = 0$ so $\phi = 0$

$q = Q\cos\omega t = (7.20 \times 10^{-4} \text{ C})\cos([105.4 \text{ rad/s}][0.0230 \text{ s}]) = -5.42 \times 10^{-4} \text{ C}$

The minus sign means that the capacitor has discharged fully and then partially charged again by the current maintained by the inductor; the plate that initially had positive charge now has negative charge and the plate that initially had negative charge now has positive charge.

e) $i = -\omega Q\sin(\omega t + \phi)$ (Eq.31-23)

$i = -(105 \text{ rad/s})(7.20 \times 10^{-4} \text{ C})\sin([105.4 \text{ rad/s}][0.0230 \text{ s}]) = -0.050 \text{ A}$

<u>or</u>

$\dfrac{1}{2}Li^2 + \dfrac{q^2}{2C} = \dfrac{Q^2}{2C}$ gives $i = \pm\sqrt{\dfrac{1}{LC}}\sqrt{Q^2 - q^2}$ (Eq.31-26)

$i = \pm(105 \text{ rad/s})\sqrt{(7.20 \times 10^{-4} \text{ C})^2 - (-5.42 \times 10^{-4} \text{ C})^2} = \pm 0.050 \text{ A}$, which checks.

f) $U_C = \dfrac{q^2}{2C} = \dfrac{(-5.42 \times 10^{-4} \text{ C})^2}{2(6.00 \times 10^{-5} \text{ F})} = 2.45 \times 10^{-3} \text{ J}$

$U_L = \frac{1}{2}Li^2 = \frac{1}{2}(1.50 \text{ H})(0.050 \text{ A})^2 = 1.87 \times 10^{-3} \text{ J}$

Note: $U_C + U_L = 2.45 \times 10^{-3}$ J $+ 1.87 \times 10^{-3}$ J $= 4.32 \times 10^{-3}$ J. This agrees with the total energy initially stored in the capacitor,

$$U = \frac{Q^2}{2C} = \frac{(7.20 \times 10^{-4} \text{ C})^2}{2(6.00 \times 10^{-5} \text{ F})} = 4.32 \times 10^{-3} \text{ J}.$$

31-29 a)

$$v_L = v_C$$
$$L\frac{di}{dt} = \frac{q}{C}$$

$$q = LC\frac{di}{dt} = (0.640 \text{ H})(3.60 \times 10^{-6} \text{ F})(2.80 \text{ A/s}) = 6.45 \times 10^{-6} \text{ C}$$

b) $v_C = \dfrac{q}{C} = \dfrac{8.50 \times 10^{-6} \text{ C}}{3.60 \times 10^{-6} \text{ F}} = 2.36$ V

$v_L = v_C = 2.36$ V

31-37 Use the equation $\omega' = \sqrt{\dfrac{1}{LC} - \dfrac{R^2}{4L^2}}$ and solve for R.

Squaring both sides gives $\omega'^2 = \dfrac{1}{LC} - \dfrac{R^2}{4L^2}$.

Set $\omega'^2 = 1/6LC$ and solve for R:

$$R = \sqrt{\frac{10L}{3C}} = \sqrt{\frac{10(0.285 \text{ H})}{3(4.60 \times 10^{-4} \text{ F})}} = 45.4 \ \Omega$$

Problems

31-43 a) $|\varepsilon| = L\left|\dfrac{di}{dt}\right|$ so $L = \dfrac{|\varepsilon|}{|di/dt|}$

$\dfrac{di}{dt} = \dfrac{\Delta i}{\Delta t}$ (since the rate of increase is constant)

So $\dfrac{di}{dt} = \dfrac{48.0 \text{ A}}{12.0 \text{ s}} = 4.00$ A/s

Then $L = \dfrac{30.0 \text{ V}}{4.00 \text{ A/s}} = 7.50$ H

b) $N\Phi_B = Li = (7.50 \text{ H})(48.0 \text{ A}) = 360 \text{ Wb}$

c) Rate at which electrical energy is being dissipated by the resistance is
$P_R = i^2 R = (48.0 \text{ A})^2(60.0 \text{ }\Omega) = 1.38 \times 10^5 \text{ W}$

Rate at which electrical energy is being stored in the magnetic field of the inductor is

$$P_L = |\varepsilon_L|i = Li|\frac{di}{dt}| = (7.50 \text{ H})(48.0 \text{ A})(4.00 \text{ A/s}) = 1.44 \times 10^3 \text{ W}.$$

The ratio is $\dfrac{P_L}{P_R} = \dfrac{1.44 \times 10^3 \text{ W}}{1.38 \times 10^5 \text{ W}} = 0.0104.$

31-47 a)

end view

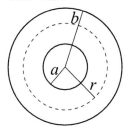

Apply Ampere's law to a circular path of radius r.
$\oint \vec{B} \cdot d\vec{l} = \mu_0 I_{\text{encl}}$

$\oint \vec{B} \cdot d\vec{l} = B(2\pi r)$

$I_{\text{encl}} = i$, the current in the inner conductor

Thus $B(2\pi r) = \mu_0 i$ and $B = \dfrac{\mu_0 i}{2\pi r}.$

b) $u = \dfrac{B^2}{2\mu_0}$

$dU = u\, dV$, where $dV = 2\pi r l\, dr$

$$dU = \frac{1}{2\mu_0}\left(\frac{\mu_0 i}{2\pi r}\right)^2 (2\pi r l)\, dr = \frac{\mu_0 i^2 l}{4\pi r}dr$$

c) $U = \int dU = \dfrac{\mu_0 i^2 l}{4\pi} \displaystyle\int_a^b \dfrac{dr}{r} = \dfrac{\mu_0 i^2 l}{4\pi}\left[\ln r\right]_a^b$

$U = \dfrac{\mu_0 i^2 l}{4\pi}(\ln b - \ln a) = \dfrac{\mu_0 i^2 l}{4\pi}\ln\left(\dfrac{b}{a}\right)$

d) Eq.(31-9): $U = \frac{1}{2}Li^2$

Part (c): $U = \dfrac{\mu_0 i^2 l}{4\pi} \ln\left(\dfrac{b}{a}\right)$

$\dfrac{1}{2} L i^2 = \dfrac{\mu_0 i^2 l}{4\pi} \ln\left(\dfrac{b}{a}\right)$

$L = \dfrac{\mu_0 l}{2\pi} \ln\left(\dfrac{b}{a}\right)$, the same result as calculated in part (d) of Problem 31-46.

31-49 Energy density in the electric field: $u_E = \frac{1}{2}\epsilon_0 E^2$ (Eq.25-11)

Energy density in the magnetic field: $u_B = B^2/2\mu_0$ (Eq.31-10)

$u_E = u_B$ so $\frac{1}{2}\epsilon_0 E^2 = B^2/2\mu_0$

$B = (\sqrt{\mu_0\epsilon_0})E =$

$(\sqrt{(4\pi \times 10^{-7} \text{ T}\cdot\text{m/A})(8.854 \times 10^{-12} \text{ C}^2/\text{N}\cdot\text{m}^2)})(650 \text{ V/m}) = 2.17 \times 10^{-6}$ T

31-51 $L = 2.50$ H, $R = 8.00$ Ω, $\varepsilon = 6.00$ V

$i = (\varepsilon/R)(1 - e^{-t/\tau})$, $\tau = L/R$

a) Eq.(31-9): $U_L = \frac{1}{2}Li^2$

$t = \tau$ so $i = (\varepsilon/R)(1 - e^{-1}) = (6.00 \text{ V}/8.00 \text{ }\Omega)(1 - e^{-1}) = 0.474$ A

Then $U_L = \frac{1}{2}Li^2 = \frac{1}{2}(2.50 \text{ H})(0.474 \text{ A})^2 = 0.281$ J

Exercise 31-23(c): $P_L = \dfrac{dU_L}{dt} = Li\dfrac{di}{dt}$

$i = \left(\dfrac{\varepsilon}{R}\right)(1 - e^{-t/\tau}); \quad \dfrac{di}{dt} = \left(\dfrac{\varepsilon}{L}\right)e^{-(R/L)t} = \dfrac{\varepsilon}{L}e^{-t/\tau}$

$P_L = L\left(\dfrac{\varepsilon}{R}(1 - e^{-t/\tau})\right)\left(\dfrac{\varepsilon}{L}e^{-t/\tau}\right) = \dfrac{\varepsilon^2}{R}(e^{-t/\tau} - e^{-2t\tau})$

$U_L = \displaystyle\int_0^\tau P_L \, dt = \dfrac{\varepsilon^2}{R}\int_0^\tau (e^{-t/\tau} - e^{-2t/\tau})\, dt = \dfrac{\varepsilon^2}{R}\left[-\tau e^{-t/\tau} + \dfrac{\tau}{2}e^{-2t/\tau}\right]_0^\tau$

$U_L = -\dfrac{\varepsilon^2}{R}\tau\left[e^{-t/\tau} - \dfrac{1}{2}e^{-2t/\tau}\right]_0^\tau = \dfrac{\varepsilon^2}{R}\tau\left[1 - \dfrac{1}{2} - e^{-1} + \dfrac{1}{2}e^{-2}\right]$

$U_L = \left(\dfrac{\varepsilon^2}{2R}\right)\left(\dfrac{L}{R}\right)(1 - 2e^{-1} + e^{-2}) = \frac{1}{2}\left(\dfrac{\varepsilon}{R}\right)^2 L(1 - 2e^{-1} + e^{-2})$

$U_L = \frac{1}{2}\left(\dfrac{6.00 \text{ V}}{8.00 \text{ }\Omega}\right)^2 (2.50 \text{ H})(0.3996) = 0.281$ J, which checks.

b) Exercise 31-23(a): The rate at which the battery supplies energy is

$$P_\varepsilon = \varepsilon i = \varepsilon \left(\frac{\varepsilon}{R}(1 - e^{-t/\tau}) \right) = \frac{\varepsilon^2}{R}(1 - e^{-t/\tau})$$

$$U_\varepsilon = \int_0^\tau P_\varepsilon \, dt = \frac{\varepsilon^2}{R} \int_0^\tau (1 - e^{-t/\tau}) dt = \frac{\varepsilon^2}{R} \left[t + \tau e^{-t/\tau} \right]_0^\tau = \left(\frac{\varepsilon^2}{R} \right)(\tau + \tau e^{-1} - \tau)$$

$$U_\varepsilon = \left(\frac{\varepsilon^2}{R} \right)\tau e^{-1} = \left(\frac{\varepsilon^2}{R} \right)\left(\frac{L}{R} \right)e^{-1} = \left(\frac{\varepsilon}{R} \right)^2 L e^{-1}$$

$$U_\varepsilon = \left(\frac{6.00 \text{ V}}{8.00 \text{ }\Omega} \right)^2 (2.50 \text{ H})(0.3679) = 0.517 \text{ J}$$

c) $P_R = i^2 R = \left(\dfrac{\varepsilon^2}{R} \right) \left(1 - e^{-t/\tau} \right)^2 = \dfrac{\varepsilon^2}{R}(1 - 2e^{-t/\tau} + e^{-2t/\tau})$

$$U_R = \int_0^\tau P_R \, dt = \frac{\varepsilon^2}{R} \int_0^\tau (1 - 2e^{-t/\tau} + e^{-2t/\tau}) dt = \frac{\varepsilon^2}{R} \left[t + 2\tau e^{-t/\tau} - \frac{\tau}{2} e^{-2t/\tau} \right]_0^\tau$$

$$U_R = \frac{\varepsilon^2}{R} \left[\tau + 2\tau e^{-1} - \frac{\tau}{2} e^{-2} - 2\tau + \frac{\tau}{2} \right] = \frac{\varepsilon^2}{R} \left[-\frac{\tau}{2} + 2\tau e^{-1} - \frac{\tau}{2} e^{-2} \right]$$

$$U_R = \left(\frac{\varepsilon^2}{2R} \right)\left(\frac{L}{R} \right)\left[-1 + 4e^{-1} - e^{-2} \right]$$

$$U_R = \left(\frac{\varepsilon}{R} \right)^2 \left(\tfrac{1}{2}L \right)\left[-1 + 4e^{-1} - e^{-2} \right] = \left(\frac{6.00 \text{ V}}{8.00 \text{ }\Omega} \right)^2 \tfrac{1}{2}(2.50 \text{ H})(0.3362) = 0.236 \text{ J}$$

d) $U_\varepsilon = U_R + U_L$ (0.517 J = 0.236 J + 0.281 J)

The energy supplied by the battery equals the sum of the enegy stored in the magnetic field of the inductor and the energy dissipated in the resistance of the inductor.

31-53 The equation is $-iR - L\dfrac{di}{dt} - \dfrac{q}{C} = 0$

Multiplying by $-i$ gives $i^2 R + Li\dfrac{di}{dt} + \dfrac{qi}{C} = 0$.

$\dfrac{d}{dt}U_L = \dfrac{d}{dt}\left(\dfrac{1}{2}Li^2 \right) = \dfrac{1}{2}L\dfrac{d}{dt}(i^2) = \dfrac{1}{2}L\left(2i\dfrac{di}{dt} \right) = Li\dfrac{di}{dt}$, the second term

$\dfrac{d}{dt}U_C = \dfrac{d}{dt}\left(\dfrac{q^2}{2C} \right) = \dfrac{1}{2C}\dfrac{d}{dt}(q^2) = \dfrac{1}{2C}(2q)\dfrac{dq}{dt} = \dfrac{qi}{C}$, the third term.

$i^2 R = P_R$, the rate at which electrical energy is dissipated in the resistance.

$\dfrac{d}{dt}U_L = P_L$, the rate at which the amount of energy stored in the inductor is changing.

$\frac{d}{dt}U_C = P_C$, the rate at which the amount of energy stored in the capacitor is changing.

The equation says that $P_R + P_L + P_C = 0$; the net rate of change of energy in the circuit is zero.

Note: At any given time one of P_C or P_L is negative. If the current and U_L are increasing the charge on the capacitor and U_C are decreasing, and vice versa.

31-55 Need $U_C = 1.60 \times 10^{-4}$ J and $V_C = 250$ V

$$U_C = \tfrac{1}{2}CV^2 \text{ so } C = \frac{2U_C}{V^2} = \frac{2(1.60 \times 10^{-4} \text{ J})}{(25.0 \text{ V})^2} = 5.12 \times 10^{-7} \text{ F} = 0.512 \text{ } \mu\text{F}$$

$$L = \frac{1}{C\omega^2} = \frac{1}{(5.12 \times 10^{-7} \text{ F})(6.40 \times 10^4 \text{ rad/s})^2} = 4.77 \times 10^{-4} \text{ H} = 0.477 \text{ mH}$$

31-57 With S closed the circuit is

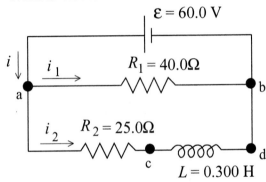

The rate of change of the current through the inductor is limited by the induced emf. Just after the switch is closed the current in the inductor has not had time to increase from zero, so $i_2 = 0$.

a) $\varepsilon - v_{ab} = 0$, so $v_{ab} = 60.0$ V

b) The voltage drops across R, as we travel through the resistor in the direction of the current, so point a is at higher potential.

c) $i_2 = 0$ so $v_{R_2} = i_2 R_2 = 0$

$\varepsilon - v_{R_2} - v_L = 0$ so $v_L = \varepsilon = 60.0$ V

d) The voltage rises when we go from b to a through the emf, so it must drop when we go from a to b through the inductor. Point c must be at higher potential than point d.

e) After the switch has been closed a long time, $\frac{di_2}{dt} \to 0$ so $v_L = 0$. Then

$\varepsilon - v_{R_2} = 0$ and $i_2 R_2 = \varepsilon$ so $i_2 = \dfrac{\varepsilon}{R_2} = \dfrac{60.0 \text{ V}}{25.0 \ \Omega} = 2.40 \text{ A}.$

The rate of change of the current through the inductor is limited by the induced emf. Just after the switch is opened again the current through the inductor hasn't had time to change and is still $i_2 = 2.40$ A. The circuit is

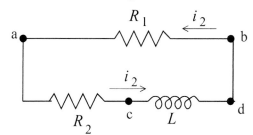

The current through R_1 is $i_2 = 2.40$ A, in the direction b to a. Thus
$v_{ab} = -i_2 R_1 = -(2.40 \text{ A})(40.0 \ \Omega)$
$v_{ab} = -96.0 \text{ V}$

f) Point where current enters resistor is at higher potential; point b is at higher potential.

g) $v_L - v_{R_1} - v_{R_2} = 0$

$v_L = v_{R_1} + v_{R_2}$

$v_{R_1} = -v_{ab} = 96.0 \text{ V};\quad v_{R_2} = i_2 R_2 = (2.40 \text{ A})(25.0 \ \Omega) = 60.0 \text{ V}$

Then $v_L = v_{R_1} + v_{R_2} = 96.0 \text{ V} + 60.0 \text{ V} = 156 \text{ V}.$

As you travel counterclockwise around the circuit in the direction of the current, the voltage drops across each resistor, so it must rise across the inductor and point d is at higher potential than point c. The current is decreasing, so the induced emf in the inductor is directed in the direction of the current. Thus, $v_{cd} = -156$ V.

h) Point d is at higher potnetial.

31-59

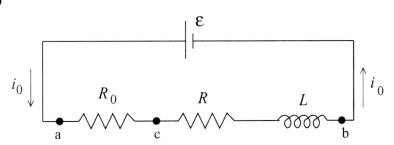

a) At time $t = 0$, $i_0 = 0$ so $v_{ac} = i_0 R_0 = 0$. By the loop rule $\varepsilon - v_{ac} - v_{cb} = 0$, so $v_{cb} = \varepsilon - v_{ac} = \varepsilon = 36.0$ V. ($i_0 R = 0$ so this potential difference of 36.0 V is across the inductor and is an induced emf produced by the changing current.)

b) After a long time $\dfrac{di_0}{dt} \to 0$ so the potential $-L\dfrac{di_0}{dt}$ across the inductor becomes zero. The loop rule gives $\varepsilon - i_0(R_0 + R) = 0$.

$$i_0 = \frac{\varepsilon}{R_0 + R} = \frac{36.0\ \text{V}}{50.0\ \Omega + 150\ \Omega} = 0.180\ \text{A}$$

$$v_{ac} = i_0 R_0 = (0.180\ \text{A})(50.0\ \Omega) = 9.0\ \text{V}$$

Thus $v_{cb} = i_0 R + L\dfrac{di_0}{dt} = (0.180\ \text{A})(150\ \Omega) + 0 = 27.0\ \text{V}$ (Note that $v_{ac} + v_{cb} = \varepsilon$.)

c) $\varepsilon - v_{ac} - v_{cb} = 0$

$$\varepsilon - iR_0 - iR - L\frac{di}{dt} = 0$$

$$L\frac{di}{dt} = \varepsilon - i(R_0 + R)$$

$$-\left(\frac{L}{R + R_0}\right)\frac{di}{dt} = -i + \frac{\varepsilon}{R + R_0}$$

$$\frac{di}{-i + \varepsilon/(R + R_0)} = -\left(\frac{R + R_0}{L}\right) dt$$

Integrate from $t = 0$, when $i = 0$, to t, when $i = i_0$:

$$\int_0^{i_0} \frac{di}{-i + \varepsilon/(R + R_0)} = \frac{R + R_0}{L} \int_0^t dt = -\ln\left[-i + \frac{\varepsilon}{R + R_0}\right]_0^{i_0} = \left(\frac{R + R_0}{L}\right) t, \text{ so}$$

$$\ln\left(-i_0 + \frac{\varepsilon}{R + R_0}\right) - \ln\left(\frac{\varepsilon}{R + R_0}\right) = -\left(\frac{R + R_0}{L}\right) t$$

$$\ln\left(\frac{-i_0 + \varepsilon/(R + R_0)}{\varepsilon/(R + R_0)}\right) = -\left(\frac{R + R_0}{L}\right) t$$

Taking exponentials of both sides gives $\dfrac{-i_0 + \varepsilon/(R + R_0)}{\varepsilon/(R + R_0)} = e^{-(R+R_0)t/L}$ and

$$i_0 = \frac{\varepsilon}{R + R_0}\left(1 - e^{-(R+R_0)t/L}\right)$$

Substituting in the numerical values gives

$$i_0 = \frac{36.0\ \text{V}}{50\ \Omega + 150\ \Omega}\left(1 - e^{-(200\ \Omega/4.00\ \text{H})t}\right) = (0.180\ \text{A})\left(1 - e^{-t/0.020\ \text{s}}\right)$$

At $t \to 0$, $i_0 = (0.180\ \text{A})(1 - 1) = 0$ (agrees with part (a)).

At $t \to \infty$, $i_0 = (0.180\ \text{A})(1 - 0) = 0.180\ \text{A}$ (agrees with part (b)).

$$v_{ac} = i_0 R_0 = \frac{\varepsilon R_0}{R + R_0}\left(1 - e^{-(R+R_0)t/L}\right) = 9.0 \text{ V}\left(1 - e^{-t/0.020 \text{ s}}\right)$$

$$v_{cb} = \varepsilon - v_{ac} = 36.0 \text{ V} - 9.0 \text{ V}\left(1 - e^{-t/0.020 \text{ s}}\right) = 9.0 \text{ V}\left(3.00 + e^{-t/0.020 \text{ s}}\right)$$

At $t \to 0$, $v_{ac} = 0$, $v_{cb} = 36.0$ V (agrees with part (a)).
At $t \to \infty$, $v_{ac} = 9.0$ V, $v_{cb} = 27.0$ V (agrees with part (b)).

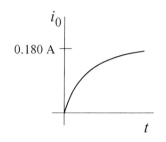
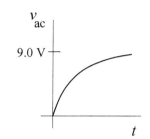
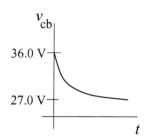

31-61 a) With switch S closed the circuit is

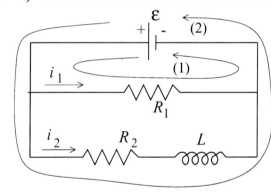

loop 1
$$\varepsilon - i_1 R_1 = 0$$
$$i_1 = \frac{\varepsilon}{R_1} \text{ (independent of } t\text{)}$$

loop (2)

$$\varepsilon - i_2 R_2 - L\frac{di_2}{dt} = 0$$

This is in the form of equation (31-12), so the solution is analogous to Eq.(31-14):

$$i_2 = \frac{\varepsilon}{R_2}\left(1 - e^{-R_2 t/L}\right)$$

b) The expressions derived in part (a) give that as $t \to \infty$, $i_1 = \frac{\varepsilon}{R_1}$, $i_2 = \frac{\varepsilon}{R_2}$

Since $\frac{di_2}{dt} \to 0$ at steady-state the inductance has no effect on the circuit.)

c) The circuit now becomes

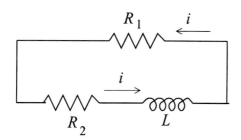

Let $t = 0$ now be when S is opened.

At $t = 0$, $i = \dfrac{\varepsilon}{R_2}$.

The loop rule applied to the single current loop gives $-i(R_1 + R_2) - L\dfrac{di}{dt} = 0$.

(Now $\dfrac{di}{dt}$ is negative.)

$L\dfrac{di}{dt} = -i(R_1 + R_2)$ gives $\dfrac{di}{i} = -\left(\dfrac{R_1 + R_2}{L}\right)dt$

Integrate from $t = 0$, when $i = I_0 = \varepsilon/R_2$, to t.

$\displaystyle\int_{I_0}^{i} \dfrac{di}{i} = -\left(\dfrac{R_1 + R_2}{L}\right)\int_0^t dt$ and $\ln\left(\dfrac{i}{I_0}\right) = -\left(\dfrac{R_1 + R_2}{L}\right)t$

Taking exponentials of both sides of this equation gives

$i = I_0 e^{-(R_1+R_2)t/L} = \dfrac{\varepsilon}{R_2}e^{-(R_1+R_2)t/L}$

d) $L = 22.0$ H

$P_{R_1} = \dfrac{V^2}{R_1} = 40.0$ W gives $R_1 = \dfrac{V^2}{P_{R_1}} = \dfrac{(120\text{ V})^2}{40.0\text{ W}} = 360\ \Omega$.

We are asked to find R_2 and ε. Use the expression derived in part (c).
$I_0 = 0.600$ A so $\varepsilon/R_2 = 0.600$ A

$i = 0.150$ A when $t = 0.080$ s, so $i = \dfrac{\varepsilon}{R_2}e^{-(R_1+R_2)t/L}$ gives

$0.150\text{ A} = (0.600\text{ A})e^{-(R_1+R_2)t/L}$
$\frac{1}{4} = e^{-(R_1+R_2)t/L}$ so $\ln 4 = (R_1 + R_2)t/L$

$R_2 = \dfrac{L\ln 4}{t} - R_1 = \dfrac{(22.0\text{ H})\ln 4}{0.080\text{ s}} - 360\ \Omega = 381.2\ \Omega - 360\ \Omega = 21.2\ \Omega$

Then $\varepsilon = (0.600\text{ A})R_2 = (0.600\text{ A})(21.2\ \Omega) = 12.7$ V.

e) Use the expressions derived in part (a). The current through the light bulb before the switch is opened is $i_1 = \dfrac{\varepsilon}{R_1} = \dfrac{12.7\text{ V}}{360\ \Omega} = 0.0353$ A

When the switch is opened the current through the light bulb jumps from 0.0353 A to 0.600 A.

CHAPTER 32
ALTERNATING CURRENT

Exercises 5, 7, 9, 13, 19, 21, 23, 25, 27, 29
Problems 33, 35, 37, 39, 49, 51, 53

Exercises

32-5 **a)** $X_L = \omega L = 2\pi f L = 2\pi(80.0 \text{ Hz})(3.00 \text{ H}) = 1510 \text{ }\Omega$

b) $X_L = 2\pi f L$ gives $L = \dfrac{X_L}{2\pi f} = \dfrac{120 \text{ }\Omega}{2\pi(80.0 \text{ Hz})} = 0.239 \text{ H}$

c) $X_C = \dfrac{1}{\omega C} = \dfrac{1}{2\pi f C} = \dfrac{1}{2\pi(80.0 \text{ Hz})(4.00 \times 10^{-6} \text{ F})} = 497 \text{ }\Omega$

d) $X_C = \dfrac{1}{2\pi f C}$ gives $C = \dfrac{1}{2\pi f X_C} = \dfrac{1}{2\pi(80.0 \text{ Hz})(120 \text{ }\Omega)} = 1.66 \times 10^{-5} \text{ F}$

32-7 $V = I X_C$ so $X_C = \dfrac{V}{I} = \dfrac{170 \text{ V}}{0.850 \text{ A}} = 200 \text{ }\Omega$

$X_C = \dfrac{1}{\omega C}$ gives $C = \dfrac{1}{2\pi f X_C} = \dfrac{1}{2\pi(60.0 \text{ Hz})(200 \text{ }\Omega)} = 1.33 \times 10^{-5} \text{ F} = 13.3 \text{ }\mu\text{F}$

32-9 **a)** $v_R = (3.80 \text{ V}) \cos[(720 \text{ rad/s})t]$

$v_R = iR$, so $i = \dfrac{v_R}{R} = \left(\dfrac{3.80 \text{ V}}{150 \text{ }\Omega}\right) \cos[(720 \text{ rad/s})t] = (0.0253 \text{ A}) \cos[(720 \text{ rad/s})t]$

b) $X_L = \omega L$
$\omega = 720 \text{ rad/s}$, $L = 0.250 \text{ H}$, so $X_L = \omega L = (720 \text{ rad/s})(0.250 \text{ H}) = 180 \text{ }\Omega$

c) If $i = I \cos \omega t$ then $v_L = V_L \cos(\omega t + 90°)$ (from Eq.32-10).
$V_L = I \omega L = I X_L = (0.02533 \text{ A})(180 \text{ }\Omega) = 4.56 \text{ V}$
$v_L = (4.56 \text{ V}) \cos[(720 \text{ rad/s})t + 90°]$
But $\cos(a + 90°) = -\sin a$ (Appendix B), so $v_L = -(4.56 \text{ V}) \sin[(720 \text{ rad/s})t]$.

32-13

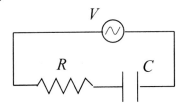

No inductor means $X_L = 0$
$R = 200 \text{ }\Omega$, $C = 6.00 \times 10^{-6} \text{ F}$,
$V = 30.0 \text{ V}$, $\omega = 250 \text{ rad/s}$

a) $X_C = \dfrac{1}{\omega C} = \dfrac{1}{(250 \text{ rad/s})(6.00 \times 10^{-6} \text{ F})} = 666.7 \ \Omega$

$Z = \sqrt{R^2 + (X_L - X_C)^2} = \sqrt{(200 \ \Omega)^2 + (666.7 \ \Omega)^2} = 696 \ \Omega$

b) $I = \dfrac{V}{Z} = \dfrac{30.0 \text{ V}}{696 \ \Omega} = 0.0431 \text{ A} = 43.1 \text{ mA}$

c) The voltage amplitude across the resistor is

$V_R = IR = (0.0431 \text{ A})(200 \ \Omega) = 8.62 \text{ V}$

The voltage amplitude across the capacitor is

$V_C = IX_C = (0.0431 \text{ A})(666.7 \ \Omega) = 28.7 \text{ V}$

d) $\tan \phi = \dfrac{X_L - X_C}{R} = \dfrac{0 - 666.7 \ \Omega}{200 \ \Omega} = -3.333$ so $\phi = -73.3°$

The phase angle is negative, so the source voltage lags behind the current.

e) The phasor diagram is qualitatively

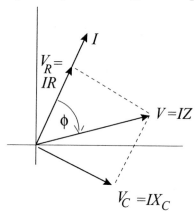

32-19 $R = 200 \ \Omega$, $L = 0.400 \text{ H}$, $C = 6.00 \times 10^{-6} \text{ F}$

a) $\omega = 400 \text{ rad/s}$

$X_L = \omega L = (400 \text{ rad/s})(0.400 \text{ H}) = 160 \ \Omega$

$X_C = \dfrac{1}{\omega C} = \dfrac{1}{(400 \text{ rad/s})(6.00 \times 10^{-6} \text{ F})} = 416.7 \ \Omega$

$Z = \sqrt{R^2 + (X_L - X_C)^2} = \sqrt{(200 \ \Omega)^2 + (160 \ \Omega - 416.7 \ \Omega)^2} = 325 \ \Omega$

$\tan \phi = \dfrac{X_L - X_C}{R} = \dfrac{160 \ \Omega - 416.7 \ \Omega}{200 \ \Omega} = -1.284$ so $\phi = -52.1°$

The phase angle is negative, so the source voltage lags behind the current.

Since $X_C > X_L$ the phasor diagram is qualitatively

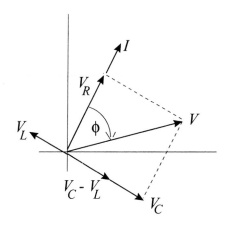

b) $\omega = 800$ rad/s

$X_L = \omega L = (800 \text{ rad/s})(0.400 \text{ H}) = 320 \; \Omega$

$$X_C = \frac{1}{\omega C} = \frac{1}{(800 \text{ rad/s})(6.00 \times 10^{-6} \text{ F})} = 208.3 \; \Omega$$

$$Z = \sqrt{R^2 + (X_L - X_C)^2} = \sqrt{(200 \; \Omega)^2 + (320 \; \Omega - 208.3 \; \Omega)^2} = 229 \; \Omega$$

$$\tan \phi = \frac{X_L - X_C}{R} = \frac{320 \; \Omega - 208.3 \; \Omega}{200 \; \Omega} = +0.5585 \text{ so } \phi = 29.2°$$

The phase angle is positive, so the source voltage leads the current.

Since $X_L > X_C$ the phasor diagram is qualitatively

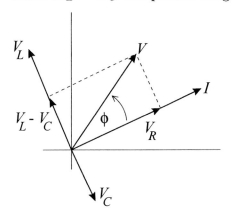

32-21 a)

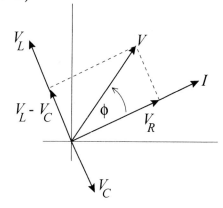

From the diagram

$$\cos\phi = \frac{V_R}{V} = \frac{IR}{IZ} = \frac{R}{Z},$$

as was to be shown.

b) $P_{\mathrm{av}} = V_{\mathrm{rms}} I_{\mathrm{rms}} \cos\phi = V_{\mathrm{rms}} I_{\mathrm{rms}} \left(\dfrac{R}{Z}\right) = \left(\dfrac{V_{\mathrm{rms}}}{Z}\right) I_{\mathrm{rms}} R.$

But $\dfrac{V_{\mathrm{rms}}}{Z} = I_{\mathrm{rms}}$, so $P_{\mathrm{av}} = I_{\mathrm{rms}}^2 R.$

32-23 a) $X_L = \omega L = 2\pi f L = 2\pi(400\ \mathrm{Hz})(0.120\ \mathrm{H}) = 301.6\ \Omega$

$$X_C = \frac{1}{\omega C} = \frac{1}{2\pi f C} = \frac{1}{2\pi(400\ \mathrm{Hz})(7.3\times10^{-6}\ \mathrm{Hz})} = 54.51\ \Omega$$

$$\tan\phi = \frac{X_L - X_C}{R} = \frac{301.6\ \Omega - 54.41\ \Omega}{240\ \Omega}, \text{ so } \phi = +45.8°.$$

The power factor is $\cos\phi = +0.697$.

b) $Z = \sqrt{R^2 + (X_L - X_C)^2} = \sqrt{(240\ \Omega)^2 + (301.6\ \Omega - 54.51\ \Omega)^2} = 344\ \Omega$

c) $V_{\mathrm{rms}} = I_{\mathrm{rms}} Z = (0.450\ \mathrm{A})(344\ \Omega) = 155\ \mathrm{V}$

d) $P_{\mathrm{av}} = I_{\mathrm{rms}} V_{\mathrm{rms}} \cos\phi = (0.450\ \mathrm{A})(155\ \mathrm{V})(0.697) = 48.6\ \mathrm{W}$

e) $P_{\mathrm{av}} = I_{\mathrm{rms}}^2 R = (0.450\ \mathrm{A})^2(240\ \Omega) = 48.6\ \mathrm{W}$

f) All the energy stored in the capacitor during one cycle of the current is released back to the circuit in another part of the cycle. There is no net dissipation of energy in the capacitor.

g) The answer is the same as for the capacitor. Energy is repeatedly being stored and released in the inductor, but no net energy is dissipated there.

32-25 $R = 150\ \Omega$, $L = 0.750$ H, $C = 0.0180\ \mu$F, $V = 150$ V

   a) At the resonance frequency $X_L = X_C$ and from $\tan\phi = \dfrac{X_L - X_C}{R}$ we have that $\phi = 0°$ and the power factor is $\cos\phi = 1$.

   b) $P_{av} = \frac{1}{2}VI\cos\phi$ (Eq.32-30)

   At the resonance frequency $Z = R$, so $I = \dfrac{V}{Z} = \dfrac{V}{R}$

   $$P_{av} = \tfrac{1}{2}V\left(\dfrac{V}{R}\right)\cos\phi = \tfrac{1}{2}\dfrac{V^2}{R} = \dfrac{1}{2}\dfrac{(150\ \text{V})^2}{150\ \Omega} = 75.0\ \text{W}$$

   c) Nothing changes in $P_{av} = \frac{1}{2}\dfrac{V^2}{R}$, so the average power is unchanged: $P_{av} = 75.0$ W. (The resonance frequency changes but since $Z = R$ at resonance the current doesn't change.)

32-27 $R = 200\ \Omega$, $L = 0.400$ H, $C = 6.00\ \mu$F

   a) The resonant angular frequency is given by Eq.(32-31)

   $$\omega_0 = \dfrac{1}{\sqrt{LC}} = \dfrac{1}{\sqrt{(0.400\ \text{H})(6.00 \times 10^{-6}\ \text{F})}} = 645\ \text{rad/s};\ f = \dfrac{\omega}{2\pi} = 103\ \text{Hz}$$

   b) At resonance $X_L = X_C$, so $\tan\phi = \dfrac{X_L - X_C}{R} = 0$ and $\phi = 0$. The source voltage and the current in the circuit are in phase.

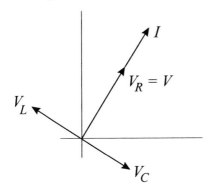

   c) $X_L = \omega L = (645\ \text{rad/s})(0.400\ \text{H}) = 258\ \Omega$

   $$X_C = \dfrac{1}{\omega C} = \dfrac{1}{(645\ \text{rad/s})(6.00 \times 10^{-6}\ \text{F})} = 258\ \Omega$$

   $X_L = X_C$ gives $Z = \sqrt{R^2 + (X_L - X_C)^2} = R = 200\ \Omega$

$$I = \frac{V}{Z} = \frac{30.0 \text{ V}}{200 \text{ }\Omega} = 0.150 \text{ A}; \quad I_{\text{rms}} = \frac{I}{\sqrt{2}} = 0.1061 \text{ A}$$

V_1 reads $V_{R,\text{rms}} = I_{\text{rms}} R = (0.1061 \text{ A})(200 \text{ }\Omega) = 21.2 \text{ V}$

V_2 reads $V_{L,\text{rms}} = I_{\text{rms}} X_L = (0.1061 \text{ A})(258 \text{ }\Omega) = 27.4 \text{ V}$

V_3 reads $V_{C,\text{rms}} = I_{\text{rms}} X_C = (0.1061 \text{ A})(258 \text{ }\Omega) = 27.4 \text{ V}$

V_4 reads $(V_L - V_C)_{\text{rms}} = I_s(X_L - X_C) = 0$. At any instant the voltages across the inductor and capacitor are equal and opposite and sum to zero.

V_5 reads $V_{\text{rms}} = 30.0 \text{ V}/\sqrt{2} = 21.2 \text{ V}$, the same as V_1.

d) $\omega_0 = \dfrac{1}{\sqrt{LC}}$ is independent of R, so $\omega_0 = 645$ rad/s and $f = 103$ Hz, the same as when $R = 200 \text{ }\Omega$.

e) At resonance $X_L = X_C$ and $Z = R$.

$$I_{\text{rms}} = \frac{V_{\text{rms}}}{Z} = \frac{V_{\text{rms}}}{R} = \frac{21.2 \text{ V}}{100 \text{ }\Omega} = 0.212 \text{ A}.$$

The current at resonance is inversely proportional to the resistance so increases by a factor of 2 when the resistance is decreased by a factor of 2.

32-29 a) Eq.(32-34): $\dfrac{V_2}{V_1} = \dfrac{N_2}{N_1}$ so $\dfrac{N_1}{N_2} = \dfrac{V_1}{V_2} = \dfrac{120 \text{ V}}{12.0 \text{ V}} = 10$

b) $I_2 = \dfrac{V_2}{R} = \dfrac{12.0 \text{ V}}{5.00 \text{ }\Omega} = 2.40 \text{ A}$

c) $P_{\text{av}} = I_2^2 R = (2.40 \text{ A})^2(5.00 \text{ }\Omega) = 28.8 \text{ W}$

d) The power drawn from the line by the transformer is the 28.8 W that is delivered by the load.

$$P_{\text{av}} = \frac{V^2}{R} \text{ so } R = \frac{V^2}{P_{\text{av}}} = \frac{(120 \text{ V})^2}{28.8 \text{ W}} = 500 \text{ }\Omega$$

And $\left(\dfrac{N_1}{N_2}\right)^2 (5.00 \text{ }\Omega) = (10)^2(5.00 \text{ }\Omega) = 500 \text{ }\Omega$, as was to be shown.

Problems

32-33 The voltage across the coil leads the curent in it by 52.3°,

so $\phi = +52.3°$

$\tan\phi = \dfrac{X_L - X_C}{R}$. But there is no capacitance in the circuit so $X_C = 0$.

Thus $\tan\phi = \dfrac{X_L}{R}$ and $X_L = R\tan\phi = (48.0~\Omega)\tan 52.3° = 62.1~\Omega$.

$X_L = \omega L = 2\pi f L$ so $L = \dfrac{X_L}{2\pi f} = \dfrac{62.1~\Omega}{2\pi(80.0~\text{Hz})} = 0.124~\text{H}$.

32-35 a) From Fig.32-2b, the rectified current is zero at the same values of t for which the sinusoidal current is zero. These times satisfy $\cos\omega t = 0$ and $\omega t = \pm\pi/2, \pm 3\pi/2, \ldots$.

The two smallest positive times are $t_1 = \pi/2\omega$, $t_2 = 3\pi/2\omega$.

b) $A = \displaystyle\int_{t_1}^{t_2} i\,dt = \int_{t_1}^{t_2} I\cos\omega t\,dt = I\left[\dfrac{1}{\omega}\sin\omega t\right]_{t_1}^{t_2} = \dfrac{I}{\omega}(\sin\omega t_2 - \sin\omega t_1)$

$\sin\omega t_1 = \sin[\omega(\pi/2\omega)] = \sin(\pi/2) = 1$
$\sin\omega t_2 = \sin[\omega(3\pi/2\omega)] = \sin(3\pi/2) = -1$

$A = \left(\dfrac{I}{\omega}\right)(1 - (-1)) = \dfrac{2I}{\omega}$

c) $I_{\text{rav}}(t_2 - t_1) = 2I/\omega$

$I_{\text{rav}} = \dfrac{2I}{\omega(t_2 - t_1)} = \dfrac{2I}{\omega(3\pi/2\omega - \pi/2\omega)} = \dfrac{2I}{\pi}$, which is Eq.(32-3).

32-37 a) Source voltage lags current so it must be that
$X_C > X_L$ and we must add an inductor in series with the circuit.

b) power factor $\cos\phi$ equals 1 so $\phi = 0$ and $X_C = X_L$.

Calculate the present value of $X_C - X_L$ to see how much more X_L is needed:
$R = Z\cos\phi = (60.0~\Omega)(0.720) = 43.2~\Omega$

$\tan\phi = \dfrac{X_L - X_C}{R}$ so $X_L - X_C = R\tan\phi$

$\cos\phi = 0.720$ gives $\phi = -43.95°$ (ϕ is negative since the voltage lags the current)
Then $X_L - X_C = R\tan\phi = (43.2~\Omega)\tan(-43.95°) = -41.64~\Omega$.

Therefore need to add $41.64~\Omega$ of X_L.

$X_L = \omega L = 2\pi f L$ and $L = \dfrac{X_L}{2\pi f} = \dfrac{41.64 \ \Omega}{2\pi(50.0 \ \text{Hz})} = 0.133 \ \text{H}$, amount of inductance to add.

32-39

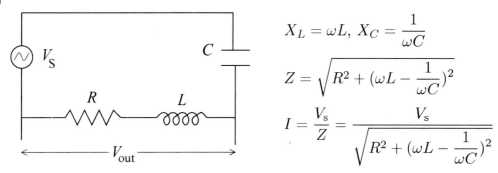

$$X_L = \omega L, \ X_C = \frac{1}{\omega C}$$

$$Z = \sqrt{R^2 + (\omega L - \frac{1}{\omega C})^2}$$

$$I = \frac{V_s}{Z} = \frac{V_s}{\sqrt{R^2 + (\omega L - \frac{1}{\omega C})^2}}$$

The voltages across the resistor and the inductor are 90° out of phase, so

$$V_{\text{out}} = \sqrt{V_R^2 + V_L^2} = I\sqrt{R^2 + X_L^2} = I\sqrt{R^2 + \omega^2 L^2} = V_s\sqrt{\frac{R^2 + \omega^2 L^2}{R^2 + (\omega L - \frac{1}{\omega C})^2}}$$

$$\frac{V_{\text{out}}}{V_s} = \sqrt{\frac{R^2 + \omega^2 L^2}{R^2 + (\omega L - \frac{1}{\omega C})^2}}$$

<u>ω small</u>

As ω gets small, $R^2 + (\omega L - \dfrac{1}{\omega C})^2 \to \dfrac{1}{\omega^2 C^2}$, $R^2 + \omega^2 L^2 \to R^2$

Therefore $\dfrac{V_{\text{out}}}{V_s} \to \sqrt{\dfrac{R^2}{(1/\omega^2 C^2)}} = \omega RC$ as ω becomes small.

<u>ω large</u>

As ω gets large, $R^2 + (\omega L - \dfrac{1}{\omega C})^2 \to R^2 + \omega^2 L^2 \to \omega^2 L^2$, $R^2 + \omega^2 L^2 \to \omega^2 L^2$

Therefore $\dfrac{V_{\text{out}}}{V_s} \to \sqrt{\dfrac{\omega^2 L^2}{\omega^2 L^2}} = 1$ as ω becomes large.

32-49 Source voltage lags current so $\phi = -54.0°$.

$X_C = 350 \ \Omega$, $R = 180 \ \Omega$, $P_{\text{av}} = 140 \ \text{W}$

a) $\tan \phi = \dfrac{X_L - X_C}{R}$

$$X_L = R \tan \phi + X_C = (180 \ \Omega) \tan(-54.0°) + 350 \ \Omega = -248 \ \Omega + 350 \ \Omega = 102 \ \Omega$$

b) $P_{\mathrm{av}} = V_{\mathrm{rms}} I_{\mathrm{rms}} \cos \phi = I_{\mathrm{rms}}^2 R$ (Exercise 32-21).

$$I_{\mathrm{rms}} = \sqrt{\frac{P_{\mathrm{av}}}{R}} = \sqrt{\frac{140 \ \mathrm{W}}{180 \ \Omega}} = 0.882 \ \mathrm{A}$$

c) $P_{\mathrm{av}} = V_{\mathrm{rms}} I_{\mathrm{rms}} \cos \phi$

$$V_{\mathrm{rms}} = \frac{P_{\mathrm{av}}}{I_{\mathrm{rms}} \cos \phi} = \frac{140 \ \mathrm{W}}{(0.882 \ \mathrm{A}) \cos(-54.0°)} = 270 \ \mathrm{V}$$

or

$$Z = \sqrt{R^2 + (X_L - X_C)^2} = \sqrt{(180 \ \Omega)^2 + (102 \ \Omega - 350 \ \Omega)^2} = 306 \ \Omega$$
$$V_{\mathrm{rms}} = I_{\mathrm{rms}} Z = (0.882 \ \mathrm{A})(306 \ \Omega) = 270 \ \mathrm{V}, \text{ which agrees.}$$

32-51 a) $V_C = I X_C$ so $I = \dfrac{V_C}{X_C} = \dfrac{360 \ \mathrm{V}}{480 \ \Omega} = 0.750 \ \mathrm{A}$

b) $V = IZ$ so $Z = \dfrac{V}{I} = \dfrac{120 \ \mathrm{V}}{0.750 \ \mathrm{A}} = 160 \ \Omega$

c) $Z^2 = R^2 + (X_L - X_C)^2$
$X_L - X_C = \pm\sqrt{Z^2 - R^2}$, so
$X_L = X_C \pm \sqrt{Z^2 - R^2} = 480 \ \Omega \pm \sqrt{(160 \ \Omega)^2 - (80.0 \ \Omega)^2} = 480 \ \Omega \pm 139 \ \Omega$
$X_L = 619 \ \Omega$ or $341 \ \Omega$

d) $X_C = \dfrac{1}{\omega C}$ and $X_L = \omega L$. At resonance, $X_C = X_L$. As the frequency is lowered below the resonance frequency X_C increases and X_L decreases. Therefore, for $\omega < \omega_0$, $X_L < X_C$. So for $X_L = 341 \ \Omega$ the angular frequency is less than the resonance angular frequency.

32-53 Consider the cycle of the repeating current that lies between $t_1 = \tau/2$ and $t_2 = 3\tau/2$. In this interval $i = \dfrac{2I_0}{\tau}(t - \tau)$.

$$I_{\mathrm{av}} = \frac{1}{t_2 - t_1} \int_{t_1}^{t_2} i \, dt = \frac{1}{\tau} \int_{\tau/2}^{3\tau/2} \frac{2I_0}{\tau}(t - \tau) \, dt = \frac{2I_0}{\tau^2} \left[\frac{1}{2}t^2 - \tau t \right]_{\tau/2}^{3\tau/2}$$

$$I_{\mathrm{av}} = \left(\frac{2I_0}{\tau^2} \right) \left(\frac{9\tau^2}{8} - \frac{3\tau^2}{2} - \frac{\tau^2}{8} + \frac{\tau^2}{2} \right) = (2I_0)\frac{1}{8}(9 - 12 - 1 + 4) = \frac{I_0}{4}(13 - 13) = 0.$$

$$I_{\text{rms}}^2 = (I^2)_{\text{av}} = \frac{1}{t_2 - t_1} \int_{t_1}^{t_2} i^2 \, dt = \frac{1}{\tau} \int_{\tau/2}^{3\tau/2} \frac{4I_0^2}{\tau^2} (t - \tau)^2 \, dt$$

$$I_{\text{rms}}^2 = \frac{4I_0^2}{\tau^3} \int_{\tau/2}^{3\tau/2} (t - \tau)^2 \, dt = \frac{4I_0^2}{\tau^3} \left[\frac{1}{3}(t - \tau)^3 \right]_{\tau/2}^{3\tau/2} = \frac{4I_0^2}{3\tau^3} \left[\left(\frac{\tau}{2} \right)^3 - \left(-\frac{\tau}{2} \right)^3 \right]$$

$$I_{\text{rms}}^2 = \frac{I_0^2}{6} [1 + 1] = \frac{1}{3} I_0^2$$

$$I_{\text{rms}} = \sqrt{I_{\text{rms}}^2} = \frac{I_0}{\sqrt{3}}.$$

CHAPTER 33
ELECTROMAGNETIC WAVES

Exercises 3, 5, 7, 9, 13, 17, 19, 23, 25
Problems 31, 33, 35, 41, 43, 45, 47

Exercises

33-3 $\omega = 2\pi f = 2\pi(6.10 \times 10^{14} \text{ Hz}) = 3.83 \times 10^{15} \text{ rad/s}$

$$k = \frac{2\pi}{\lambda} = \frac{2\pi f}{c} = \frac{\omega}{c} = \frac{3.83 \times 10^{15} \text{ rad}/s}{3.00 \times 10^8 \text{ m/s}} = 1.28 \times 10^7 \text{ rad/s}$$

$B_{\text{max}} = 5.80 \times 10^{-4} \text{ T}$

$E_{\text{max}} = cB_{\text{max}} = (3.00 \times 10^8 \text{ m/s})(5.80 \times 10^{-4} \text{ T}) = 1.74 \times 10^5 \text{ V/m}$

$\vec{B}$ is along the y-axis. $\vec{E} \times \vec{B}$ is in the direction of propagation (the $+z$-direction).
From this we can deduce the direction of $\vec{E}$.

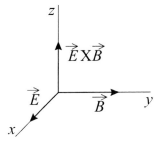

$\vec{E}$ is along the x-axis.

The equations are in the form of Eqs.(33-17):

$\vec{E} = E_{\text{max}}\hat{i} \, \sin(\omega t - kz) =$
$(1.74 \times 10^5 \text{ V/m})\hat{i} \, \sin[(3.83 \times 10^{15} \text{ rad/s})t - (1.28 \times 10^7 \text{ rad/m})z]$
$\vec{B} = B_{\text{max}}\hat{j} \, \sin(\omega t - kz) =$
$(5.80 \times 10^{-4} \text{ T})\hat{j} \, \sin[(3.83 \times 10^{15} \text{ rad/s})t - (1.28 \times 10^7 \text{ rad/m})z]$

33-5 **a)** The equation for the electric field contains the factor $\sin(\omega t - ky)$
so the wave is traveling in the $+y$-direction.

b) $\vec{E}(y,t) = -(3.10 \times 10^5 \text{ V/m})\hat{k} \, \sin[(2.65 \times 10^{12} \text{ rad/s})t - ky]$
Comparing to Eq.(33-16) gives $\omega = 2.65 \times 10^{12} \text{ rad/s}$

$$\omega = 2\pi f = \frac{2\pi c}{\lambda} \text{ so } \lambda = \frac{2\pi c}{\omega} = \frac{2\pi(2.998 \times 10^8 \text{ m/s})}{(2.65 \times 10^{12} \text{ rad/s}} = 7.11 \times 10^{-4} \text{ m}$$

c)

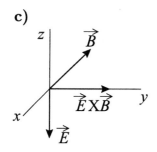

$\vec{E} \times \vec{B}$ must be in the +y-direction (the direction in which the wave is traveling). When $\vec{E}$ is in the $-z$-direction then $\vec{B}$ must be in the $-x$-direction.

$$k = \frac{2\pi}{\lambda} = \frac{\omega}{c} = \frac{2.65 \times 10^{12} \text{ rad/s}}{2.998 \times 10^8 \text{ m/s}} = 8.84 \times 10^3 \text{ rad/m}$$

$$E_{max} = 3.10 \times 10^5 \text{ V/m}$$

Then $B_{max} = \dfrac{E_{max}}{c} = \dfrac{3.10 \times 10^5 \text{ V/m}}{2.998 \times 10^8 \text{ m/s}} = 1.03 \times 10^{-3} \text{ T}$

Using Eq.(33-16) and the fact that $\vec{B}$ is in the $-\hat{i}$ direction when $\vec{E}$ is in the $-\hat{k}$ direction,

$$\vec{B} = -(1.03 \times 10^{-3} \text{ T})\hat{i} \sin[(2.65 \times 10^{12} \text{ rad/s})t - (8.84 \times 10^3 \text{ rad/m})y]$$

33-7 **a)** $c = f\lambda$ so $\lambda = \dfrac{c}{f} = \dfrac{2.998 \times 10^8 \text{ m/s}}{830 \times 10^3 \text{ Hz}} = 361 \text{ m}$

b) $k = \dfrac{2\pi}{\lambda} = \dfrac{2\pi \text{ rad}}{361 \text{ m}} = 0.0174 \text{ rad/m}$

c) $\omega = 2\pi f = (2\pi)(830 \times 10^3 \text{ Hz}) = 5.22 \times 10^6 \text{ rad/s}$

d) Eq.(33-4): $E = cB = (2.998 \times 10^8 \text{ m/s})(4.82 \times 10^{-11} \text{ T}) = 0.0144 \text{ V/m}$

33-9 **a)** $E_{max} = cB_{max}$ so

$$B_{max} = \frac{E_{max}}{c} = \frac{0.0900 \text{ V/m}}{2.998 \times 10^8 \text{ m/s}} = 3.00 \times 10^{-10} \text{ T}$$

b) $I = S_{av} = \dfrac{E_{max}B_{max}}{2\mu_0} = \dfrac{(0.0900 \text{ V/m})(3.00 \times 10^{-10} \text{ T})}{2(4\pi \times 10^{-7} \text{ T} \cdot \text{m/A})} = 1.074 \times 10^{-5} \text{ W/m}^2$

I gives the power per unit area. As in Example 33-3 surround the antenna with a sphere of radius $R = 25.0 \times 10^3$ m. The upper half of this sphere has area $A = 2\pi R^2$. All the power radiated passes through this surface, so the power per unit area at this surface is given by $I = P/(2\pi R^2)$.

Thus $P = 2\pi R^2 I = 2\pi(25.0 \times 10^3 \text{ m})^2(1.074 \times 10^{-5} \text{ W/m}^2) = 4.22 \times 10^4 \text{ W} = 42.2$ kW.

c) The discussion in part (b) shows that since $P = I2\pi R^2$, IR^2 is constant, so $I_1 R_1^2 = I_2 R_2^2$.

But $I = \dfrac{E_{\max}^2}{2\mu_0 c}$, so $E_{\max,1}^2 R_1^2 = E_{\max,2}^2 R_2^2$ and $E_{\max,1} R_1 = E_{\max,2} R_2$

$$R_2 = R_1 \left(\frac{E_{\max,1}}{E_{\max,2}} \right) = 25.0 \text{ km} \left(\frac{0.0900 \text{ V/m}}{0.0300 \text{ V/m}} \right) = 75.0 \text{ km}$$

33-13 a) By Eq.(33-27) the average momentum density is $\dfrac{dp}{dV} = \dfrac{S_{av}}{c^2} = \dfrac{I}{c^2}$

$$\frac{dp}{dV} = \frac{0.78 \times 10^3 \text{ W/m}^2}{(2.998 \times 10^8 \text{ m/s})^2} = 8.7 \times 10^{-15} \text{ kg/m}^2 \cdot \text{s}$$

b) By Eq.(33-28) the average momentum flow rate per unit area is

$$\frac{S_{av}}{c} = \frac{I}{c} = \frac{0.78 \times 10^3 \text{ W/m}^2}{2.998 \times 10^8 \text{ m/s}} = 2.6 \times 10^{-6} \text{ Pa}$$

33-17 $K = 1.74,\quad K_m = 1.23,\quad B_{\max} = 3.80 \times 10^{-9} \text{ T}$

$E_{\max} = vB_{\max}$

By Eq.(33-32) $v = \dfrac{c}{\sqrt{KK_m}}$, so

$$E_{\max} = \frac{cB_{\max}}{\sqrt{KK_m}} = \frac{(2.998 \times 10^8 \text{ m/s})(3.80 \times 10^{-9} \text{ T})}{\sqrt{(1.74)(1.23)}} = 0.779 \text{ V/m}$$

33-19 a) $\lambda = \dfrac{v}{f} = \dfrac{2.17 \times 10^8 \text{ m/s}}{5.70 \times 10^{14} \text{ Hz}} = 3.81 \times 10^{-7} \text{ m}$

b) $\lambda = \dfrac{c}{f} = \dfrac{2.998 \times 10^8 \text{ m/s}}{5.70 \times 10^{14} \text{ Hz}} = 5.26 \times 10^{-7} \text{ m}$

c) $n = \dfrac{c}{v} = \dfrac{2.998 \times 10^8 \text{ m/s}}{2.17 \times 10^8 \text{ m/s}} = 1.38$

d) $n = \sqrt{KK_m} \approx \sqrt{K}$ so $K = n^2 = (1.38)^2 = 1.90$

33-23 a) The distance between adjacent nodal planes of $\vec{B}$ is $\lambda/2$.

There is an antinodal plane of $\vec{B}$ midway between any two adjacent nodal planes, so the distance between a nodal plane and an adjacent nodal plane is $\lambda/4$.

$$\lambda = \frac{v}{f} = \frac{2.10 \times 10^8 \text{ m/s}}{1.20 \times 10^{10} \text{ Hz}} = 0.0175 \text{ m}$$

$$\frac{\lambda}{4} = \frac{0.0175 \text{ m}}{4} = 4.38 \times 10^{-3} \text{ m} = 4.38 \text{ mm}$$

b) The nodal planes of $\vec{E}$ are at $x = 0$, $\lambda/2$, λ. $3\lambda/2,\ldots$, so the antinodal planes of $\vec{E}$ are at $x = \lambda/4$, $3\lambda/4$, $5\lambda/4,\ldots$. The nodal planes of $\vec{B}$ are at $x = \lambda/4$, $3\lambda/4$, $5\lambda/4,\ldots$, so the antinodal planes of $\vec{B}$ are at $\lambda/2$, λ, $3\lambda/2,\ldots$.

The distance between adjacent antinodal planes of $\vec{E}$ and antinodal planes of $\vec{B}$ is therefore $\lambda/4 = 4.38$ mm.

c) From Eqs.(33-40) and (33-41) the distance between adjacent nodal planes of $\vec{E}$ and $\vec{B}$ is $\lambda/4 = 4.38$ mm.

33-25 a) By Eq.(33-41) we see that the nodal planes of the $\vec{B}$ field are a distance $\lambda/2$ apart, so $\lambda/2 = 3.55$ mm and $\lambda = 7.10$ mm.

b) By Eq.(33-40) we see that the nodal planes of the $\vec{E}$ field are also a distance $\lambda/2 = 3.55$ mm apart.

c) $v = f\lambda = (2.20 \times 10^{10} \text{ Hz})(7.10 \times 10^{-3} \text{ m}) = 1.56 \times 10^8 \text{ m/s}$.

Problems

33-31 Eq.(33-12): $\dfrac{\partial E_y}{\partial x} = -\dfrac{\partial B_z}{\partial t}$

Taking $\dfrac{\partial}{\partial t}$ of both sides of this equation gives

$$\frac{\partial^2 E_y}{\partial x \partial t} = -\frac{\partial^2 B_z}{\partial t^2}$$

Eq.(33-14) says $-\dfrac{\partial B_z}{\partial x} = \epsilon_0 \mu_0 \dfrac{\partial E_y}{\partial t}$.

Taking $\dfrac{\partial}{\partial x}$ of both sides of this equation gives

$$-\frac{\partial^2 B_z}{\partial x^2} = \epsilon_0 \mu_0 \frac{\partial^2 E_y}{\partial t \partial x}, \text{ so } \frac{\partial^2 E_y}{\partial t \partial x} = -\frac{1}{\epsilon_0 \mu_0} \frac{\partial^2 B_z}{\partial x^2}$$

But $\dfrac{\partial^2 E_y}{\partial x \partial t} = \dfrac{\partial^2 E_y}{\partial t \partial x}$ (The order in which the partial derivatives are taken doesn't change the result.)

So $-\dfrac{\partial^2 B_z}{\partial t^2} = -\dfrac{1}{\epsilon_0 \mu_0} \dfrac{\partial^2 B_z}{\partial x^2}$

$\dfrac{\partial^2 B_z}{\partial x^2} = \epsilon_0 \mu_0 \dfrac{\partial^2 B_z}{\partial t^2}$, as was to be shown.

33-33 **a)** I gives the energy flow per unit time per unit area:

$I = \dfrac{1}{A} \dfrac{dU}{dt}$ and thus $\dfrac{dU}{dt} = AI$

$I = \dfrac{E_{max}^2}{2\mu_0 c} = \dfrac{(0.0280 \text{ V/m})^2}{2(4\pi \times 10^{-7} \text{ T} \cdot \text{m/A})(2.998 \times 10^8 \text{ m/s})} = 1.04 \times 10^{-6} \text{ W/m}^2$

Then $\dfrac{dU}{dt} = AI = (5.00 \times 10^{-4} \text{ m}^2)(1.04 \times 10^{-6} \text{ W/m}^2) = 5.20 \times 10^{-10} \text{ W}.$

The energy incident on the mirror in 1.00 s is

$(5.20 \times 10^{-10} \text{ W})(1.00 \text{ s}) = 5.20 \times 10^{-10} \text{ J}.$

b) The light is reflected by the mirror, so the average pressure is

$\dfrac{2I}{c} = \dfrac{2(1.04 \times 10^{-6} \text{ W/m}^2)}{2.998 \times 10^8 \text{ m/s}} = 6.94 \times 10^{-15} \text{ Pa}$

c) Surround the light bulb with a spherical surface of radius $R = 3.20$ m and surface area $A = 4\pi R^2$. All the power radiated by the bulb passes through this surface, so $I = \dfrac{P}{A} = \dfrac{P}{4\pi R^2}.$

$P = 4\pi R^2 I = 4\pi (3.20 \text{ m})^2 (1.04 \times 10^{-6} \text{ W/m}^2) = 1.34 \times 10^{-4} \text{ W}.$

33-35 **a)** The intensity is power per unit area:

$I = \dfrac{P}{A} = \dfrac{3.20 \times 10^{-3} \text{ W}}{\pi (1.25 \times 10^{-3} \text{ m})^2} = 652 \text{ W/m}^2.$

$I = \dfrac{E_{max}^2}{2\mu_0 c}$, so $E_{max} = \sqrt{2\mu_0 c I}$

$E_{max} = \sqrt{2(4\pi \times 10^{-7} \text{ T} \cdot \text{m/A})(2.998 \times 10^8 \text{ m/s})(652 \text{ W/m}^2)} = 701 \text{ V/m}$

$B_{max} = \dfrac{E_{max}}{c} = \dfrac{701 \text{ V/m}}{2.998 \times 10^8 \text{ m/s}} = 2.34 \times 10^{-6} \text{ T}$

b) The energy density in the electric field is $u_E = \frac{1}{2}\epsilon_0 E^2$.

$E = E_{max}\sin(\omega t - kx)$ and the average value of $\sin^2(\omega t - kx)$ is $\frac{1}{2}$. The average energy density in the electric field then is

$$u_{E,av} = \frac{1}{4}\epsilon_0 E_{max}^2 = \frac{1}{4}(8.854 \times 10^{-12}\ \text{C}^2/\text{N}\cdot\text{m}^2)(701\ \text{V/m})^2 = 1.09 \times 10^{-6}\ \text{J/m}^3.$$

The energy density in the magnetic field is $u_B = \dfrac{B^2}{2\mu_0}$.

The average value is $u_{B,av} = \dfrac{B_{max}^2}{4\mu_0} = \dfrac{(2.34 \times 10^{-6}\ \text{T})^2}{4(4\pi \times 10^{-7}\ \text{T}\cdot\text{m/A})} = 1.09 \times 10^{-6}\ \text{J/m}^3.$

c) The total energy in this length of beam is the total energy density $u_{av} = u_{E,av} + u_{B,av} = 2.18 \times 10^{-6}\ \text{J/m}^3$ times the volume of this part of the beam.

Thus $U = u_{av}LA = (2.18 \times 10^{-6}\ \text{J/m}^3)(1.00\ \text{m})\pi(1.25 \times 10^{-3}\ \text{m})^2 = 1.07 \times 10^{-11}\ \text{J}.$

This quantity can also be calculated as the power output times the time it takes the light to travel $L = 1.00$ m:

$$U = P\left(\frac{L}{c}\right) = (3.20 \times 10^{-3}\ \text{W})\left(\frac{1.00\ \text{m}}{2.998 \times 10^8\ \text{m/s}}\right) = 1.07 \times 10^{-11}\ \text{J, which}$$

checks.

33-41 a) The direction of $\vec{E}$ is parallel to the axis of the cylinder, in the direction of the current. From Eq.26-7, $E = \rho J = \rho I/\pi a^2$. ($E$ is uniform across the cross section of the conductor.)

b) Cross-sectional view of the conductor; take the current to be coming out of the page.

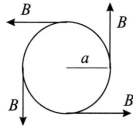

Apply Ampere's law to a circle of radius a.
$\oint \vec{B} \cdot d\vec{l} = B(2\pi a)$
$I_{encl} = I$

$\oint \vec{B} \cdot d\vec{l} = \mu_0 I_{encl}$ gives $B(2\pi a) = \mu_0 I$ and

$$B = \frac{\mu_0 I}{2\pi a}$$

The direction of $\vec{B}$ is counterclockwise around the circle.

c)

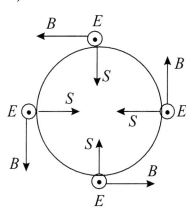

The direction of $\vec{S} = \dfrac{1}{\mu_0}\vec{E} \times \vec{B}$

is radially inward.

$$S = \frac{1}{\mu_0}EB = \frac{1}{\mu_0}\left(\frac{\rho I}{\pi a^2}\right)\left(\frac{\mu_0 I}{2\pi a}\right)$$

$$S = \frac{\rho I^2}{2\pi^2 a^3}$$

d) Since S is constant over the surface of the conductor, the rate of energy flow P is given by S times the surface area of a length l of the conductor:

$$P = SA = S(2\pi a l) = \frac{\rho I^2}{2\pi^2 a^3}(2\pi a l) = \frac{\rho l I^2}{\pi a^2}.$$

But $R = \dfrac{\rho l}{\pi a^2}$, so the result from the Poynting vector is $P = RI^2$. This agrees with $P_R = I^2 R$, the rate at which electrical energy is being dissipated by the resistance of the wire.

Since $\vec{S}$ is radially inward at the surface of the wire and has magnitude equal to the rate at which electrical energy is being dissipated in the wire, this energy can be thought of as entering through the cylindrical sides of the conductor.

33-43 The magnitude of the induced emf is given by Faraday's law:

$$|\varepsilon| = |\frac{d\Phi_B}{dt}|.$$

$\Phi_B = B\pi R^2$, where $R = 0.0900$ m is the radius of the loop. (This assumes that the magnetic field is uniform across the loop, an excellent approximation.)

$$|\varepsilon| = \pi R^2 |\frac{dB}{dt}|$$

$B = B_{max}\sin(\omega t - kx)$ so $\dfrac{dB}{dt} = B_{max}\omega\cos(\omega t - kx)$

The maximum value of $\dfrac{dB}{dt}$ is $B_{max}\omega$, so $|\varepsilon|_{max} = \pi R^2 B_{max}\omega$.

$R = 0.0900$ m, $\omega = 2\pi f = 2\pi(95.0 \times 10^6 \text{ Hz}) = 5.97 \times 10^8$ rad/s

Calculate the intensity I at this distance from the source, and from that the magnetic field amplitude B_{max}:

$$I = \frac{P}{4\pi r^2} = \frac{55.0 \times 10^3 \text{ W}}{4\pi (2.50 \times 10^3 \text{ m})^2} = 7.00 \times 10^{-4} \text{ W/m}^2.$$

$$I = \frac{E_{\text{max}}^2}{2\mu_0 c} = \frac{(cB_{\text{max}})^2}{2\mu_0 c} = \frac{c}{2\mu_0} B_{\text{max}}^2$$

Thus $B_{\text{max}} = \sqrt{\dfrac{2\mu_0 I}{c}} =$

$$\sqrt{\frac{2(4\pi \times 10^{-7} \text{ T} \cdot \text{m/A})(7.00 \times 10^{-4} \text{ W/m}^2)}{2.998 \times 10^8 \text{ m/s}}} = 2.42 \times 10^{-9} \text{ T}.$$

Then $|\varepsilon|_{\text{max}} = \pi R^2 B_{\text{max}} \omega = \pi (0.0900 \text{ m})^2 (2.42 \times 10^{-9} \text{ T})(5.97 \times 10^8 \text{ rad/s}) = 0.0368 \text{ V}.$

33-45 The average time rate of change of the momentum of the light emitted by the flashlight (the momentum flow rate in the beam) is $\dfrac{dp}{dt} = \dfrac{AI}{c}$, where A is the area of the beam and I is the intensity. But $IA = P$ is the power output of the flashlight, so $\dfrac{dp}{dt} = \dfrac{P}{c}$.

By Newton's second law $\dfrac{dp}{dt}$ equals the force F exerted by the light on the astronaut, so $F_{\text{av}} = P/c$. Then $F = ma$ gives $ma = P/c$.

$$a = \frac{P}{mc} = \frac{120 \text{ W}}{(150 \text{ kg})(2.998 \times 10^8 \text{ m/s})} = 2.67 \times 10^{-9} \text{ m/s}^2$$

Now that we have the acceleration we can use a constant-acceleration kinematic equation to find the time: $x - x_0 = v_{0x} t + \frac{1}{2} a_x t^2$.

$$v_{0x} = 0 \text{ so } t = \sqrt{\frac{2(x - x_0)}{a}} = \sqrt{\frac{2(16.0 \text{ m})}{2.67 \times 10^{-9} \text{ m/s}^2}} = 1.10 \times 10^5 \text{ s} = 30.6 \text{ h}.$$

33-47 **a)** The gravitational force is $F_{\text{g}} = G\dfrac{mM}{r^2}$. The mass of the dust particle is $m = \rho V = \rho \frac{4}{3}\pi R^3$. Thus $F_{\text{g}} = \dfrac{4\rho G \pi M R^3}{3r^2}.$

b) For a totally absorbing surface $p_{\text{rad}} = \dfrac{I}{c}.$

If L is the power output of the sun, the intensity of the solar radiation a distance r from the sun is $I = \dfrac{L}{4\pi r^2}.$

Thus $p_{rad} = \dfrac{L}{4\pi cr^2}$.

The force F_{rad} that corresponds to p_{rad} is in the direction of propagation of the radiation, so $F_{rad} = p_{rad}A_\perp$, where $A_\perp = \pi R^2$ is the component of area of the particle perpendicular to the radiation propagation direction. Thus

$$F_{rad} = \left(\frac{L}{4\pi cr^2}\right)(\pi R^2) = \frac{LR^2}{4cr^2}.$$

c) $F_g = F_{rad}$

$$\frac{4\rho G\pi M R^3}{3r^2} = \frac{LR^2}{4cr^2}$$

$$\left(\frac{4\rho G\pi M}{3}\right)R = \frac{L}{4c}$$

$$R = \frac{3L}{16c\rho G\pi M}$$

$$R = \frac{3(3.9 \times 10^{26}\ \text{W})}{16(2.998 \times 10^8\ \text{m/s})(3000\ \text{kg/m}^3)(6.673 \times 10^{-11}\ \text{N}\cdot\text{m}^2/\text{kg}^2)\pi(1.99 \times 10^{30}\ \text{kg})}$$

$R = 1.9 \times 10^{-7}\ \text{m} = 0.19\ \mu\text{m}$.

The gravitation force and the radiation force both have a r^{-2} dependence on the distance from the sun, so this distance divides out in the calculation of R.

d) $\dfrac{F_{rad}}{F_g} = \left(\dfrac{LR^2}{4cr^2}\right)\left(\dfrac{3r^2}{4\rho G\pi m R^3}\right) = \dfrac{3L}{16c\rho G\pi M R}$.

If $R < 0.20\ \mu\text{m}$ then $F_{rad} > F_g$ and the radiation force will drive the particles out of the solar system.

CHAPTER 34
THE NATURE AND PROPAGATION OF LIGHT

Exercises 1, 5, 7, 11, 13, 15, 21
Problems 29, 31, 33, 35, 37, 39, 41, 43, 49

Exercises

34-1 **a)** Eq.(34-1): $n = \dfrac{c}{v}$ so $v = \dfrac{c}{n} = \dfrac{2.998 \times 10^8 \text{ m/s}}{1.47} = 2.04 \times 10^8 \text{ m/s}$

b) Eq.(34-5): $\lambda = \dfrac{\lambda_0}{n} = \dfrac{650 \text{ nm}}{1.47} = 442 \text{ nm}$

34-5 **a)**

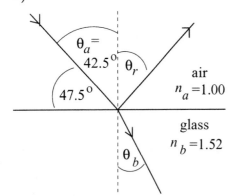

$\theta_r = \theta_a = 42.5°$
The reflected ray makes an angle of
$90.0° - \theta_r = 47.5°$ with the
surface of the glass.

b) $n_a \sin \theta_a = n_b \sin \theta_b$, where the angles are measured from the normal to the interface.

$$\sin \theta_b = \frac{n_a \sin \theta_a}{n_b} = \frac{(1.00)(\sin 42.5°)}{1.66} = 0.4070$$

$\theta_b = 24.0°$

The refracted ray makes an angle of $90.0° - \theta_b = 66.0°$ with the surface of the glass.

34-7

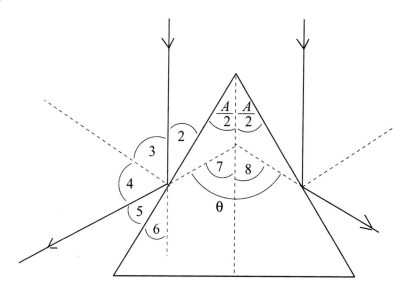

Angle θ is angle 7 + angle 8. We are asked to show that $\theta = 2A$. By symmetry angle 7 = angle 8, so we need to show that angle 7 = A.

From the sketch we see that angle 5 + angle 6 = angle 7 and that angle 2 = $A/2$. But angle 2 = angle 6, so angle 6 = $A/2$.

The law of reflection $\theta_a = \theta_r$ says that angle 3 = angle 4. But then angle 2 = angle 5. Thus angle 5 = $A/2$.

Then angle 7 = angle 5 + angle 6 = $A/2 + A/2 = A$, which completes the proof.

34-11 Use the critical angle to find the index of refraction of the liquid.

Total internal reflection requires that the light be incident on the material with the larger n, in this case the liquid. Apply $n_a \sin\theta_a = n_b \sin\theta_b$ with a = liquid and b = air, so $n_a = n_{\text{liq}}$ and $n_b = 1.0$.

$\theta_a = \theta_{\text{crit}}$ when $\theta_b = 90°$, so $n_{\text{liq}} \sin\theta_{\text{crit}} = (1.0)\sin 90°$

$$n_{\text{liq}} = \frac{1}{\sin\theta_{\text{crit}}} = \frac{1}{\sin 42.5°} = 1.48.$$

a) $n_a \sin\theta_a = n_b \sin\theta_b$ (a = liquid, b = air)

$$\sin\theta_b = \frac{n_a \sin\theta_a}{n_b} = \frac{(1.48)\sin 35.0°}{1.0} = 0.8489 \text{ and } \theta_b = 58.1°$$

b) Now $n_a \sin\theta_a = n_b \sin\theta_b$ with a = air, b = liquid

$$\sin\theta_b = \frac{n_a \sin\theta_a}{n_b} = \frac{(1.0)\sin 35.0°}{1.48} = 0.3876 \text{ and } \theta_b = 22.8°$$

34-13 **a)** Define the index of refraction of air for sound waves to be 1.00.

Then $n_{\text{water}} = \dfrac{v_{\text{air}}}{v_{\text{water}}}$ (Eq.34-1), so $n_{\text{water}} = \dfrac{344 \text{ m/s}}{1320 \text{ m/s}} = 0.2606.$

$n_{\text{water}} < n_{\text{air}}$; air has the larger index of refraction for sound waves since sound travels slower in air than it does in water.

b) $n_a \sin\theta_a = n_b \sin\theta_b$

$\theta_a = \theta_{\text{crit}}$ when $\sin\theta_b = 1$

so $\sin\theta_{\text{crit}} = \dfrac{n_b \sin 90^\circ}{n_a} = \dfrac{n_{\text{water}}}{n_{\text{air}}} = \dfrac{0.2606}{1.00} = 0.2606.$

$\theta_{\text{crit}} = 15.1^\circ$

c) For total internal reflection the wave must be traveling in the material with the larger index of refraction, so the sound wave must be traveling in air.

34-15 a) Reflected beam completely linearly polarized implies that

the angle of incidence equals the polarizing angle, so $\theta_p = 54.5^\circ$.

Then $\tan\theta_p = \dfrac{n_b}{n_a}$ gives $n_{\text{glass}} = n_{\text{air}} \tan\theta_p = (1.00)\tan 54.5^\circ = 1.40.$

b) $n_a \sin\theta_a = n_b \sin\theta_b$

$\sin\theta_b = \dfrac{n_a \sin\theta_a}{n_b} = \dfrac{(1.00)\sin 54.5^\circ}{1.40} = 0.5815$ and $\theta_b = 35.5^\circ$

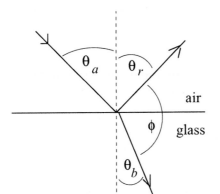

air

glass

Note: $\phi = 180.0^\circ - \theta_r - \theta_b$ and $\theta_r = \theta_a$.
Thus $\phi = 180.0^\circ - 54.5^\circ - 35.5^\circ = 90.0^\circ$;
the reflected ray and the refracted ray
are perpendicular to each other.

34-21 a)

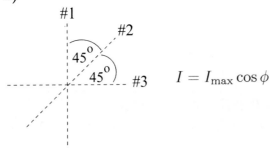

$I = I_{\text{max}} \cos\phi$

After the first filter the intensity is $I_1 = \frac{1}{2}I_0$ and the light is linearly polarized along the axis of the first polarizer.

After the second filter the intensity is

$I_2 = I_1 \cos^2 \phi = (\frac{1}{2}I_0)(\cos 45.0°)^2 = 0.250 I_0$

and the light is linearly polarized along the axis of the second polarizer.

After the third filter the intensity is

$I_3 = I_2 \cos^2 \phi = 0.250 I_0 (\cos 45.0°)^2 = 0.125 I_0$

and the light is linearly polarized along the axis of the third polarizer.

b)

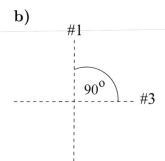

After the first filter the intensity is $I_1 = \frac{1}{2}I_0$ and the light is linearly polarized along the axis of the first polarizer.

After the next filter the intensity is $I_3 = I_1 \cos^2 \phi = (\frac{1}{2}I_0)(\cos 90.0°)^2 = 0$. No light is passed.

Problems

34-29

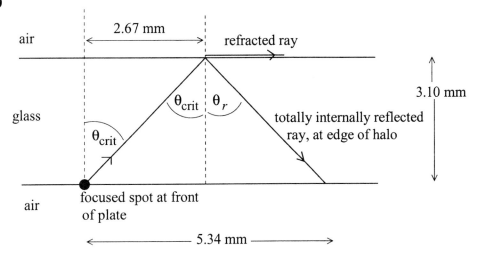

From the distances given in the sketch, $\tan \theta_{\text{crit}} = \dfrac{2.67 \text{ mm}}{3.10 \text{ mm}} = 0.8613; \ \theta_{\text{crit}} = 40.7°$.

Apply Snell's law to the total internal reflection to find the refractive index of the glass:

$$n_a \sin \theta_a = n_b \sin \theta_b$$

$$n_{\text{glass}} \sin \theta_{\text{crit}} = 1.00 \sin 90°$$

$$n_{\text{glass}} = \frac{1}{\sin \theta_{\text{crit}}} = \frac{1}{\sin 40.7°} = 1.53$$

34-31 Before the liquid is poured in

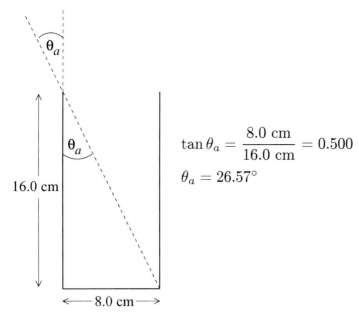

$$\tan \theta_a = \frac{8.0 \text{ cm}}{16.0 \text{ cm}} = 0.500$$

$$\theta_a = 26.57°$$

After the liquid is poured in, θ_a is the same and the refracted ray passes through the center of the bottom of the glass:

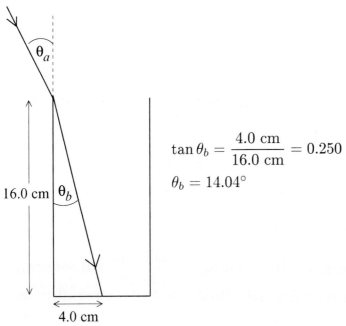

$$\tan \theta_b = \frac{4.0 \text{ cm}}{16.0 \text{ cm}} = 0.250$$

$$\theta_b = 14.04°$$

Then can use Snell's law to find n_b, the refractive index of the liquid:

$n_a \sin \theta_a = n_b \sin \theta_b$

$$n_b = \frac{n_a \sin \theta_a}{\sin \theta_b} = \frac{(1.00)(\sin 26.57°)}{\sin 14.04°} = 1.84$$

34-33

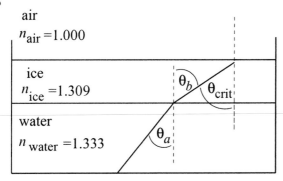

We want to find the incident angle θ_a at the water-ice interface that causes the incident angle at the ice-air interface to be the critical angle.

ice-air interface:

$n_{\text{ice}} \sin \theta_{\text{crit}} = 1.0 \sin 90°$

$n_{\text{ice}} \sin \theta_{\text{crit}} = 1.0$ so $\sin \theta_{\text{crit}} = \dfrac{1}{n_{\text{ice}}}$

But from the diagram we see that $\theta_b = \theta_{\text{crit}}$, so $\sin \theta_b = \dfrac{1}{n_{\text{ice}}}$.

water-ice interface:

$n_{\text{w}} \sin \theta_a = n_{\text{ice}} \sin \theta_b$

But $\sin \theta_b = \dfrac{1}{n_{\text{ice}}}$ so $n_w \sin \theta_a = 1.0$.

$\sin \theta_a = \dfrac{1}{n_{\text{w}}} = \dfrac{1}{1.333} = 0.7502$ and $\theta_a = 48.6°$.

Note: This is the critical angle for a water-air interface; the answer would be the same if the ice layer wasn't there!

34-35 Apply Snell's law to the refraction of each ray as it emerges from the glass. The angle of incidence equals the angle $A = 25.0°$.

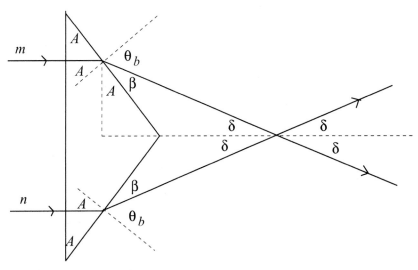

$n_a \sin \theta_a = n_b \sin \theta_b$

$n_{\text{glass}} \sin 25.0° = 1.00 \sin \theta_b$

$\sin \theta_b = n_{\text{glass}} \sin 25.0°$

$\sin \theta_b = 1.66 \sin 25.0° = 0.7015$

$\theta_b = 44.55°$

$\beta = 90.0° - \theta_b = 45.45°$

Then $\delta = 90.0° - A - \beta = 90.0° - 25.0° - 45.45° = 19.55°$.

The angle between the two rays is $2\delta = 39.1°$.

34-37 $n_a \sin \theta_a = n_b \sin \theta_b$

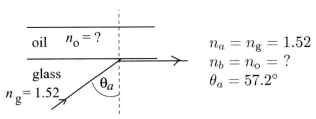

$n_a = n_g = 1.52$
$n_b = n_o = ?$
$\theta_a = 57.2°$

Maximum n_o for total internal reflection means $\theta_b = 90°$

$(1.52) \sin 57.2° = n_o \sin 90°$

$n_o = (1.52) \sin 57.2° = 1.28$

34-39 a)

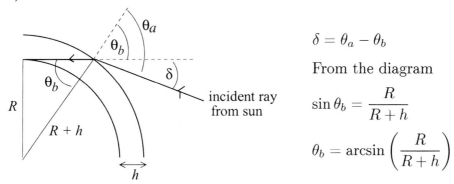

$\delta = \theta_a - \theta_b$

From the diagram

$\sin \theta_b = \dfrac{R}{R+h}$

$\theta_b = \arcsin\left(\dfrac{R}{R+h}\right)$

Apply Snell's law to the refraction that occurs at the top of the atmosphere:

$n_a \sin \theta_a = n_b \sin \theta_b$

(a = vacuum of space, refractive index 1.0; b = atmosphere, refractive index n)

$\sin \theta_a = n \sin \theta_b = n\left(\dfrac{R}{R+h}\right)$ so $\theta_a = \arcsin\left(\dfrac{nR}{R+h}\right)$

$\delta = \theta_a - \theta_b = \arcsin\left(\dfrac{nR}{R+h}\right) - \arcsin\left(\dfrac{R}{R+h}\right)$

b) $\dfrac{R}{R+h} = \dfrac{6.38 \times 10^6 \text{ m}}{6.38 \times 10^6 \text{ m} + 20 \times 10^3 \text{ m}} = 0.99688$

$\dfrac{nR}{R+h} = 1.0003(0.99688) = 0.99718$

$\theta_b = \arcsin\left(\dfrac{R}{R+h}\right) = 85.47°$

$\theta_b = \arcsin\left(\dfrac{nR}{R+h}\right) = 85.70°$

$\delta = \theta_a - \theta_b = 85.70° - 85.47° = 0.23°$

This is about the same as the angular radius of the sun.

34-41 a)

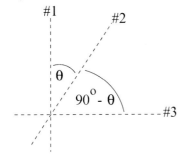

1st filter: $I_1 = \frac{1}{2} I_0$
2nd filter: $I_2 = I_1 (\cos \theta)^2$
$\qquad = \frac{1}{2} I_0 \cos^2 \theta$
3rd filter: $I_3 = I_2 (\cos(90° - \theta))^2$
$\qquad = \frac{1}{2} I_0 \cos^2 \theta \cos^2(90° - \theta)$

$\cos(90° - \theta) = \cos(\theta - 90°) = \sin \theta$ (Using the trig identities in Appendix B.)

Therefore, $I_3 = \frac{1}{2} I_0 \cos^2 \theta \sin^2 \theta$.

But $\cos\theta \sin\theta = \frac{1}{2}\sin 2\theta$ (Appendix B again), so $I_2 = \frac{1}{8}I_0(\sin 2\theta)^2$

b) I_3 maximum implies $\sin 2\theta = 1$ and $\theta = 45°$.

34-43 a) The light travels a distance $\sqrt{h_1^2 + x^2}$ in traveling from point A to the interface. Along this path the speed of the light is v_1, so the time it takes to travel this distance is $t_1 = \dfrac{\sqrt{h_1^2 + x^2}}{v_1}$.

The light travels a distance $\sqrt{h_2^2 + (l-x)^2}$ in traveling from the interface to point B. Along this path the speed of the light is v_2, so the time it takes to travel this distance is $t_2 = \dfrac{\sqrt{h_2^2 + (l-x)^2}}{v_2}$.

The total time to go from A to B is $t = t_1 + t_2 = \dfrac{\sqrt{h_1^2 + x^2}}{v_1} + \dfrac{\sqrt{h_2^2 + (l-x)^2}}{v_2}$.

b) $\dfrac{dt}{dx} = \dfrac{1}{v_1}(\dfrac{1}{2})(h_1^2 + x^2)^{-1/2}(2x) + \dfrac{1}{v_2}(\dfrac{1}{2})(h_2^2 + (l-x)^2)^{-1/2}2(l-x)(-1) = 0$

$$\dfrac{x}{v_1\sqrt{h_1^2 + x^2}} = \dfrac{l-x}{v_2\sqrt{h_2^2 + (l-x)^2}}$$

Multiplying both sides by c gives $\dfrac{c}{v_1}\dfrac{x}{\sqrt{h_1^2 + x^2}} = \dfrac{c}{v_2}\dfrac{l-x}{\sqrt{h_2^2 + (l-x)^2}}$

$\dfrac{c}{v_1} = n_1$ and $\dfrac{c}{v_2} = n_2$ (Eq.34-1)

From Fig.34-40, $\sin\theta_1 = \dfrac{x}{\sqrt{h_1^2 + x^2}}$ and $\sin\theta_2 = \dfrac{l-x}{\sqrt{h_2^2 + (l-x)^2}}$.

So $n_1 \sin\theta_1 = n_2 \sin\theta_2$, which is Snell's law.

34-49 a)

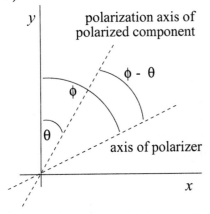

The polarizer passes $\frac{1}{2}$ of the intensity of the unpolarized component, independent of ϕ. Out of the intensity I_p of the polarized component the polarizer passes intensity $I_p \cos^2(\phi - \theta)$, where $\phi - \theta$ is the angle between the plane of polarization and the axis of the polarizer.

The total transmitted intensity is $I = \frac{1}{2} I_0 + I_p \cos^2(\phi - \theta)$. This is maximum when $\theta = \phi$ and from the table of data this occurs for ϕ between $30°$ and $40°$, say at $35°$ and $\theta = 35°$.

Alternatively, the total transmitted intensity is minimum when $\phi - \theta = 90°$ and from the data this occurs for $\phi = 125°$. Thus $\theta = \phi - 90° = 125° - 90° = 35°$, in agreement with the above.

b) $I = \frac{1}{2} I_0 + I_p \cos^2(\phi - \theta)$

Use data at two values of ϕ to determine the two constants I_0 and I_p. Use data where the I_p term is large ($\phi = 30°$) and where it is small ($\phi = 130°$) to have the greatest sensitivity to both I_0 and I_p:

$\phi = 30°$ gives $24.8 \text{ W/m}^2 = \frac{1}{2} I_0 + I_p \cos^2(30° - 35°)$

$24.8 \text{ W/m}^2 = 0.500 I_0 + 0.9924 I_p$

$\phi = 130°$ gives $5.2 \text{ W/m}^2 = \frac{1}{2} I_0 + I_p \cos^2(130° - 35°)$

$5.2 \text{ W/m}^2 = 0.500 I_0 + 0.0076 I_p$

Subtracting the second equation from the first gives $19.6 \text{ W/m}^2 = 0.9848 I_p$ and $I_p = 19.9 \text{ W/m}^2$.

And then $I_0 = 2(5.2 \text{ W/m}^2 - 0.0076(19.9 \text{ W/m}^2)) = 10.1 \text{ W/m}^2$.

CHAPTER 35
GEOMETRIC OPTICS

Exercises 3, 5, 13, 15, 19, 21, 25, 27, 29, 35
Problems 37, 39, 41, 43, 49, 53, 55, 57, 59, 61, 63, 65, 67, 71, 75

Exercises

35-3 Plane mirror: $s = -s'$ (Eq.35-1) and $m = y'/y = -s'/s = +1$ (Eq.35-2).

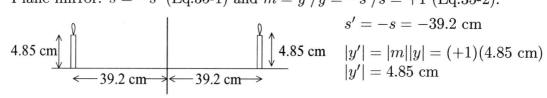

$$s' = -s = -39.2 \text{ cm}$$
$$|y'| = |m||y| = (+1)(4.85 \text{ cm})$$
$$|y'| = 4.85 \text{ cm}$$

The image is 39.2 cm to the right of the mirror and is 4.85 cm tall.

35-5 Concave means R and f are positive; $R = +22.0$ cm; $f = R/2 = +11.0$ cm

a)

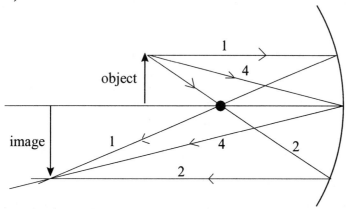

Three principal rays, numbered as in Sect.35-4 are shown.

The principal ray diagram shows that the image is real, inverted, and enlarged.

b) $\dfrac{1}{s} + \dfrac{1}{s'} = \dfrac{1}{f}$

$$\frac{1}{s'} = \frac{1}{f} - \frac{1}{s} = \frac{s - f}{sf} \text{ so } s' = \frac{sf}{s - f} = \frac{(16.5 \text{ cm})(11.0 \text{ cm})}{16.5 \text{ cm} - 11.0 \text{ cm}} = +33.0 \text{ cm}$$

$s' > 0$ so real image, 33.0 cm to left of mirror vertex

$$m = -\frac{s'}{s} = -\frac{33.0 \text{ cm}}{16.5 \text{ cm}} = -2.00 \ (m < 0 \text{ means inverted image})$$

$$|y'| = |m||y| = 2.00(0.600 \text{ cm}) = 1.20 \text{ cm}$$

The image is 33.0 cm to the left of the mirror vertex. It is real, inverted, and is 1.20 cm tall (enlarged). The calculation agrees with the image characterization from the principal ray diagram.

35-13 Convex means R and f are negative; $R = -20.0$ cm; $f = R/2 = -10.0$ cm

a) $s = 9.00$ cm, so $|s| < |f|$

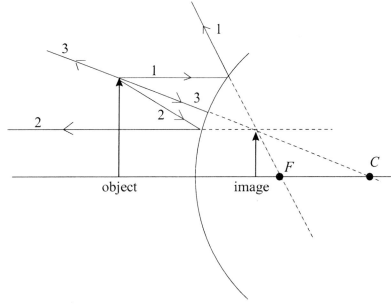

Three principal rays, numbered as in Sect. 35-4, are shown.

The principal ray diagram shows that the image is virtual, erect, reduced in size, and located closer to the mirror than the focal point.

b) $\dfrac{1}{s} + \dfrac{1}{s'} = \dfrac{1}{f}$

$\dfrac{1}{s'} = \dfrac{1}{f} - \dfrac{1}{s} = \dfrac{s - f}{sf}$ so $s' = \dfrac{sf}{s - f} = \dfrac{(12.0 \text{ cm})(-10.0 \text{ cm})}{12.0 \text{ cm} - (-10.0 \text{ cm})} = -5.45$ cm

$s' < 0$ so virtual image, 5.45 cm to right of mirror vertex

$m = -\dfrac{s'}{s} = -\dfrac{-5.45 \text{ cm}}{12.0 \text{ cm}} = +0.454$ ($m > 0$ means erect image)

$|y'| = |m||y| = (+0.454)(9.00 \text{ mm}) = 4.09$ mm

The image is 5.45 cm to the right of the mirror vertex, closer to the mirror than the focal point. It is 4.09 mm tall (reduced), erect, and virtual. The calculation agrees with the image characterization from the principal ray diagram.

35-15

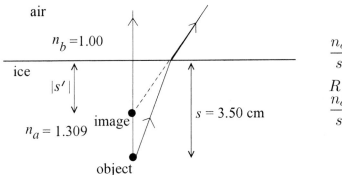

$\dfrac{n_a}{s} + \dfrac{n_b}{s'} = \dfrac{n_b - n_a}{R}$;

$R \to \infty$ (flat surface), so

$\dfrac{n_a}{s} + \dfrac{n_b}{s'} = 0$

$$s' = -\frac{n_b s}{n_a} = -\frac{(1.00)(3.50 \text{ cm})}{1.309} = -2.67 \text{ cm}$$

The apparent depth is 2.67 cm.

35-19

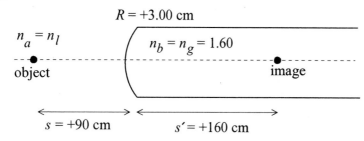

Let n_l be the refractive index of the liquid. s, s', and R are all positive.

$$\frac{n_a}{s} + \frac{n_b}{s'} = \frac{n_b - n_a}{R}$$

$$\frac{n_l}{90.0 \text{ cm}} + \frac{1.60}{160 \text{ cm}} = \frac{1.60 - n_l}{3.00 \text{ cm}}$$

Multiplying the equation by 1440 cm gives $16.0 n_l + 14.4 = 768 - 480 n_l$

$496 n_l = 753.6$ and $n_l = 1.52$.

35-21

$$R = -4.00 \text{ cm}$$

$y = 1.50 \text{ mm}$ $n_a = 1.00$

$s = +24.0 \text{ cm}$ $n_b = 1.60$

$$\frac{n_a}{s} + \frac{n_b}{s'} = \frac{n_b - n_a}{R}$$

$$\frac{1.00}{24.0 \text{ cm}} + \frac{1.60}{s'} = \frac{1.60 - 1.00}{-4.00 \text{ cm}}$$

Multiplying by 24.0 cm gives $1.00 + \dfrac{38.4}{s'} = -3.60$

$$\frac{38.4 \text{ cm}}{s'} = -4.60 \text{ and } s' = -\frac{38.4 \text{ cm}}{4.60} = -8.35 \text{ cm}$$

Eq.(35-12): $m = -\dfrac{n_a s'}{n_b s} = -\dfrac{(1.00)(-8.35 \text{ cm})}{(1.60)(+24.0 \text{ cm})} = +0.217$

$|y'| = |m||y| = (0.217)(1.50 \text{ mm}) = 0.326 \text{ mm}$

The image is virtual ($s' < 0$) and is 8.35 cm to the left of the vertex. The image is erect ($m > 0$) and is 0.326 mm tall.

35-25 $f = +7.00$ cm ($f > 0$ since the lens is converging)

$m = -\dfrac{s'}{s}$; image is erect implies $m > 0$ so must have $s' < 0$ (image is virtual)

$m = \dfrac{y'}{y} = \dfrac{+1.30 \text{ cm}}{0.400 \text{ cm}} = +3.25$

$s' = -ms = -(3.25)s$

Using this in $\dfrac{1}{s} + \dfrac{1}{s'} = \dfrac{1}{f}$ gives

$\dfrac{1}{s} + \dfrac{1}{(-3.25 \text{ cm})s} = \dfrac{1}{7.00 \text{ cm}}$

$s = +4.85 \text{ cm}$

$s' = -(3.25)s = -(3.25)(4.85 \text{ cm}) = -15.8 \text{ cm}$

The object is 4.85 cm to the left of the lens. $s' < 0$ so the image is 15.8 cm to the left of the lens and is virtual.

35-27

$R_1 = +5.00 \text{ cm}$
$R_2 = +3.50 \text{ cm}$
$n = 1.48$

$\dfrac{1}{f} = (n-1)\left(\dfrac{1}{R_1} - \dfrac{1}{R_2}\right) = (0.48)\left(\dfrac{1}{5.00 \text{ cm}} - \dfrac{1}{3.50 \text{ cm}}\right); \ f = -24.31 \text{ cm}$

$s = 18.0 \text{ cm}$

$\dfrac{1}{s} + \dfrac{1}{s'} = \dfrac{1}{f}$, so $\dfrac{1}{s'} = \dfrac{1}{f} - \dfrac{1}{s} = \dfrac{s-f}{sf}$

$s' = \dfrac{sf}{s-f} = \dfrac{(18.0 \text{ cm})(-24.31 \text{ cm})}{18.0 \text{ cm} - (-24.31 \text{ cm})} = -10.3 \text{ cm}$

$s' < 0$ means that the image is virtual and is 10.3 cm to the left of the lens, on the same side of the lens as the object.

35-29

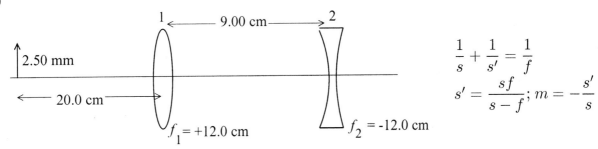

image formed by lens #1

$s_1' = \dfrac{s_1 f}{s_1 - f} = \dfrac{(20.0 \text{ cm})(12.0 \text{ cm})}{20.0 \text{ cm} - 12.0 \text{ cm}} = +30.0 \text{ cm};$

$$m_1 = -\frac{s_1'}{s_1} = -1.50$$

This image serves as the object for the second lens. The image is 30.0 cm to the right of lens #1 so is 30.0 cm − 9.00 cm = 21.0 cm to the right of lens #2. So for lens #2, $s_2 = -21.0$ cm.

image formed by lens #2

$$s_2' = \frac{s_2 f}{s_2 - f} = \frac{(-21.0 \text{ cm})(-12.0 \text{ cm})}{-21.0 \text{ cm} + 12.0 \text{ cm}} = -28.0 \text{ cm};$$

$$m_2 = -\frac{s_2'}{s_2} = -1.33$$

a) $s_2' = -28.0$ cm so the final image is 28.0 cm to the left of lens #2 and 28.0 cm − 9.00 cm = 19.0 cm to the left of lens #1.

b) $s_2' < 0$ so the final image is virtual.

c) $m_{\text{tot}} = m_1 m_2 = (-1.50)(-1.33) = +2.00$

$y' = m_{\text{tot}} y = (2.00)(2.50 \text{ mm}) = 5.00$ mm (height of final image).

$m_{\text{tot}} > 0$ so final image is erect (with respect to original object)

($m_1 < 0$ so lens #1 inverts the image. But $m_2 < 0$ also, so lens #2 also inverts the image, making it erect.)

35-35 $f = -48.0$ cm

virtual image 17.0 cm from lens so $s' = -17.0$ cm

$$\frac{1}{s} + \frac{1}{s'} = \frac{1}{f}, \text{ so } \frac{1}{s'} = \frac{1}{f} - \frac{1}{s} = \frac{s - f}{sf}$$

$$s = \frac{s' f}{s' - f} = \frac{(-17.0 \text{ cm})(-48.0 \text{ cm})}{-17.0 \text{ cm} - (-48.0 \text{ cm})} = +26.3 \text{ cm}$$

$$m = -\frac{s'}{s} = -\frac{-17.0 \text{ cm}}{+26.3 \text{ cm}} = +0.646$$

$$m = \frac{y'}{y} \text{ so } |y| = \frac{|y'|}{|m|} = \frac{8.00 \text{ mm}}{0.646} = 12.4 \text{ mm}$$

Virtual image, real object ($s > 0$) so image and object are on same side of lens.

$m > 0$ so image is erect with respect to the object. The height of the object is 12.4 mm.

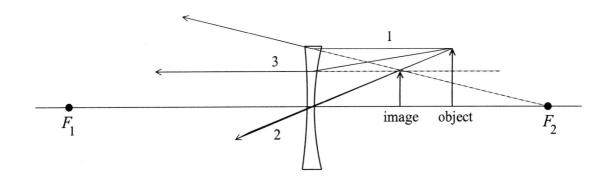

Problems

35-37

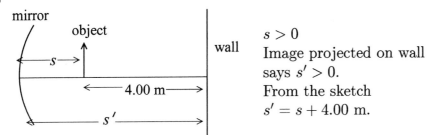

$s > 0$
Image projected on wall
says $s' > 0$.
From the sketch
$s' = s + 4.00$ m.

$m = -\dfrac{s'}{s}$ and $s > 0$, $s' > 0$ says $m < 0$.

$|m| = \dfrac{|y'|}{|y|} = 2.25$ so $m = -2.25$

Then $m = -\dfrac{s'}{s}$ gives $-2.25 = -\dfrac{s'}{s}$ and $s' = 2.25s$.

Using this in the first equation gives $2.25s = s + 4.00$ m and $s = \dfrac{4.00\text{ m}}{1.25} = 3.20$ m.

$s' = 2.25s = (2.25)(3.20\text{ m}) = 7.20$ m

The mirror should be 7.20 m from the wall.

$\dfrac{1}{s} + \dfrac{1}{s'} = \dfrac{2}{R}$ so $\dfrac{2}{R} = \dfrac{s + s'}{ss'}$

$R = 2\left(\dfrac{ss'}{s + s'}\right) = 2\dfrac{(3.20\text{ m})(7.20\text{ m})}{3.20\text{ m} + 7.20\text{ m}} = 4.43$ m.

Note that R is calculated to be positive, which is the correct sign for a concave mirror.

35-39 For a plane mirror $s' = -s$. $v = \dfrac{ds}{dt}$ and $v' = \dfrac{ds'}{dt}$, so $v' = -v$.

The velocities of the object and image relative to the mirror are equal in magnitude and opposite in direction. Thus both you and your image are receding from the mirror surface at 2.40 m/s, in opposite directions. Your image is therefore moving at 4.80 m/s relative to you.

35-41 **a)** Image is to be formed on screen so is real image; $s' > 0$. Mirror to screen distance is 8.00 m, so $s' = +800$ cm.

$$m = -\frac{s'}{s} < 0 \text{ since both } s \text{ and } s' \text{ are positive.}$$

$$|m| = \frac{|y'|}{|y|} = \frac{36.0 \text{ cm}}{0.600 \text{ cm}} = 60.0 \text{ and } m = -60.0.$$

Then $m = -\dfrac{s'}{s}$ gives $s = -\dfrac{s'}{m} = -\dfrac{800 \text{ cm}}{-60.0} = +13.3$ cm.

b) $\dfrac{1}{s} + \dfrac{1}{s'} = \dfrac{2}{R}$, so $\dfrac{2}{R} = \dfrac{s + s'}{ss'}$

$$R = 2\left(\frac{ss'}{s + s'}\right) = 2\left(\frac{(13.3 \text{ cm})(800 \text{ cm})}{800 \text{ cm} + 13.3 \text{ cm}}\right) = 26.2 \text{ cm.}$$

Note that R is calculated to be positive, which is correct for a concave mirror.

35-43 **a)** convex implies $R < 0$; $R = -24.0$ cm; $f = R/2 = -12.0$ cm

$$\frac{1}{s} + \frac{1}{s'} = \frac{1}{f}, \text{ so } \frac{1}{s'} = \frac{1}{f} - \frac{1}{s} = \frac{s - f}{sf}$$

$$s' = \frac{sf}{s - f} = \frac{(-12.0 \text{ cm})s}{s + 12.0 \text{ cm}}$$

s is negative, so write as $s = -|s|$; $s' = +\dfrac{(12.0 \text{ cm})|s|}{12.0 \text{ cm} - |s|}$.

Thus $s' > 0$ (real image) for $|s| < 12.0$ cm. Since s is negative this means -12.0 cm $< s < 0$. A real image is formed if the virtual object is closer to the mirror than the focus.

b) $m = -\dfrac{s'}{s}$; real image implies $s' > 0$; virtual object implies $s < 0$. Thus $m > 0$ and the image is erect.

c)

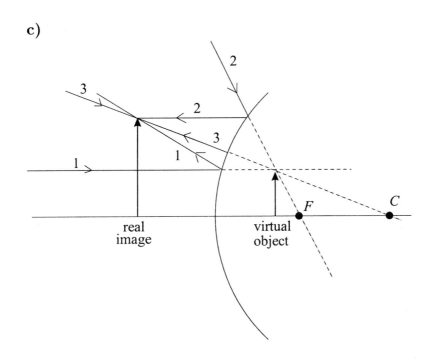

35-49 convex mirror means $R < 0$ so $R = -1.25$ m; $f = R/2 = -0.625$ m

Find the relation between s and s':

$$\frac{1}{s} + \frac{1}{s'} = \frac{1}{f}, \text{ so } \frac{1}{s'} = \frac{1}{f} - \frac{1}{s} = \frac{s-f}{sf} \text{ and } s' = \frac{sf}{s-f}$$

The speed of the image is $v' = \dfrac{ds'}{dt} = \dfrac{ds'}{ds}\dfrac{ds}{dt}$.

$$\frac{ds'}{ds} = \frac{f}{s-f} - \frac{sf}{(s-f)^2} = \frac{(s-f)f - sf}{(s-f)^2} = -\frac{f^2}{(s-f)^2} = -\left(\frac{f}{s-f}\right)^2$$

$\dfrac{ds}{dt} = v$, the speed of the object.

Thus $v' = -\left(\dfrac{f}{s-f}\right)^2 v$. (The quantity $\left(\dfrac{f}{s-f}\right)^2$ is always positive so the minus sign means that the object and image move in opposite directions.)

a) $s = 10.0$ m

$$v' = -\left(\frac{-0.625 \text{ m}}{10.0 \text{ m} - (-0.625 \text{ m})}\right)^2 (2.50 \text{ m/s}) = -0.00865 \text{ m/s} = -8.65 \text{ mm/s}$$

b) $s = 2.0$ m

$$v' = -\left(\frac{-0.625 \text{ m}}{2.0 \text{ m} - (-0.625 \text{ m})}\right)^2 (2.50 \text{ m/s}) = -0.142 \text{ m/s} = -14.2 \text{ cm/s}$$

35-53

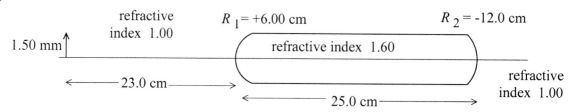

a) image formed by refraction at first surface (left end of rod):

$s = +23.0$ cm; $n_a = 1.00$; $n_b = 1.60$; $R = +6.00$ cm

$$\frac{n_a}{s} + \frac{n_b}{s'} = \frac{n_b - n_a}{R}$$

$$\frac{1}{23.0 \text{ cm}} + \frac{1.60}{s'} = \frac{1.60 - 1.00}{6.00 \text{ cm}}$$

$$\frac{1.60}{s'} = \frac{1}{10.0 \text{ cm}} - \frac{1}{23.0 \text{ cm}} = \frac{23 - 10}{230 \text{ cm}} = \frac{13}{230 \text{ cm}}$$

$$s' = 1.60 \left(\frac{230 \text{ cm}}{13} \right) = +28.3 \text{ cm}; \text{ image is 28.3 cm to right of first vertex.}$$

This image serves as the object for the refraction at the second surface (right end of rod). It is 28.3 cm − 25.0 cm = 3.3 cm to the right of the second vertex. For the second surface $s = -3.3$ cm (virtual object).

b) Object is on side of outgoing light, so is a virtual object.

c) Image formed by refraction at second surface (right end of rod):

$s = -3.3$ cm; $n_a = 1.60$; $n_b = 1.00$; $R = -12.0$ cm

$$\frac{n_a}{s} + \frac{n_b}{s'} = \frac{n_b - n_a}{R}$$

$$\frac{1.60}{-3.3 \text{ cm}} + \frac{1.00}{s'} = \frac{1.00 - 1.60}{-12.0 \text{ cm}}$$

$s' = +1.9$ cm; $s' > 0$ so image is 1.9 cm to right of vertex at right-hand end of rod.

d) $s' > 0$ so final image is real.

magnification for first surface:

$$m = -\frac{n_a s'}{n_b s} = -\frac{(1.00)(+28.3 \text{ cm})}{(1.60)(+23.0 \text{ cm})} = -0.769$$

magnification for second surface:

$$m = -\frac{n_a s'}{n_b s} = -\frac{(1.60)(+1.9 \text{ cm})}{(1.00)(-3.3 \text{ cm})} = +0.92$$

The overall magnification is $m_{\text{tot}} = m_1 m_2 = (-0.769)(+0.92) = -0.71$

$m_{\text{tot}} < 0$ so final image is inverted with respect to the original object.

e) $y' = m_{\text{tot}} y = (-0.71)(1.50 \text{ mm}) = -1.06 \text{ mm}$

The final image has a height of 1.06 mm.

35-55 Image is formed by refraction at the front surface of the sphere.

Let n_g be the index of refraction of the glass.

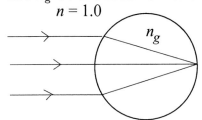

$n = 1.0$

n_g

$s = \infty$

$s' = +2r$, where r is the radius of the sphere

$n_a = 1.00$, $n_b = n_g$, $R = +r$

$$\frac{n_a}{s} + \frac{n_b}{s'} = \frac{n_b - n_a}{R}$$

$$\frac{1}{\infty} + \frac{n_g}{2r} = \frac{n_g - 1.00}{r}$$

$$\frac{n_g}{2r} = \frac{n_g}{r} - \frac{1}{r}; \quad \frac{n_g}{2r} = \frac{1}{r} \text{ and } n_g = 2.00$$

35-57

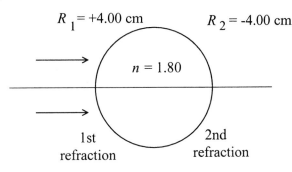

$R_1 = +4.00$ cm $R_2 = -4.00$ cm

$n = 1.80$

1st refraction   2nd refraction

The image formed by the first refraction serves as the object for the second refraction.

The two vertexes are a distance $2R = 8.00$ cm apart.

1st refraction

$n_a = 1.00$; $n_b = 1.80$; $R = +4.00$ cm; $s = \infty$ (parallel rays)

$$\frac{n_a}{s} + \frac{n_b}{s'} = \frac{n_b - n_a}{R}$$

$$\frac{1.00}{\infty} + \frac{1.80}{s'} = \frac{1.80 - 1.00}{+4.00 \text{ cm}}$$

$$\frac{1.80}{s'} = \frac{0.8}{4.00 \text{ cm}} \text{ and } s' = (1.80)\left(\frac{4.00 \text{ cm}}{0.80}\right) = 9.00 \text{ cm}$$

$s' > 0$ means the first image is 9.00 cm to the right of the first vertex, so is 1.00 cm to the right of the second vertex. This image serves as a virtual object for the second refraction, with $s = -1.00$ cm.

2nd refraction

$n_a = 1.80$; $n_b = 1.00$; $R = -4.00$ cm; $s = -1.00$ cm

$$\frac{n_a}{s} + \frac{n_b}{s'} = \frac{n_b - n_a}{R}$$

$$\frac{1.80}{-1.00 \text{ cm}} + \frac{1.00}{s'} = \frac{1.00 - 1.80}{-4.00 \text{ cm}}$$

$$\frac{1.00}{s'} = \frac{0.8}{4.00 \text{ cm}} + \frac{1.80}{1.00 \text{ cm}} \quad \text{and} \quad s' = 0.50 \text{ cm}$$

$s' > 0$ so this the final image is 0.50 cm to the right of the second vertex, or 4.50 cm from the center of the sphere.

35-59 a)

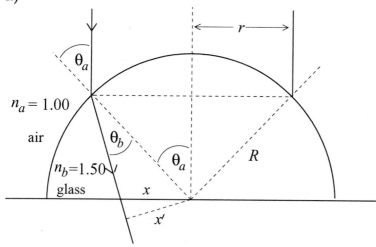

The width of the incident beam is exaggerated in the sketch, to make it easier to draw.

The diameter of the circle of light formed on the table is $2x$. Note the two right triangles containing the angles θ_a and θ_b.

$r = 0.190$ cm is the radius of the incident beam.

$R = 12.0$ cm is the radius of the glass hemisphere.

θ_a and θ_b small imply $x \approx x'$; $\sin \theta_a = \dfrac{r}{R}$, $\sin \theta_b = \dfrac{x'}{R} \approx \dfrac{x}{R}$

Snell's law: $n_a \sin \theta_a = n_b \sin \theta_b$

Using the above expressions for $\sin \theta_a$ and $\sin \theta_b$ gives $n_a \dfrac{r}{R} = n_b \dfrac{x}{R}$

$n_a r = n_b x$ so $x = \dfrac{n_a r}{n_b} = \dfrac{1.00(0.190 \text{ cm})}{1.50} = 0.1267$ cm

The diameter of the circle on the table is $2x = 2(0.1267 \text{ cm}) = 0.253$ cm.

b) R divides out of the expression; the result for the diameter of the spot is independent of the radius R of the hemisphere.

35-61

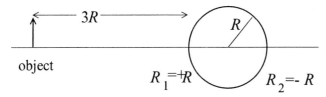

Thin-walled glass means the glass has no effect on the light rays. The problem is that of refraction by a sphere of water surrounded by air.

First refraction (air→water):

$n_a = 1.00$ (air); $n_b = 4/3$ (water); $s_1 = 3R$; $R_1 = +R$

$$\frac{n_a}{s_1} + \frac{n_b}{s_1'} = \frac{n_b - n_a}{R_1}$$

$$\frac{1}{3R} + \frac{4/3}{s_1'} = \frac{4/3 - 1.00}{R}$$

$$\frac{4}{3s_1'} = \frac{1}{3R} - \frac{1}{3R} = 0 \text{ and } s_1' = \infty \text{ (parallel rays)}$$

Second refraction (water→air):

$n_a = 4/3$ (water); $n_b = 1.00$ (air); $s_2 = -\infty$; $R_2 = -R$

$$\frac{n_a}{s_2} + \frac{n_b}{s_2'} = \frac{n_b - n_a}{R_2}$$

$$\frac{4/3}{-\infty} + \frac{1.00}{s_2'} = \frac{1.00 - 4/3}{-R}$$

$$\frac{1.00}{s_2'} = \frac{1.00}{3R} \text{ and } s_2' = +3R$$

The final image is $3R$ to the right of the second surface so is $4R$ from the center of the sphere, on the opposite side from the object.

35-63

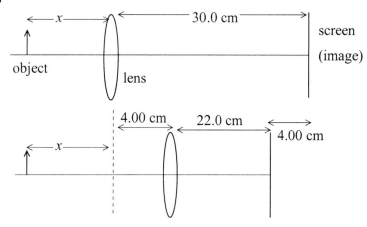

$s = x;\ s' = 30.0$ cm

$$\frac{1}{s} + \frac{1}{s'} = \frac{1}{f}$$

$$\frac{1}{x} + \frac{1}{30.0\text{ cm}} = \frac{1}{f}$$

$s = x+4.00$ cm; $s' = 22.0$ cm

$$\frac{1}{s} + \frac{1}{s'} = \frac{1}{f}$$

$$\frac{1}{x + 4.00\text{ cm}} + \frac{1}{22.0\text{ cm}} = \frac{1}{f}$$

Equate these two expressions for $1/f$:

$$\frac{1}{x} + \frac{1}{30.0\text{ cm}} = \frac{1}{x + 4.00\text{ cm}} + \frac{1}{22.0\text{ cm}}$$

$$\frac{1}{x} - \frac{1}{x + 4.00\text{ cm}} = \frac{1}{22.0\text{ cm}} - \frac{1}{30.0\text{ cm}}$$

$$\frac{x + 4.00\text{ cm} - x}{x(x + 4.00\text{ cm})} = \frac{30.0 - 22.0}{660\text{ cm}} \text{ and } \frac{4.00\text{ cm}}{x(x + 4.00\text{ cm})} = \frac{8}{660\text{ cm}}$$

$x^2 + (4.00\text{ cm})x - 330\text{ cm}^2 = 0$ and $x = \frac{1}{2}(-4.00 \pm \sqrt{16.0 + 4(330)})$ cm

x must be positive so $x = \frac{1}{2}(-4.00 + 36.55)$ cm $= 16.28$ cm

Then $\dfrac{1}{x} + \dfrac{1}{30.0\text{ cm}} = \dfrac{1}{f}$ and $\dfrac{1}{f} = \dfrac{1}{16.28\text{ cm}} + \dfrac{1}{30.0\text{ cm}}$

$f = +10.6$ cm $(f > 0$; the lens is converging)

35-65

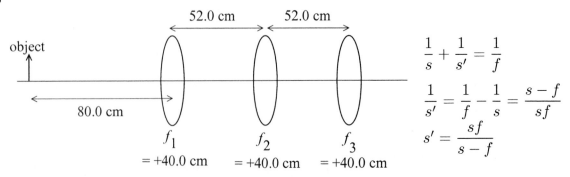

$$\frac{1}{s} + \frac{1}{s'} = \frac{1}{f}$$

$$\frac{1}{s'} = \frac{1}{f} - \frac{1}{s} = \frac{s - f}{sf}$$

$$s' = \frac{sf}{s - f}$$

<u>lens #1</u>

$s = +80.0$ cm; $f = +40.0$ cm

$$s' = \frac{sf}{s-f} = \frac{(+80.0 \text{ cm})(+40.0 \text{ cm})}{+80.0 \text{ cm} - 40.0 \text{ cm}} = +80.0 \text{ cm}$$

The image formed by the first lens is 80.0 cm to the right of the first lens, so it is 80.0 cm − 52.0 cm = 28.0 cm to the right of the second lens.

lens #2

$s = -28.0$ cm; $f = +40.0$ cm

$$s' = \frac{sf}{s-f} = \frac{(-28.0 \text{ cm})(+40.0 \text{ cm})}{-28.0 \text{ cm} - 40.0 \text{ cm}} = +16.47 \text{ cm}$$

The image formed by the second lens is 16.47 cm to the right of the second lens, so it is 52.0 cm − 16.47 cm = 35.53 cm to the left of the third lens.

lens #3

$s = +35.53$ cm; $f = +40.0$ cm

$$s' = \frac{sf}{s-f} = \frac{(+35.53 \text{ cm})(+40.0 \text{ cm})}{+35.53 \text{ cm} - 40.0 \text{ cm}} = -318 \text{ cm}$$

The final image is 318 cm to the left of the third lens, so it is 318 cm − 52 cm − 52 cm − 80 cm = 134 cm to the left of the object.

35-67 a)

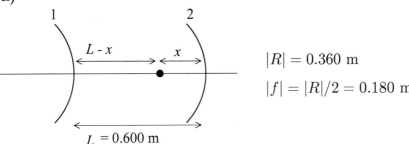

$|R| = 0.360$ m

$|f| = |R|/2 = 0.180$ m

$L = 0.600$ m

Image formed by convex mirror (mirror #1):

convex means $f_1 = -0.180$ m; $s_1 = L - x$

$$s_1' = \frac{s_1 f_1}{s_1 - f_1} = \frac{(L-x)(-0.180 \text{ m})}{L - x + 0.180 \text{ m}} = -(0.180 \text{ m})\left(\frac{0.600 \text{ m} - x}{0.780 \text{ m} - x}\right) < 0$$

The image is $(0.180 \text{ m})\left(\dfrac{0.600 \text{ m} - x}{0.780 \text{ m} - x}\right)$ to the left of mirror #1 so is

$$0.600 \text{ m} + (0.180 \text{ m})\left(\frac{0.600 \text{ m} - x}{0.780 \text{ m} - x}\right) =$$

$$\frac{0.576 \text{ m}^2 - (0.780 \text{ m})x}{0.780 \text{ m} - x} \text{ to the left of mirror \#2.}$$

Image formed by concave mirror (mirror #2):

concave implies $f_2 = +0.180$ m

$$s_2 = \frac{0.576\ \text{m}^2 - (0.780\ \text{m})x}{0.780\ \text{m} - x}$$

Rays return to the source implies $s_2' = x$.

Using these expressions in $s_2 = \dfrac{s_2' f_2}{s_2' - f_2}$ gives

$$\frac{0.576\ \text{m}^2 - (0.780\ \text{m})x}{0.780\ \text{m} - x} = \frac{(0.180\ \text{m})x}{x - 0.180\ \text{m}}$$

$$0.600x^2 - (0.576\ \text{m})x + 0.10368\ \text{m}^2 = 0$$

$$x = \tfrac{1}{1.20}\left(0.576 \pm \sqrt{(0.576)^2 - 4(0.600)(0.10368)}\right)\ \text{m} = \tfrac{1}{1.20}(0.576 \pm 0.288)\ \text{m}$$

$x = 0.72$ m (imposible; can't have $x > L = 0.600$ m) or $x = 0.24$ m.

b)

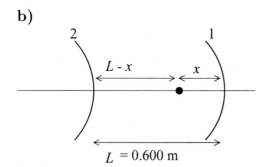

Image formed by concave mirror (mirror #1):

concave means $f_1 = +0.180$ m; $s_1 = x$

$$s_1' = \frac{s_1 f_1}{s_1 - f_1} = \frac{(0.180\ \text{m})x}{x - 0.180\ \text{m}}$$

The image is $\dfrac{(0.180\ \text{m})x}{x - 0.180\ \text{m}}$ to the left of mirror #1, so

$$s_2 = 0.600\ \text{m} - \frac{(0.180\ \text{m})x}{x - 0.180\ \text{m}} = \frac{(0.420\ \text{m})x - 0.108\ \text{m}^2}{x - 0.180\ \text{m}}$$

Image formed by convex mirror (mirror #2):

convex means $f_2 = -0.180$ m

rays return to the source means $s_2' = L - x = 0.600\ \text{m} - x$

$\dfrac{1}{s} + \dfrac{1}{s'} = \dfrac{1}{f}$ gives

$$\frac{x - 0.180\ \text{m}}{(0.420\ \text{m})x - 0.108\ \text{m}^2} + \frac{1}{0.600\ \text{m} - x} = -\frac{1}{0.180\ \text{m}}$$

$$\frac{x - 0.180\text{ m}}{(0.420\text{ m})x - 0.108\text{ m}^2} = -\left(\frac{0.780\text{ m} - x}{0.108\text{ m}^2 - (0.180\text{ m})x}\right)$$

$0.600x^2 - (0.576\text{ m})x + 0.1036\text{ m}^2 = 0$

This is the same quadratic equation as obtained in part (a), so again $x = 0.24$ m.

35-71

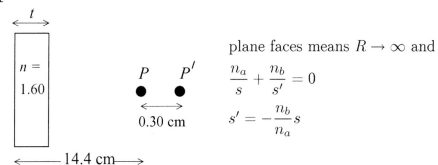

plane faces means $R \to \infty$ and

$$\frac{n_a}{s} + \frac{n_b}{s'} = 0$$

$$s' = -\frac{n_b}{n_a}s$$

refraction at first (left-hand) surface of the piece of glass:

The rays converging toward point P constitute a virtual object for this surface, so $s = -14.4$ cm.

$n_a = 1.00$, $n_b = 1.60$

$$s' = -\frac{1.60}{1.00}(-14.4\text{ cm}) = +23.0\text{ cm}$$

This image is 23.0 cm to the right of the first surface so is a distance $23.0\text{ cm} - t$ to the right of the second surface. This image serves as a virtual object for the second surface.

refraction at the second (right-hand) surface of the piece of glass:

The image is at P' so $s' = 14.4\text{ cm} + 0.30\text{ cm} - t = 14.7\text{ cm} - t$.

$s = -(23.0\text{ cm} - t)$; $n_a = 1.60$; $n_b = 1.00$

$s' = -\dfrac{n_b}{n_a}s$ gives $14.7\text{ cm} - t = -\left(\dfrac{1.00}{1.60}\right)(-[23.0\text{ cm} - t])$

$14.7\text{ cm} - t = +14.4\text{ cm} - 0.625t$

$0.375t = 0.30\text{ cm}$ and $t = 0.80\text{ cm}$

35-75

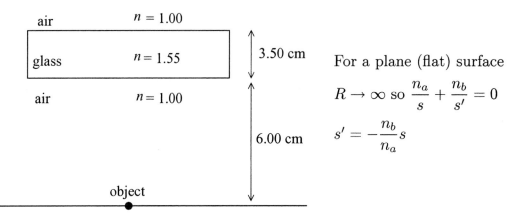

For a plane (flat) surface

$$R \to \infty \text{ so } \frac{n_a}{s} + \frac{n_b}{s'} = 0$$

$$s' = -\frac{n_b}{n_a}s$$

First refraction (air→ glass):

$n_a = 1.00$; $n_b = 1.55$; $s = 6.00$ cm

$$s' = -\frac{n_b}{n_a}s = -\frac{1.55}{1.00}(6.00 \text{ cm}) = -9.30 \text{ cm}$$

The image is 9.30 cm below the lower surface of the glass, so is 9.30 cm + 3.50 cm = 12.8 cm below the upper surface.

Second refraction (glass→ air):

$n_a = 1.55$; $n_b = 1.00$; $s = +12.8$ cm

$$s' = -\frac{n_b}{n_a}s = -\frac{1.00}{1.55}(12.8 \text{ cm}) = -8.26 \text{ cm}$$

The image of the page is 8.26 cm below the top surface of the glass plate and therefore 9.50 cm - 8.26 cm = 1.24 cm above the page.

CHAPTER 36
OPTICAL INSTRUMENTS

Exercises 1, 7, 9, 11, 15, 17, 19, 21, 27
Problems 29, 31, 35, 39, 41

Exercises

36-1 **a)** We need $m = -\dfrac{24 \times 10^{-3} \text{ m}}{160 \text{ m}} = -1.5 \times 10^{-4}$. Alternatively,

$$m = -\frac{36 \times 10^{-3} \text{ m}}{240 \text{ m}} = -1.5 \times 10^{-4}.$$

$s \gg f$ so $s' \approx f$

Then $m = -\dfrac{s'}{s} = -\dfrac{f}{s} = -1.5 \times 10^{-4}$ and

$f = (1.5 \times 10^{-4})(600 \text{ m}) = 0.090 \text{ m} = 90 \text{ mm}.$

A smaller f means a smaller s' and a smaller m, so with $f = 85$ mm the object's image nearly fills the picture area.

b) We need $m = -\dfrac{36 \times 10^{-3} \text{ m}}{9.6 \text{ m}} = -3.75 \times 10^{-3}.$

Then, as in part (a), $\dfrac{f}{s} = 3.75 \times 10^{-3}$ and

$f = (40.0 \text{ m})(3.75 \times 10^{-3}) = 0.15 \text{ m} = 150 \text{ mm}.$

Therefore use the 135 mm lens.

36-7 **a)** $f/4$ lens means f-number $= 4$

f-number $= \dfrac{f}{D}$ (Eq.36-1) so $D = \dfrac{f}{f\text{-number}} = \dfrac{300 \text{ m}}{4} = 75$ mm

b) $f/8$ lens means f-number $= 8$; $D = \dfrac{f}{f\text{-number}} = \dfrac{300 \text{ m}}{8} = 37.5$ mm

D is smaller by a factor of 2. The aperture area is smaller by a factor of $2^2 = 4$. So need an exposure time larger by a factor of 4; $4(1/250)$ s $= (1/62.5)$ s.

36-9 **a)** $f = 0.200$ m; $s' = 5.00$ m; $s = ?$

$$\frac{1}{s} + \frac{1}{s'} = \frac{1}{f}, \text{ so } \frac{1}{s} = \frac{1}{f} - \frac{1}{s'} = \frac{s' - f}{s'f}$$

$$s = \frac{s'f}{s' - f} = \frac{(5.00 \text{ m})(0.200 \text{ m})}{5.00 \text{ m} - 0.200 \text{ m}} = 0.208 \text{ m} = 20.8 \text{ cm}$$

b) $f = 0.200$ m; $s = 0.100$ m; $s' =?$

$$\frac{1}{s} + \frac{1}{s'} = \frac{1}{f}$$

$$s' = \frac{sf}{s - f} = \frac{(0.100 \text{ m})(0.200 \text{ m})}{0.100 \text{ m} - 0.200 \text{ m}} = -0.200 \text{ m}$$

$s' < 0$ means the image is virtual and a virtual image cannot be projected onto a screen. So no, the screen cannot be moved to achieve focus.

36-11 a) $\frac{1}{f} = +2.75$ diopters means $f = +\dfrac{1}{2.75}$ m $= +0.3636$ m (converging lens)

The purpose of the corrective lens is to take an object 25 cm from the eye and form a virtual image at the eye's near point.

$f = 36.36$ cm; $s = 25$ cm; $s' =?$

$$\frac{1}{s} + \frac{1}{s'} = \frac{1}{f} \text{ so}$$

$$s' = \frac{sf}{s - f} = \frac{(25 \text{ cm})(36.36 \text{ cm})}{25 \text{ cm} - 36.36 \text{ cm}} = -80.0 \text{ cm}$$

The eye's near point is 80.0 cm from the eye.

b) $\frac{1}{f} = -1.30$ diopters means $f = -\dfrac{1}{1.30}$ m $= -0.7692$ m (diverging lens)

The purpose of the corrective lens is to take an object at infinity and form a virtual image of it at the eye's far point.

$f = -76.92$ cm; $s = \infty$; $s' =?$

$$\frac{1}{s} + \frac{1}{s'} = \frac{1}{f} \text{ and } s = \infty \text{ says } \frac{1}{s'} = \frac{1}{f} \text{ and } s = f = -76.9 \text{ cm}$$

The eye's far point is 76.9 cm from the eye.

36-15 a) $f = 8.00$ cm; $s' = -25.0$ cm; $s =?$

$$\frac{1}{s} + \frac{1}{s'} = \frac{1}{f}, \text{ so } \frac{1}{s} = \frac{1}{f} - \frac{1}{s'} = \frac{s' - f}{s' f}$$

$$s = \frac{s' f}{s' - f} = \frac{(-25.0 \text{ cm})(+8.00 \text{ cm})}{-25.0 \text{ cm} - 8.00 \text{ cm}} = +6.06 \text{ cm}$$

b) $m = -\dfrac{s'}{s} = -\dfrac{-25.0 \text{ cm}}{6.06 \text{ cm}} = +4.125$

$|m| = \dfrac{|y'|}{|y|}$ so $|y'| = |m||y| = (4.125)(1.00 \text{ mm}) = 4.12$ mm

36-17 The image formed by a simple magnifier is virtual and $m = -\dfrac{s'}{s}$ is positive.

Thus $m = +6.50$. $m = -\dfrac{s'}{s}$ gives that $s' = -6.50s$.

Use this in $\dfrac{1}{s} + \dfrac{1}{s'} = \dfrac{1}{f}$: $\quad \dfrac{1}{s} + \dfrac{1}{-6.50s} = \dfrac{1}{4.00 \text{ cm}}$

$s = 3.38$ cm and $s' = -22.0$ cm

The flea is 3.38 cm from the lens and the image is 22.0 cm from the lens, on the same side of the lens as the flea.

36-19 a)

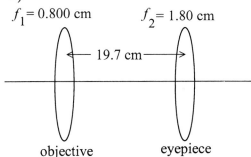

Final image is at ∞ so the object for the eyepiece is at its focal point. But the object for the eyepiece is the image of the objective so the image formed by the objective is 19.7 cm $-$ 1.80 cm $=$ 17.9 cm to the right of the lens.

Image formation by the objective:

$f = 0.800$ cm; $s' = 17.9$ cm; $s = ?$

$\dfrac{1}{s} + \dfrac{1}{s'} = \dfrac{1}{f}$, so $\dfrac{1}{s} = \dfrac{1}{f} - \dfrac{1}{s'} = \dfrac{s' - f}{s'f}$

$s = \dfrac{s'f}{s' - f} = \dfrac{(17.9 \text{ cm})(+0.800 \text{ cm})}{17.9 \text{ cm} - 0.800 \text{ cm}} = +8.37$ mm

b) $m_1 = -\dfrac{s'}{s} = -\dfrac{17.9 \text{ cm}}{0.837 \text{ cm}} = -21.4$

The magnification of the linear magnification of the objective is 21.4.

c) $M = m_1 M_2$

$M_2 = \dfrac{25 \text{ cm}}{f_2} = \dfrac{25 \text{ cm}}{1.80 \text{ cm}} = 13.9$

$M = m_1 M_2 = (-21.4)(13.9) = -297$

36-21 $f_1 = 95.0$ cm (objective); $f_2 = 15.0$ cm (eyepiece)

a) Eq.(36-5) $M = -\dfrac{f_1}{f_2} = -\dfrac{95.0 \text{ cm}}{15.0 \text{ cm}} = -6.33$

b) $s = 3.00 \times 10^3$ m

$s' = f_1 = 95.0$ cm (since s is very large, $s' \approx f$)

$$m = -\frac{s'}{s} = -\frac{0.950 \text{ m}}{3.00 \times 10^3 \text{ m}} = -3.167 \times 10^{-4}$$

$|y'| = |m||y| = (3.167 \times 10^{-4})(60.0 \text{ m}) = 0.0190 \text{ m} = 1.90 \text{ cm}$

c) The angular size of the object for the eyepiece is

$$\theta = \frac{0.0190 \text{ m}}{0.950 \text{ m}} = 0.0200 \text{ rad.}$$

(Note that this is also the angular size of the object for the objective: $\theta = \dfrac{60.0 \text{ m}}{3.00 \times 10^3 \text{ m}} = 0.0200$ rad. For a thin lens the object and image have the same angular size and the image of the objective is the object for the eyepiece.)

$M = \dfrac{\theta'}{\theta}$ (Eq.36-5) so the angular size of the image is

$\theta' = M\theta = -(6.33)(0.0200 \text{ rad}) = -0.127$ rad

(The minus sign shows that the final image is inverted.)

36-27 For a thin lens the image has the same angular size as the object. The object is very far from the lens so the image is formed at the focal plane and $s' = f$.

Thus $\dfrac{y'}{s'} = 0.014° = 2.44 \times 10^{-4}$ rad.

$y' = (18 \text{ m})(2.44 \times 10^{-4} \text{ rad}) = 4.4$ mm

Problems

36-29 **a)** $f = 35.0 \times 10^{-3}$ m

$$|m| = \frac{|y'|}{|y|} = \frac{(3/4)(36.0 \times 10^{-3} \text{ m})}{22.7 \text{ m}} = 1.189 \times 10^{-3}$$

Image on film means image is real. $s > 0$, $s' > 0$ and $m = -\dfrac{s'}{s} < 0$, so

$m = -1.189 \times 10^{-3}$

$s' = -ms = -(-1.189 \times 10^{-3})s = +(1.189 \times 10^{-3})s$

Use this in $\dfrac{1}{s} + \dfrac{1}{s'} = \dfrac{1}{f}$, so $\dfrac{1}{s} + \dfrac{1}{(1.189 \times 10^{-3})s} = \dfrac{1}{f}$

$\dfrac{842}{s} = \dfrac{1}{f}$ so $s = 842f = 842(35.0 \times 10^{-3} \text{ m}) = 29.5$ m

b) Fill the viewfinder frame means $|m| = \dfrac{36.0 \times 10^{-3} \text{ m}}{22.7 \text{ m}} = 1.586 \times 10^{-3}$.

Thus $s' = (1.586 \times 10^{-3})s$.

$$\frac{1}{s} + \frac{1}{s'} = \frac{1}{f} \text{ gives } \frac{1}{s} + \frac{1}{(1.586 \times 10^{-3})s} = \frac{1}{f}$$

$$\frac{631.5}{s} = \frac{1}{f} \text{ so } s = 631.5f = 631.5(35.0 \times 10^{-3} \text{ m}) = 22.1 \text{ m}$$

36-31 a)

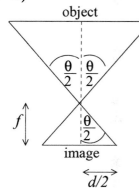

object

$\frac{\theta}{2}$ $\frac{\theta}{2}$

f

$\frac{\theta}{2}$

image

$d/2$

From the sketch, $\tan(\theta/2) = \dfrac{d/2}{f} = \dfrac{d}{2f}$

$$\frac{\theta}{2} = \arctan\left(\frac{d}{2f}\right)$$

$$\theta = 2\arctan\left(\frac{d}{2f}\right)$$

b) $d = \sqrt{(24 \text{ mm})^2 + (36 \text{ mm})^2} = 43.3 \text{ mm}$

$f = 28 \text{ mm}$ gives $\theta = 2\arctan\left(\dfrac{43.3 \text{ mm}}{2(28 \text{ mm})}\right) = 75°$ (Fig.36-2d says 75°)

$f = 105 \text{ mm}$ gives $\theta = 2\arctan\left(\dfrac{43.3 \text{ mm}}{2(105 \text{ mm})}\right) = 23°$ (Fig.36-2d says 25°)

$f = 300 \text{ mm}$ gives $\theta = 2\arctan\left(\dfrac{43.3 \text{ mm}}{2(300 \text{ mm})}\right) = 8.3°$ (Fig.36-2d says 8°)

36-35 The generalization of Eq.(36-3) is $M = \dfrac{\text{near point}}{f}$, so $f = \dfrac{\text{near point}}{M}$.

a) age 10, near point = 7 cm

$$f = \frac{7 \text{ cm}}{2.0} = 3.5 \text{ cm}$$

b) age 30, near point = 14 cm

$$f = \frac{14 \text{ cm}}{2.0} = 7.0 \text{ cm}$$

c) age 60, near point = 200 cm

$$f = \frac{200 \text{ cm}}{2.0} = 100 \text{ cm}$$

d) $f = 3.5$ cm (from part (a)) and near point = 200 cm (for 60-year-old)

$$M = \frac{200 \text{ cm}}{3.5 \text{ cm}} = 57$$

e) No. The reason $f = 3.5$ cm gives a larger M for a 60-year-old than for a 10-year-old is that the eye of the older person can't focus on as close of an object as the younger person can. The unaided eye of the 60-year-old must view a much smaller angular size, and that is why the same f gives a much larger M. The angular size of the image depends only on f, so it is the same for the two ages.

36-39 a)

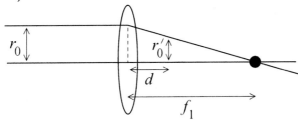

From similar triangles in the sketch,

$$\frac{r_0}{f_1} = \frac{r_0'}{f_1 - d}$$

Thus $r_0' = \left(\dfrac{f_1 - d}{f_1}\right) r_0,$

as was to be shown.

b) The image at the focal point of the first lens, a distance f_1 to the right of the first lens, serves as the object for the second lens. The image is a distance $f_1 - d$ to the right of the second lens, so $s_2 = -(f_1 - d) = d - f_1$.

$$s_2' = \frac{s_2 f_2}{s_2 - f_2} = \frac{(d - f_1) f_2}{d - f_1 - f_2}$$

$f_2 < 0$ so $|f_2| = -f_2$ and $s_2' = \dfrac{(f_1 - d)|f_2|}{|f_2| - f_1 + d}$, as was to be shown.

c)

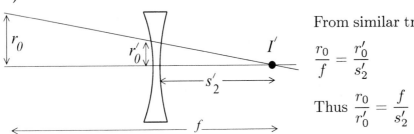

From similar triangles in the sketch,

$$\frac{r_0}{f} = \frac{r_0'}{s_2'}$$

Thus $\dfrac{r_0}{r_0'} = \dfrac{f}{s_2'}$

From the results of part (a), $\dfrac{r_0}{r_0'} = \dfrac{f_1}{f_1 - d}$.

Combining the two results gives $\dfrac{f_1}{f_1 - d} = \dfrac{f}{s_2'}$

$$f = s_2' \left(\frac{f_1}{f_1 - d}\right) = \frac{(f_1 - d)|f_2| f_1}{(|f_2| - f_1 + d)(f_1 - d)} = \frac{f_1 |f_2|}{|f_2| - f_1 + d}, \text{ as was to be shown.}$$

d) $f = \dfrac{f_1 |f_2|}{|f_2| - f_1 + d}$

$f_1 = 12.0$ cm, $|f_2| = 18.0$ cm, so $f = \dfrac{216 \text{ cm}^2}{6.0 \text{ cm} + d}$

$d = 0$ gives $f = 36.0$ cm; maximum f

$d = 4.0$ gives $f = 21.6$ cm; minimum f

$$f = 30.0 \text{ cm says } 30.0 \text{ cm} = \frac{216 \text{ cm}^2}{6.0 \text{ cm} + d}$$

$6.0 \text{ cm} + d = 7.2 \text{ cm and } d = 1.2 \text{ cm}$

36-41 $f_1 = 8.00$ mm (objective); $f_2 = 7.50$ cm (eyepiece)

a) The total magnification of the final image is $m_{\text{tot}} = m_1 m_2$.

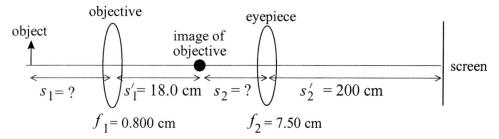

Find the object distace s_1 for the objective:

$s_1' = +18.0$ cm, $f_1 = 0.800$ cm, $s_1 = ?$

$$\frac{1}{s_1} + \frac{1}{s_1'} = \frac{1}{f_1}, \text{ so } \frac{1}{s_1} = \frac{1}{f_1} - \frac{1}{s_1'} = \frac{s_1' - f_1}{s_1' f_1}$$

$$s_1 = \frac{s_1' f_1}{s_1' - f_1} = \frac{(18.0 \text{ cm})(0.800 \text{ cm})}{18.0 \text{ cm} - 0.800 \text{ cm}} = 0.8372 \text{ cm}$$

Find the object distance s_2 for the eyepiece:

$s_2' = +200$ cm, $f_2 = 7.50$ cm, $s_2 = ?$

$$\frac{1}{s_2} + \frac{1}{s_2'} = \frac{1}{f_2}$$

$$s_2 = \frac{s_2' f_2}{s_2' - f_2} = \frac{(200 \text{ cm})(7.50 \text{ cm})}{200 \text{ cm} - 7.50 \text{ cm}} = 7.792 \text{ cm}$$

Now we can calculate the magnification for each lens:

$$m_1 = -\frac{s_1'}{s_1} = -\frac{18.0 \text{ cm}}{0.8372 \text{ cm}} = -21.50$$

$$m_2 = -\frac{s_2'}{s_2} = -\frac{200 \text{ cm}}{7.792 \text{ cm}} = -25.67$$

$m_{\text{tot}} = m_1 m_2 = (-21.50)(-25.67) = 552.$

b) From the sketch we can see that the distance between the two lenses is

$s_1' + s_2 = 18.0 \text{ cm} + 7.792 \text{ cm} = 25.8 \text{ cm}.$

CHAPTER 37
INTERFERENCE

Exercises 5, 7, 9, 11, 13, 15, 19, 21, 25, 27, 31
Problems 35, 37, 43, 47, 49, 51, 53

Exercises

37-5

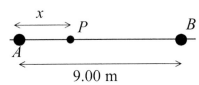

The distance of point P from each coherent source is $r_A = x$ and $r_B = 9.00$ m $- x$.

The path difference is $r_B - r_A = 9.00$ m $- 2x$. For constructive interence this path difference is an integer multiple of the wavelength:

$$r_B - r_A = m\lambda, \ m = 0, \pm 1, \pm 2, \ldots$$

$$\lambda = \frac{c}{f} = \frac{2.998 \times 10^8 \text{ m/s}}{120 \times 10^6 \text{ Hz}} = 2.50 \text{ m}$$

Thus 9.00 m $- 2x = m(2.50$ m$)$ and $x = \dfrac{9.00 \text{ m} - m(2.50 \text{ m})}{2} = 4.50$ m $- (1.25$ m$)x$.

x must lie in the range 0 to 9.00 m since P is said to be between the two antennas.

$m = 0$ gives $x = 4.50$ m

$m = +1$ gives $x = 4.50$ m $- 1.25$ m $= 3.25$ m

$m = +2$ gives $x = 4.50$ m $- 2.50$ m $= 2.00$ m

$m = +3$ gives $x = 4.50$ m $- 3.75$ m $= 0.75$ m

$m = -1$ gives $x = 4.50$ m $+ 1.25$ m $= 5.75$ m

$m = -2$ gives $x = 4.50$ m $+ 2.50$ m $= 7.00$ m

$m = -3$ gives $x = 4.50$ m $+ 3.75$ m $= 8.25$ m

All other vlues of m give values of x out of the allowed range.

Constuctive interference will occur for $x = 0.75$ m, 2.00 m, 3.25 m, 4.50 m, 5.75 m, 7.00 m, and 8.25 m.

37-7 For destructive interference the path difference $r_2 - r_1$ must equal a half-integer number of wavelengths: $r_2 - r_1 = (m + \frac{1}{2})\lambda$, $m = 0, 1, 2, \ldots$

$$\lambda = \frac{r_2 - r_1}{m + \frac{1}{2}} = \frac{2040 \text{ nm}}{m + \frac{1}{2}}$$

$m = 0$ gives $\lambda = 4080$ nm; $m = 1$ gives $\lambda = 1360$ nm; $m = 2$ gives $\lambda = 816$ nm; $m = 3$ gives $\lambda = 583$ nm; $m = 4$ gives $\lambda = 453$ nm; $m = 5$ gives $\lambda = 371$ nm; $\ldots$

Of these wavelengths, 583 nm and 453 nm are in the visible region.

37-9 The positions of the bright fringes are given by Eq.(37-6): $y_m = R(m\lambda/d)$.
For each fringe the adjacent fringe is located at $y_{m+1} = R(m+1)\lambda/d$.
The separation between adjacent fringes is $\Delta y = y_{m+1} - y_m = R\lambda/d$.

$$\lambda = \frac{d\,\Delta y}{R} = \frac{(0.460 \times 10^{-3} \text{ m})(2.82 \times 10^{-3} \text{ m})}{2.20 \text{ m}} = 5.90 \times 10^{-7} \text{ m} = 590 \text{ nm}$$

37-11 The dark lines correspond to destructive interference and hence are located
by Eq.(37-5):

$$d\sin\theta = (m+\tfrac{1}{2})\lambda \text{ so } \sin\theta = \frac{(m+\frac{1}{2})\lambda}{d}, \; m = 0, \pm 1, \pm 2, \ldots$$

1st dark line is for $m = 0$

2nd dark line is for $m = 1$ and
$$\sin\theta_1 = \frac{3\lambda}{2d} = \frac{3(500 \times 10^{-9} \text{ m})}{2(0.450 \times 10^{-3} \text{ m})} = 1.667 \times 10^{-3} \text{ and } \theta_1 = 1.667 \times 10^{-3} \text{ rad}$$

3rd dark line is for $m = 2$ and
$$\sin\theta_2 = \frac{5\lambda}{2d} = \frac{5(500 \times 10^{-9} \text{ m})}{2(0.450 \times 10^{-3} \text{ m})} = 2.778 \times 10^{-3} \text{ and } \theta_2 = 2.778 \times 10^{-3} \text{ rad}$$

(Note that θ_1 and θ_2 are small so that the approximation $\theta \approx \sin\theta \approx \tan\theta$ is valid.)
The distance of each dark line from the center of the central bright band is given
by $y_m = R\tan\theta$, where $R = 0.850$ m is the distance to the screen.
$\tan\theta \approx \theta$ so $y_m = R\theta_m$
$y_1 = R\theta_1 = (0.750 \text{ m})((1.667 \times 10^{-3} \text{ rad}) = 1.25 \times 10^{-3} \text{ m}$
$y_2 = R\theta_2 = (0.750 \text{ m})((2.778 \times 10^{-3} \text{ rad}) = 2.08 \times 10^{-3} \text{ m}$
$\Delta y = y_2 - y_1 = 2.08 \times 10^{-3} \text{ m} - 1.25 \times 10^{-3} \text{ m} = 0.83 \text{ mm}$

37-13 **a)** The minima are located at angles θ given by $d\sin\theta = (m+\tfrac{1}{2})\lambda$.
The first minimum corresponds to $m = 0$, so

$$\sin\theta = \frac{\lambda}{2d} = \frac{660 \times 10^{-9} \text{ m}}{2(0.260 \times 10^{-3} \text{ m})} = 1.27 \times 10^{-3} \text{ and } \theta = 1.27 \times 10^{-3} \text{ rad}$$

The distance on the screen is
$y = R\tan\theta = (0.700 \text{ m})\tan(1.27 \times 10^{-3} \text{ rad}) = 0.889 \text{ mm}.$

b) Eq.(37-15) given the intensity I as a function of the position y on the screen:
$$I = I_0 \cos^2\left(\frac{\pi dy}{\lambda R}\right).$$

$$I = \tfrac{1}{2}I_0 \text{ says } \cos^2\left(\frac{\pi dy}{\lambda R}\right) = \tfrac{1}{2}$$

$$\cos\left(\frac{\pi dy}{\lambda R}\right) = \frac{1}{\sqrt{2}} \text{ so } \frac{\pi dy}{\lambda R} = \frac{\pi}{4} \text{ rad}$$

$$y = \frac{\lambda R}{4d} = \frac{(660 \times 10^{-9} \text{ m})(0.700 \text{ m})}{4((0.260 \times 10^{-3} \text{ m})} = 0.444 \text{ mm}$$

37-15 **a)** The minima are located by $d\sin\theta = (m + \tfrac{1}{2})\lambda$. The first minimum is for

$$m = 0 \text{ so } \sin\theta = \frac{\lambda}{2d} = \frac{550 \times 10^{-9} \text{ m}}{2(0.130 \times 10^{-3} \text{ m})} = 2.115 \times 10^{-3} \text{ and}$$

$$\theta = 2.115 \times 10^{-3} \text{ rad}$$

(θ is small so $\theta \approx \sin\theta \approx \tan\theta$)

The distance on the screen is $y_1 = R\tan\theta \approx R\theta$, where R is the distance to the screen.

$$y_1 = R\theta_1 = (0.900 \text{ m})(2.115 \times 10^{-3} \text{ rad}) = 1.90 \times 10^{-3} \text{ m} = 1.90 \text{ mm.}$$

b) Since θ is small we can use Eq.(37-15): $I = I_0 \cos^2\left(\frac{\pi dy}{\lambda R}\right)$.

We want I for $y = \tfrac{1}{2}y_1 = 0.950 \times 10^{-3}$ m.

$$I = (4.00 \times 10^{-6} \text{ W/m}^2)\left(\cos\left[\frac{\pi(0.130 \times 10^{-3} \text{ m})(0.950 \times 10^{-3} \text{ m})}{(550 \times 10^{-9} \text{ m})(0.900 \text{ m})}\right]\right)^2 =$$

$$2.01 \times 10^{-6} \text{ W/m}^2.$$

37-19

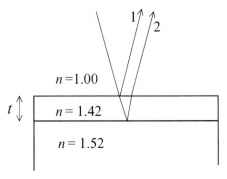

Both rays (1) and (2) undergo a 180° phase change on reflection, so there is no net phase difference introduced and the condition for destructive interference is $2t = (m + \tfrac{1}{2})\lambda$.

$$t = \frac{(m + \tfrac{1}{2})\lambda}{2}; \text{ thinnest film says } m = 0 \text{ so } t = \frac{\lambda}{4}$$

$$\lambda = \frac{\lambda_0}{1.42} \text{ and } t = \frac{\lambda_0}{4(1.42)} = \frac{650 \times 10^{-9} \text{ m}}{4(1.42)} = 1.14 \times 10^{-7} \text{ m} = 114 \text{ nm}$$

37-21

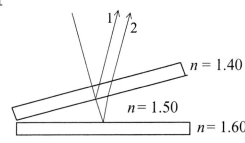

When ray (1) reflects off the top of the wedge of silicone grease it undergoes a 180° phase change ($n = 1.40 < n = 1.50$).

When ray (2) reflects off the top of the lower plate it undergoes a 180° phase change ($n = 1.50 < n = 1.60$).

There is no net phase difference introduced by the phase changes on reflection and the condition for an interference minimum is $2t = (m + \frac{1}{2})\lambda$, where λ is the wavelength in the silicone grease, $\lambda = \lambda_0/n = 500 \times 10^{-9}$ m$/1.50 = 333 \times 10^{-9}$ m.

As in Example 37-4 (Fig.37-11), $\dfrac{t}{x} = \dfrac{h}{l}$ and $t = \dfrac{hx}{l}$.

Then $\dfrac{2hx}{l} = (m + \dfrac{1}{2})\lambda$.

$$x_m = (m + \tfrac{1}{2})\frac{\lambda l}{2h}; \qquad x_{m+1} = (m + \tfrac{3}{2})\frac{\lambda l}{2h}$$

The spacing between adjacent dark fringes is

$$\Delta x = x_{m+1} - x_m = \frac{\lambda l}{2h} = \frac{(333 \times 10^{-9}\ \text{m})(0.10\ \text{m})}{2(0.020 \times 10^{-3}\ \text{m})}$$

$$\Delta x = 0.83 \times 10^{-3}\ \text{m} = 0.83\ \text{mm}.$$

37-25 a)

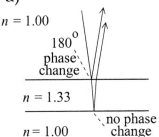

There is a 180° phase change when the light is reflected from the outside surface of the bubble and no phase change when the light is reflected from the inside surface.

Thus the reflections produce a net 180° phase difference and for there to be constructive interference the path difference $2t$ must correspond to a half-integer number of wavelengths to compensate for the $\lambda/2$ shift due to the reflections. Hence the condition for constructive interference is

$2t = (m + \frac{1}{2})(\lambda_0/n)$, $m = 0, 1, 2, \ldots$

Here λ_0 is the wavelength in air and (λ_0/n) is the wavelength in the bubble, where the path difference occurs.

$$\lambda_0 = \frac{2tn}{m + \frac{1}{2}} = \frac{2(290\ \text{nm})(1.33)}{m + \frac{1}{2}} = \frac{771.4\ \text{nm}}{m + \frac{1}{2}}$$

for $m = 0$, $\lambda = 1543$ nm; for $m = 1$, $\lambda = 514$ nm; for $m = 2$, $\lambda = 308$ nm; $\ldots$

Only 514 nm is in the visible region; the color for this wavelength is green.

b) $\lambda_0 = \dfrac{2tn}{m + \frac{1}{2}} = \dfrac{2(340 \text{ nm})(1.33)}{m + \frac{1}{2}} = \dfrac{904.4 \text{ nm}}{m + \frac{1}{2}}$

for $m = 0$, $\lambda = 1809$ nm;   for $m = 1$, $\lambda = 603$ nm;   for $m = 2$, $\lambda = 362$ nm; ...

Only 603 nm is in the visible region; the color for this wavelength is orange.

37-27 Eq.(37-19): $y = m(\lambda/2) = 1800(633 \times 10^{-9} \text{ m})/2 = 5.70 \times 10^{-4} \text{ m} = 0.570 \text{ mm}$

37-31 By conservation of linear momentum, since the initial momentum is zero, the nucleus and the photon must have equal and opposite momenta.

Find the recoil speed of the nucleus:

$K = \frac{1}{2}mv^2$ and $v = \sqrt{\dfrac{2K}{m}} = \sqrt{\dfrac{2(2.75 \times 10^{-19} \text{ J})}{2.27 \times 10^{-25} \text{ kg}}} = 1557 \text{ m/s}.$

The momentum of the recoiling nucleus is

$p = mv = (2.27 \times 10^{-25} \text{ kg})(1557 \text{ m/s}) = 3.53 \times 10^{-22} \text{ kg·m/s}.$

The gamma ray photon must have $p = 3.53 \times 10^{-22}$ kg·m/s.

$E = pc = (3.53 \times 10^{-22} \text{ kg} \cdot \text{m/s})(2.998 \times 10^8 \text{ m/s}) = 1.06 \times 10^{-13} \text{ J}$

$p = \dfrac{h}{\lambda}$ so $\lambda = \dfrac{h}{p} = \dfrac{6.626 \times 10^{-34} \text{ J/s}}{3.53 \times 10^{-22} \text{ kg} \cdot \text{m/s}} = 1.88 \times 10^{-12} \text{ m}$

$c = f\lambda$ so $f = \dfrac{c}{\lambda} = \dfrac{2.998 \times 10^8 \text{ m/s}}{1.88 \times 10^{-12} \text{ m}} = 1.59 \times 10^{20} \text{ Hz}$

Problems

37-35 The only effect of the water is to change the wavelength to

$\lambda = \lambda_0/n = \dfrac{500 \times 10^{-9} \text{ m}}{1.333} = 375 \times 10^{-9} \text{ m}$

$\theta_1 \approx \sin\theta_1 = \dfrac{3\lambda}{2d} = \dfrac{3(375 \times 10^{-9} \text{ m})}{2(0.450 \times 10^{-3} \text{ m})} = 1.250 \times 10^{-3} \text{ rad}$

$\theta_2 \approx \sin\theta_2 = \dfrac{5\lambda}{2d} = \dfrac{5(375 \times 10^{-9} \text{ m})}{2(0.450 \times 10^{-3} \text{ m})} = 2.083 \times 10^{-3} \text{ rad}$

$y_1 \approx R\theta_1 = (0.750 \text{ m})(1.250 \times 10^{-3} \text{ rad}) = 9.38 \times 10^{-4} \text{ m}$

$y_2 \approx R\theta_2 = (0.750 \text{ m})(2.083 \times 10^{-3} \text{ rad}) = 1.562 \times 10^{-3} \text{ m}$

$\Delta y = y_2 - y_1 = 1.562 \times 10^{-3} \text{ m} - 9.38 \times 10^{-4} \text{ m} = 6.24 \times 10^{-4} \text{ m} = 0.62 \text{ mm}$

37-37 **a)** There must be destructive interference between the sound waves from the two speakers.

b) The change in path length must be $\lambda/2$, so $\lambda/2 = 0.398$ m and $\lambda = 0.796$ m.

$$v = f\lambda \text{ so } f = \frac{v}{\lambda} = \frac{340 \text{ m/s}}{0.796 \text{ m}} = 427 \text{ Hz}$$

c) The change in path length must equal λ to go from one point of constructive interference to the next, so the speakers must be moved 0.796 m.

37-43 The dark lines are located by $d\sin\theta = (m + \frac{1}{2})\lambda$.

First dark line is for $m = 0$ and $d\sin\theta_1 = \lambda/2$.

$$\sin\theta_1 = \frac{\lambda}{2d} = \frac{550 \times 10^{-9} \text{ m}}{2(1.80 \times 10^{-6} \text{ m})} = 0.1528 \text{ and } \theta_1 = 8.789°.$$

Second dark line is for $m = 1$ and $d\sin\theta_2 = 3\lambda/2$.

$$\sin\theta_2 = \frac{3\lambda}{2d} = 3\left(\frac{550 \times 10^{-9} \text{ m}}{2(1.80 \times 10^{-6} \text{ m})}\right) = 0.4583 \text{ and } \theta_2 = 27.28°.$$

$$y_1 = R\tan\theta_1 = (0.350 \text{ m})\tan 8.789° = 0.0541 \text{ m}$$

$$y_2 = R\tan\theta_2 = (0.350 \text{ m})\tan 27.28° = 0.1805 \text{ m}$$

The distance between the lines is $\Delta y = y_2 - y_1 = 0.1805 \text{ m} - 0.0541 \text{ m} = 0.126 \text{ m} = 12.6$ cm.

37-47 a)

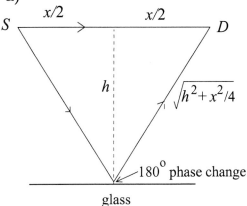

path difference $= 2\sqrt{h^2 + x^2/4} - x = \sqrt{4h^2 + x^2} - x$

Since there is a 180° phase change for the reflected ray, the condition for constructive interference is

path difference $= (m + \frac{1}{2})\lambda$ (constructive interference)

and the condition for destructive interference is

path difference $= m\lambda$ (destructive interference)

b) Constructive interference: $(m + \frac{1}{2})\lambda = \sqrt{4h^2 + x^2} - x$

$$\lambda = \frac{\sqrt{4h^2 + x^2} - x}{m + \frac{1}{2}}$$

Longest λ is for $m = 0$ and then

$$\lambda = 2(\sqrt{4h^2 + x^2} - x) = 2(\sqrt{4(0.24 \text{ m})^2 + (0.14 \text{ m})^2} - 0.14 \text{ m}) = 0.72 \text{ m}$$

37-49

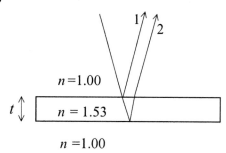

Ray (1) undergoes a 180° phase change on reflection at the top surface of the glass.

Ray (2) has no phase change on reflection from the lower surface of the glass.

a) Intensified in the reflected beam means constructive interference for the reflected light. The reflections produce a net phase difference of 180°, so the condition for constructive interference is $2t = (m + \frac{1}{2})\lambda$, $m = 0, 1, 2, \ldots$ and $\lambda = 2t/(m + \frac{1}{2})$.
λ is the wavelength in the glass plate: $\lambda = \lambda_0/n$.

$$\frac{\lambda_0}{n} = \frac{2t}{m + \frac{1}{2}} \text{ and } \lambda_0 = \frac{2tn}{m + \frac{1}{2}} = \frac{2(0.485 \times 10^{-6} \text{ m})(1.53)}{m + \frac{1}{2}} = \frac{1484 \text{ nm}}{m + \frac{1}{2}}.$$

$m = 0$ gives $\lambda_0 = 2968$ nm; $m = 1$ gives $\lambda_0 = 989$ nm; $m = 2$ gives $\lambda_0 = 594$ nm; $m = 3$ gives $\lambda_0 = 424$ nm; $m = 4$ gives $\lambda_0 = 330$ nm; ...

The wavelengths 424 nm and 594 nm are within the limits of the visible spectrum.

b) Intensified in the transmitted light means destructive interference in the reflected light. The condition for destructive interference between the light reflected at the top and at the bottom of the oil film is $2t = m\lambda_0/n$.

$$\lambda_0 = \frac{2tn}{m} = \frac{2(0.485 \times 10^{-6} \text{ m})(1.53)}{m} = \frac{1484 \text{ nm}}{m}$$

$m = 1$ gives $\lambda_0 = 1484$ nm; $m = 2$ gives $\lambda_0 = 742$ nm; $m = 3$ gives $\lambda_0 = 495$ nm; $m = 4$ gives $\lambda_0 = 371$ nm; ...

The only wavelength within the visible spectrum is 495 nm.

37-51 This problem deals with Newton's rings (Sect.37-5). The interference is between rays reflecting from the top and bottom edges of the air that is between the lens and the plate.

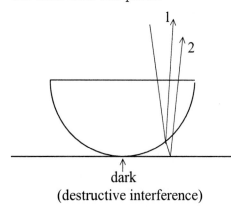

dark
(destructive interference)

Ray (1) does not undergo any phase change on reflection.

Ray (2) does undergo a 180° phase change on reflection.

Hence the path difference $2t$, where t is the thickness of the air wedge, must satisfy $2t = (m + \frac{1}{2})\lambda$, $m = 0, 1, 2, \ldots$ for constructive interference

Second bright ring means $m = 1$ and $t = \dfrac{3\lambda}{4} = \dfrac{3(580 \times 10^{-9} \text{ m})}{4} = 4.35 \times 10^{-7}$ m.

Now must relate t to the diameter of the ring:

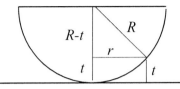

The radius of the ring is r.
$$r^2 + (R - t)^2 = R^2$$

$r = \sqrt{R^2 - (R - t)^2} = \sqrt{R^2 - R^2 + 2Rt - t^2} = \sqrt{2Rt - t^2}$

But $R = 0.952$ m $\gg t$, so we can neglect t^2 relative to $2Rt$:

$r = \sqrt{2Rt} = \sqrt{2(0.952 \text{ m})(4.35 \times 10^{-7} \text{ m})} = 0.910 \times 10^{-3}$ m

This is the radius. The diameter of the ring is $2r = 1.82 \times 10^{-3}$ m $= 1.82$ mm.

37-53 **a)** The wavelength in the glass is decreased by a factor of $1/n$, so for light through the upper slit a shorter path is needed to produce the same phase at the screen. Therefore, the interference pattern is shifted downward on the screen.

b) At a point on the screen located by the angle θ the difference in path length is $d \sin \theta$. This introduces a phase difference of $\phi = \left(\dfrac{2\pi}{\lambda_0}\right)(d \sin \theta)$, where λ_0 is the wavelength of the light in air or vacuum.

In the thickness L of glass the number of wavelengths is $\dfrac{L}{\lambda} = \dfrac{nL}{\lambda_0}$. A corresponding length L of the path of the ray through the lower slit, in air, contains L/λ_0 wavelenths. The phase difference this introduces is $\phi = 2\pi \left(\dfrac{nL}{\lambda_0} - \dfrac{L}{\lambda_0}\right)$ and $\phi = 2\pi(n - 1)(L/\lambda_0)$.

The total phase difference is the sum of these two,

$\left(\dfrac{2\pi}{\lambda_0}\right)(d \sin \theta) + 2\pi(n - 1)(L/\lambda_0) = (2\pi/\lambda_0)(d \sin \theta + L(n - 1))$.

Eq.(37-10) then gives $I = I_0 \cos^2 \left[\left(\dfrac{\pi}{\lambda_0}\right)(d \sin \theta + L(n - 1))\right]$.

c) Maxima means $\cos \phi/2 = \pm 1$ and $\phi/2 = m\pi$, $m = 0, \pm 1, \pm 2, \ldots$

$(\pi/\lambda_0)(d \sin \theta + L(n - 1)) = m\pi$

$d \sin \theta + L(n - 1) = m\lambda_0$

$\sin \theta = \dfrac{m\lambda_0 - L(n - 1)}{d}$

CHAPTER 38
DIFFRACTION

Exercises 1, 7, 9, 11, 13, 15, 17, 19, 21, 27, 29, 35, 37
Problems 39, 41, 43, 49, 55, 57

Exercises

38-1 The minima are located by Eq.(38-2): $\sin\theta = \dfrac{m\lambda}{a}$, $m = \pm 1, \pm 2, \ldots$

First minimum means $m = 1$ and $\sin\theta_1 = \lambda/2$ and $\lambda = a\sin\theta_1$.

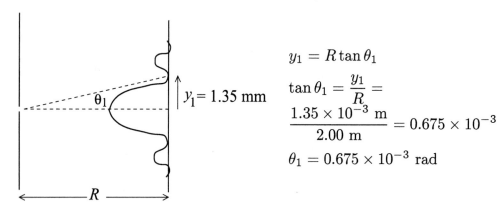

$$y_1 = R\tan\theta_1$$

$$\tan\theta_1 = \frac{y_1}{R} =$$

$$\frac{1.35 \times 10^{-3}\ \text{m}}{2.00\ \text{m}} = 0.675 \times 10^{-3}$$

$$\theta_1 = 0.675 \times 10^{-3}\ \text{rad}$$

$$\lambda = a\sin\theta_1 = (0.750 \times 10^{-3}\ \text{m})\sin(0.675 \times 10^{-3}\ \text{rad}) = 506\ \text{nm}$$

38-7 a)

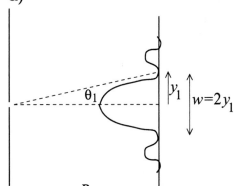

The first minimum is located by

$$\sin\theta_1 = \frac{\lambda}{a} =$$

$$\frac{633 \times 10^{-9}\ \text{m}}{0.350 \times 10^{-3}\ \text{m}} = 1.809 \times 10^{-3}$$

$$\theta_1 = 1.809 \times 10^{-3}\ \text{rad}$$

$$y_1 = R\tan\theta_1 = (3.00\ \text{m})\tan(1.809 \times 10^{-3}\ \text{rad}) = 5.427 \times 10^{-3}\ \text{m}$$
$$w = 2y_1 = 2(5.427 \times 10^{-3}\ \text{m}) = 1.09 \times 10^{-2}\ \text{m} = 10.9\ \text{mm}$$

b)

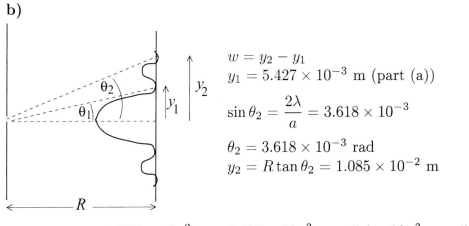

$$w = y_2 - y_1$$
$$y_1 = 5.427 \times 10^{-3} \text{ m (part (a))}$$

$$\sin \theta_2 = \frac{2\lambda}{a} = 3.618 \times 10^{-3}$$

$$\theta_2 = 3.618 \times 10^{-3} \text{ rad}$$
$$y_2 = R \tan \theta_2 = 1.085 \times 10^{-2} \text{ m}$$

$$w = y_2 - y_1 = 1.085 \times 10^{-2} \text{ m} - 5.427 \times 10^{-3} \text{ m} = 5.4 \times 10^{-3} \text{ m} = 5.4 \text{ mm}$$

38-9   a)

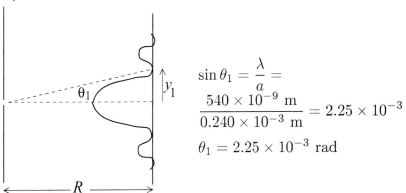

$$\sin \theta_1 = \frac{\lambda}{a} =$$
$$\frac{540 \times 10^{-9} \text{ m}}{0.240 \times 10^{-3} \text{ m}} = 2.25 \times 10^{-3}$$

$$\theta_1 = 2.25 \times 10^{-3} \text{ rad}$$

$$y_1 = R \tan \theta_1 = (3.00 \text{ m}) \tan(2.225 \times 10^{-3} \text{ rad}) = 6.75 \times 10^{-3} \text{ m} = 6.75 \text{ mm}$$

b) Midway between the center of the central maximum and the first minimum implies $y = \frac{1}{2}(6.75 \text{ mm}) = 3.375 \times 10^{-3}$ m.

$$\tan \theta = \frac{y}{R} = \frac{3.375 \times 10^{-3} \text{ m}}{3.00 \text{ m}} = 1.125 \times 10^{-3}; \theta = 1.125 \times 10^{-3} \text{ rad}$$

The phase angle β at this point on the screen is

$$\beta = \left(\frac{2\pi}{\lambda}\right) a \sin \theta = \frac{2\pi}{540 \times 10^{-9} \text{ m}} (0.240 \times 10^{-3} \text{ m}) \sin(1.125 \times 10^{-3} \text{ rad}) = \pi.$$

Then $I = I_0 \left(\dfrac{\sin \beta/2}{\beta/2}\right)^2 = (6.00 \times 10^{-6} \text{ W/m}^2) \left(\dfrac{\sin \pi/2}{\pi/2}\right)^2$

$$I = \left(\frac{4}{\pi^2}\right) (6.00 \times 10^{-6} \text{ W/m}^2) = 2.43 \times 10^{-6} \text{ W/m}^2.$$

38-11 Eq.(38-6): $\beta = \left(\dfrac{2\pi}{\lambda}\right) a \sin \theta$

$$\lambda = \left(\frac{2\pi}{\beta}\right) a \sin\theta = \left(\frac{2\pi}{(\pi/2)\ \text{rad}}\right)(0.320 \times 10^{-3}\ \text{m}) \sin 0.24° = 5.4\ \mu\text{m}$$

38-13 a) $\theta = 3.25°$, $\beta = 56.0$ rad, $a = 0.105 \times 10^{-3}$ m

Eq.(38-6); $\beta = \left(\dfrac{2\pi}{\lambda}\right) a \sin\theta$

$$\lambda = \frac{2\pi a \sin\theta}{\beta} = \frac{2\pi(0.105 \times 10^{-3}\ \text{m}) \sin 3.25°}{56.0\ \text{rad}} = 668\ \text{nm}$$

b) $I = I_0 \left(\dfrac{\sin\beta/2}{\beta/2}\right)^2 = I_0 \left(\dfrac{4}{\beta^2}\right)(\sin(\beta/2))^2 = I_0 \dfrac{4}{(56.0\ \text{rad})^2}[\sin(28.0\ \text{rad})]^2 = 9.36 \times 10^{-5} I_0$

38-15 a) The interference fringes (maxima) are located by $d\sin\theta = m\lambda$,

with $m = 0, \pm 1, \pm 2, \ldots$. The intensity I in the diffraction pattern is given by $I = I_0 \left(\dfrac{\sin\beta/2}{\beta/2}\right)^2$, with $\beta = \left(\dfrac{2\pi}{\lambda}\right) a \sin\theta$.

We want $m = \pm 3$ in the first equation to give θ that makes $I = 0$ in the second equation.

$d\sin\theta = m\lambda$ gives $\beta = \left(\dfrac{2\pi}{\lambda}\right) a \left(\dfrac{3\lambda}{d}\right) = 2\pi(3a/d)$.

$I = 0$ says $\dfrac{\sin\beta/2}{\beta/2} = 0$ so $\beta = 2\pi$ and then $2\pi = 2\pi(3a/d)$ and $(d/a) = 3$.

b) Fringes $m = 0, \pm 1, \pm 2$ are within the central diffraction maximum and the $m = \pm 3$ fringes coincide with the first diffraction minimum.

Find the value of m for the fringes that coincide with the second diffraction minimum:

Second minimum implies $\beta = 4\pi$.

$\beta = \left(\dfrac{2\pi}{\lambda}\right) a \sin\theta = \left(\dfrac{2\pi}{\lambda}\right) a \left(\dfrac{m\lambda}{d}\right) = 2\pi m(a/d) = 2\pi(m/3)$

Then $\beta = 4\pi$ says $4\pi = 2\pi(m/3)$ and $m = 6$.

Therefore the $m = +4$ and $m = +5$ fringes are contained within the first diffraction maximum on one side of the central maximum; two fringes.

38-17 a) If the slits are very narrow then the central maximum of the diffraction pattern for each slit completely fills the screen and the intensity distribution is given solely by the two-slit interference. The maxima are given by

$d\sin\theta = m\lambda$ so $\sin\theta = m\lambda/d$

1st order maximum: $m = 1$, so $\sin\theta = \dfrac{\lambda}{d} = \dfrac{580 \times 10^{-9} \text{ m}}{0.530 \times 10^{-3} \text{ m}} = 1.094 \times 10^{-3}$;
$\theta = 0.0627°$

2nd order maximum: $m = 2$, so $\sin\theta = \dfrac{2\lambda}{d} = 2.188 \times 10^{-3}$; $\theta = 0.125°$

b) The intensity is given by Eq.(38-12): $I = I_0 \cos^2(\phi/2)\left(\dfrac{\sin\beta/2}{\beta/2}\right)^2$

$\phi = \left(\dfrac{2\pi d}{\lambda}\right)\sin\theta = \left(\dfrac{2\pi d}{\lambda}\right)\left(\dfrac{m\lambda}{d}\right) = 2\pi m$, so $\cos^2(\phi/2) = \cos^2(m\pi) = 1$

(Since the angular positions in part (a) correspond to interference maxima.)

$\beta = \left(\dfrac{2\pi a}{\lambda}\right)\sin\theta =$

$\left(\dfrac{2\pi a}{\lambda}\right)\left(\dfrac{m\lambda}{d}\right) = 2\pi m(a/d) = m2\pi\left(\dfrac{0.320 \text{ mm}}{0.530 \text{ mm}}\right) = m(3.794 \text{ rad})$

1st order maximum: $m = 1$, so $I = I_0(1)\left(\dfrac{\sin(3.794/2) \text{ rad}}{(3.794/2) \text{ rad}}\right)^2 = 0.249 I_0$

2nd order maximum: $m = 2$, so $I = I_0(1)\left(\dfrac{\sin 3.794 \text{ rad}}{3.794 \text{ rad}}\right)^2 = 0.0256 I_0$

38-19

(i)

$\phi = \pi/2$

There is destructive interference between the light through slits 1 and 3 and between 2 and 4.

(ii)

$\phi = \pi$

There is destructive interference between the light through slits 1 and 2 and between 3 and 4.

(iii)

$\phi = 3\pi/2$

There is destructive interference between the light through slits 1 and 3 and between

2 and 4.

38-21 a) The interference <u>maxima</u> are at angles θ_m given by $\sin \theta_m = \dfrac{m\lambda}{d}$.

The diffraction <u>minima</u> are at angles θ_n given by $\sin \theta_n = \dfrac{n\lambda}{a}$.

The mth interference maximum coincides with the nth diffraction minimum if $m(\lambda/d) = n(\lambda/a)$, or $m = n(d/a)$.

If $m = 3$ coincides with $n = 1$, then $3 = d/a$ and $a = d/3 = 0.840 \text{ nm}/3 = 0.280 \text{ nm}$.

b) With $d/a = 3$ the equation in part (a) becomes $m = 3n$. n takes on the values 1, 2, 3, 4, ..., so the other interference maxima which are missing because they coincide with a diffraction minimum are $m = 6, 9, 12, \ldots$.

c) The wavelength λ does not appear in the equation $m = n(d/a)$ from part (a) so changing the wavelength will not affect which interference maxima are missing; $m = 3, 6, 9, 12, \ldots$ will still be the ones missing.

38-27 4000 lines/cm means 4.00×10^5 lines/m

The slit spacing is $d = \dfrac{1}{4.00 \times 10^5}$ m $= 2.50 \times 10^{-6}$ m.

a) The line positions are given by $\sin \theta = m\lambda/d$.

first order means $m = 1$

$\sin \theta_\alpha = \dfrac{\lambda_\alpha}{d} = \dfrac{656 \times 10^{-9} \text{ m}}{2.50 \times 10^{-6} \text{ m}} = 0.2624$ and $\theta_\alpha = 15.21°$.

$\sin \theta_\beta = \dfrac{\lambda_\beta}{d} = \dfrac{486 \times 10^{-9} \text{ m}}{2.50 \times 10^{-6} \text{ m}} = 0.1944$ and $\theta_\beta = 11.21°$.

The angular separation is $\theta_\alpha - \theta_\beta = 15.21° - 11.21° = 4.0°$.

b) second order means $m = 2$

$\sin \theta_\alpha = 2\lambda_\alpha/d = 0.5248$ and $\theta_\alpha = 31.65°$

$\sin \theta_\beta = 2\lambda_\beta/d = 0.3888$ and $\theta_\beta = 22.88°$

The angular separation is $\theta_\alpha - \theta_\beta = 31.65° - 22.88° = 8.8°$.

Note that the angular separation is larger in higher orders.

38-29 The maxima occur at angles θ given by Eq.(38-16):

$2d \sin \theta = m\lambda$, where d is the spacing between adjacent atomic planes.

second order says $m = 2$

$d = \dfrac{m\lambda}{2 \sin \theta} = \dfrac{2(0.0850 \times 10^{-9} \text{ m})}{2 \sin 21.5°} = 2.32 \times 10^{-10}$ m $= 0.232$ nm

38-35 The angular size of the first dark ring is given by $\sin \theta_1 = 1.22\lambda/D$ (Eq.38-17).

$$\sin\theta_1 = 1.22\left(\frac{620 \times 10^{-9} \text{ m}}{7.4 \times 10^{-6} \text{ m}}\right) = 0.1022; \quad \theta_1 = 0.1024 \text{ rad}$$

The radius of the Airy disk (central bright spot) is $r = (4.5 \text{ m})\tan\theta_1 = 0.462$ m. The diameter is $2r = 0.92$ m $= 92$ cm.

38-37 Resolved by Rayleigh's criterion means angular separation θ of the objects equals $1.22\lambda/D$. The angular separation θ of the objects is

$$\theta = \frac{250 \times 10^3 \text{ m}}{5.93 \times 10^{11} \text{ m}}, \text{ where } 5.93 \times 10^{11} \text{ m is the distance from earth to Jupiter.}$$

Thus $\theta = 4.216 \times 10^{-7}$.

Then $\theta = 1.22\dfrac{\lambda}{D}$ and $D = \dfrac{1.22\lambda}{\theta} = \dfrac{1.22(500 \times 10^{-9} \text{ m})}{4.216 \times 10^{-7}} = 1.45$ m

Problems

38-39 a) $I = I_0\left(\dfrac{\sin\beta/2}{\beta/2}\right)^2$,where $\beta = \left(\dfrac{2\pi}{\lambda}\right)a\sin\theta$

$I = \frac{1}{2}I_0$ so $\dfrac{\sin\beta/2}{\beta/2} = \dfrac{1}{\sqrt{2}}$

Let $x = \beta/2$; the equation for x is $\dfrac{\sin x}{x} = \dfrac{1}{\sqrt{2}} = 0.7071$.

Use trial and error to find the value of x that is a solution to this equation.

x	$(\sin x)/x$
1.0 rad	0.841
1.5 rad	0.665
1.2 rad	0.777
1.4 rad	0.7039
1.39 rad	0.7077; thus $x = 1.39$ rad and $\beta = 2x = 2.78$ rad

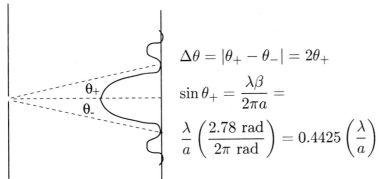

$$\Delta\theta = |\theta_+ - \theta_-| = 2\theta_+$$

$$\sin\theta_+ = \frac{\lambda\beta}{2\pi a} =$$

$$\frac{\lambda}{a}\left(\frac{2.78 \text{ rad}}{2\pi \text{ rad}}\right) = 0.4425\left(\frac{\lambda}{a}\right)$$

(i) For $\dfrac{a}{\lambda} = 2$, $\sin\theta_+ = 0.4425(\frac{1}{2}) = 0.2212$; $\theta_+ = 12.78°$; $\Delta\theta = 2\theta_+ = 25.6°$

(ii) For $\dfrac{a}{\lambda} = 5$, $\sin\theta_+ = 0.4425(\frac{1}{5}) = 0.0885$; $\theta_+ = 5.077°$; $\Delta\theta = 2\theta_+ = 10.2°$

(iii) For $\dfrac{a}{\lambda} = 10$, $\sin\theta_+ = 0.4425(\frac{1}{10}) = 0.04425$; $\theta_+ = 2.536°$; $\Delta\theta = 2\theta_+ = 5.1°$

b) $\sin\theta_0 = \dfrac{\lambda}{a}$ locates the first minimum

(i) For $\dfrac{a}{\lambda} = 2$, $\sin\theta_0 = \frac{1}{2}$; $\theta_0 = 30.0°$; $2\theta_0 = 60.0°$

(ii) For $\dfrac{a}{\lambda} = 5$, $\sin\theta_0 = \frac{1}{5}$; $\theta_0 = 11.54°$; $2\theta_0 = 23.1°$

(iii) For $\dfrac{a}{\lambda} = 10$, $\sin\theta_0 = \frac{1}{10}$; $\theta_0 = 5.74°$; $2\theta_0 = 11.5°$

Note: Either definition of the width shows that the central maximum gets narrower as the slit gets wider.

38-41 **a)** The angular position of the first minimum is given by $a\sin\theta = m\lambda$ (Eq.38-2), with $m = 1$.

$$\sin\theta = \frac{\lambda}{a} = \frac{540 \times 10^{-9}\ \text{m}}{0.360 \times 10^{-3}\ \text{m}} = 1.50 \times 10^{-3};\ \theta = 1.50 \times 10^{-3}\ \text{rad}$$

The distance y_1 from the center of the central maximum to the first minimum is given by $y_1 = x\tan\theta = (1.20\ \text{m})\tan(1.50 \times 10^{-3}\ \text{rad}) = 1.80 \times 10^{-3}\ \text{m} = 1.80\ \text{mm}$. (Note that θ is small enough for $\theta \approx \sin\theta \approx \tan\theta$, and Eq.(38-3) applies.)

b) From part (a) of Problem 38-39, $I = \frac{1}{2}I_0$ when $\beta = 2.78$ rad.

$$\beta = \left(\frac{2\pi}{\lambda}\right) a\sin\theta \ \text{(Eq.(38-6))},\ \text{so} \ \sin\theta = \frac{\beta\lambda}{2\pi a}.$$

$$y = x\tan\theta \approx x\sin\theta \approx \frac{\beta\lambda x}{2\pi a} = \frac{(2.78\ \text{rad})(540 \times 10^{-9}\ \text{m})(1.20\ \text{m})}{2\pi(0.360 \times 10^{-3}\ \text{m})} =$$

$7.96 \times 10^{-4}\ \text{m} = 0.796\ \text{mm}$

38-43 The phase difference between adjacent slits is $\phi = \dfrac{2\pi d}{\lambda}\sin\theta \approx \dfrac{2\pi d\theta}{\lambda}$

when θ is small and $\sin\theta \approx \theta$. Thus $\theta = \dfrac{\lambda\phi}{2\pi d}$.

A principal maximum occurs when $\phi = \phi_{max} = m2\pi$, where m is an integer, since then all the phasors add. The first minima on either side of the mth principal maximum occurs when $\phi = \phi_{min}^{\pm} = m2\pi \pm (2\pi/N)$ and the phasor diagram for N slits forms a closed loop and the resultant phasor is zero.

The angular position of a principal maximum is $\theta = \left(\dfrac{\lambda}{2\pi d}\right)\phi_{max}$. The angular position of the adjacent minimum is $\theta_{min}^{\pm} = \left(\dfrac{\lambda}{2\pi d}\right)\phi_{min}^{\pm}$.

$$\theta^+_{\min} = \left(\frac{\lambda}{2\pi d}\right)\left(\phi_{\max} + \frac{2\pi}{N}\right) = \theta + \left(\frac{\lambda}{2\pi d}\right)\left(\frac{2\pi}{N}\right) = \theta + \frac{\lambda}{Nd}$$

$$\theta^-_{\min} = \left(\frac{\lambda}{2\pi d}\right)\left(\phi_{\max} - \frac{2\pi}{N}\right) = \theta - \frac{\lambda}{Nd}$$

The angular width of the principal maximum is $\theta^+_{\min} - \theta^-_{\min} = \dfrac{2\lambda}{Nd}$, as was to be shown.

38-49 The condition for an intensity maximum is $d\sin\theta = m\lambda$, $m = 0, \pm 1, \pm 2, \ldots$ Third order means $m = 3$.

6500 lines/cm so 6.50×10^5 lines/m and $d = \dfrac{1}{6.50 \times 10^5}$ m $= 1.538 \times 10^{-6}$ m

The longest observable wavelength is the one that gives $\theta = 90°$ and hence $\sin\theta = 1$,

$$\lambda = \frac{d\sin\theta}{m} = \frac{(1.538 \times 10^{-6}\text{ m})(1)}{3} = 5.13 \times 10^{-7}\text{ m} = 513\text{ nm}$$

38-55 Rayleigh's criterion says that the two objects are resolved if the center of one diffraction pattern coincides with the first minimum of the other.

By Eq.(38-2) the angular position of the first minimum relative to the center of the central maximum is $\sin\theta = \lambda/a$, where a is the slit width. Hence if the objects are resolved according to Rayleigh's criterion, the angular separation between centers of the images of the two objects must be at least λ/a.

But as discussed in Example 38-8, the angular separation of the image points equals the angular separation of the object points. The angular separation of the object points is y/s, where $y = 2.50$ m is the linear separation of the two points and s is their distance from the observer.

Thus $\dfrac{y}{s} = \dfrac{\lambda}{a}$ and $s = \dfrac{ya}{\lambda} = \dfrac{(2.50\text{ m})(0.350 \times 10^{-3}\text{ m})}{600 \times 10^{-9}\text{ m}} = 1.46$ km.

38-57 Resolved by Rayleigh's criterion means the angular separation θ of the objects is given by $\theta = 1.22\lambda/D$. $\theta = y/s$, where $y = 75.0$ m is the distance between the two objects and s is their distance from the astronaut (her altitude).

$$\frac{y}{s} = 1.22\frac{\lambda}{D}$$

$$s = \frac{yD}{1.22\lambda} = \frac{(75.0\text{ m})(4.00 \times 10^{-3}\text{ m})}{1.22(500 \times 10^{-9}\text{ m})} = 4.92 \times 10^5\text{ m} = 492\text{ km}$$

CHAPTER 39

RELATIVITY

Exercises 1, 3, 5, 9, 13, 19, 21, 23, 25, 27, 29, 35, 37, 41
Problems 43, 45, 49, 51, 55, 57, 59

Exercises

39-1

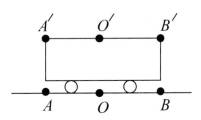

Simultaneous to observer on train means light pulses from A' and B' arrive at O' at the same time. To observer at O light from A' has a shorter distance to travel than light from B' so O will conclude that the pulse from $A(A')$ started before the pulse at $B(B')$. To observer at O bolt A appeared to strike first.

39-3 $\Delta t = 4.00 \text{ h} = 4.00 \text{ h} (3600 \text{ s}/1 \text{ h}) = 1.44 \times 10^4 \text{ s}$

The elapsed time for the clock on the plane is Δt_0.

$$\Delta t = \frac{\Delta t_0}{\sqrt{1 - u^2/c^2}} \text{ and } \Delta t_0 = \Delta t \sqrt{1 - u^2/c^2}$$

$\frac{u}{c}$ small so $\sqrt{1 - u^2/c^2} = (1 - u^2/c^2)^{1/2} \approx 1 - \frac{1}{2}\frac{u^2}{c^2}$; thus $\Delta t_0 = \Delta t \left(1 - \frac{1}{2}\frac{u^2}{c^2}\right)$

The difference in the clock readings is

$\Delta t - \Delta t_0 = \frac{1}{2}\frac{u^2}{c^2}\Delta t = \frac{1}{2}\left(\frac{250 \text{ m/s}}{2.998 \times 10^8 \text{ m/s}}\right)^2 (1.44 \times 10^4 \text{ s}) = 5.01 \times 10^{-9} \text{ s}.$ The clock on the plane has the shorter elapsed time.

39-5 **a)** $\Delta t_0 = 2.60 \times 10^{-8} \text{ s}$; $\Delta t = 4.20 \times 10^{-7} \text{ s}$

$$\Delta t = \frac{\Delta t_0}{\sqrt{1 - u^2/c^2}} \text{ says } 1 - \frac{u^2}{c^2} = \left(\frac{\Delta t_0}{\Delta t}\right)^2$$

$\frac{u}{c} = \sqrt{1 - \left(\frac{\Delta t_0}{\Delta t}\right)^2} = \sqrt{1 - \left(\frac{2.60 \times 10^{-8} \text{ s}}{4.20 \times 10^{-7} \text{ s}}\right)^2} = 0.998$; $u = 0.998c$

b) The speed in the laboratory frame is $u = 0.998c$; the time measured in this frame is Δt, so the distance as measured in this frame is

$d = u\Delta t = (0.998)(2.998 \times 10^8 \text{ m/s})(4.20 \times 10^{-7} \text{ s}) = 126$ m

39-9 **a)** The distance measured in the earth's frame is the proper length $l_0 = 55.0 \times 10^3$ m.

$l = l_0\sqrt{1 - u^2/c^2} = (55.0 \times 10^3 \text{ m})\sqrt{1 - (0.9860c/c)^2} = 9.17 \times 10^3 \text{ m} = 9.17$ km

b) Use the lifetime measured in the muon's frame as the time of travel to calculate the distance traveled as measured in that frame.

$d = u\,\Delta t = (0.9860)(2.998 \times 10^8 \text{ m/s})(2.20 \times 10^{-6} \text{ s}) = 650 \text{ m} = 0.650$ km

The muon's original height as measured in the muon's frame (part (A)) is 9.17 km, so the fraction is $\dfrac{0.650 \text{ km}}{9.17 \text{ km}} = 0.0709$.

c) $\Delta t_0 = 2.20 \times 10^{-6}$ s; $\Delta t = ?$

$\Delta t = \dfrac{\Delta t_0}{\sqrt{1 - u^2/c^2}} = \dfrac{2.20 \times 10^{-6} \text{ s}}{\sqrt{1 - (0.9860c/c)^2}} = 1.32 \times 10^{-5} \text{ s} = 13.2 \ \mu\text{s}$

Use the lifetime in the earth's frame to find the distance traveled in that frame:

$d = u\,\Delta t = (0.9860)(2.998 \times 10^8 \text{ m/s})(1.32 \times 10^{-5} \text{ s}) = 3.90 \times 10^3 \text{ m} = 3.90$ km

The fraction is $\dfrac{3.90 \text{ km}}{55.0 \text{ km}} = 0.0709$, the same fraction as in the muon's frame.

39-13 $l = l_0\sqrt{1 - u^2/c^2}$. The length measured when the spacecraft is moving is $l = 74.0$ m; l_0 is the length measured in a frame at rest relative to the spacecraft.

$l_0 = \dfrac{l}{\sqrt{1 - u^2/c^2}} = \dfrac{74.0 \text{ m}}{\sqrt{1 - (0.600c/c)^2}} = 92.5$ m

39-19 Use the Lorentz velocity transformation equation, Eq.(39-23): $v' = \dfrac{v - u}{1 - uv/c^2}$.

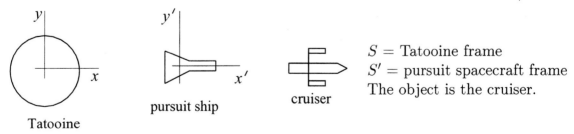

S = Tatooine frame
S' = pursuit spacecraft frame
The object is the cruiser.

With the coordinates shown, each ship is moving in the positive coordinate direction in the Tatooine frame.

u is the velocity of the pursuit spacecraft relative to Tatooine; $u = +0.800c$

$v = +0.600c$

$v' = ?$ (velocity of the cruiser relative to the pursuit spacecraft)

$$v' = \frac{v - u}{1 - uv/c^2} = \frac{0.600c - 0.800c}{1 - (0.800c)(0.600c)/c^2} = \frac{-0.200c}{0.520} = -0.385c$$

The crusier is moving toward the pursuit spacecraft with a speed of $0.385c = 1.15 \times 10^8$ m/s.

39-21

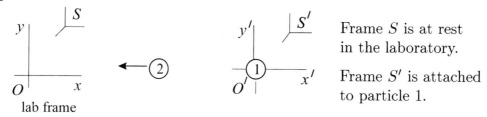

lab frame

Frame S is at rest in the laboratory.

Frame S' is attached to particle 1.

u is the speed of S' relative to S; this is the speed of particle 1 as measured in the laboratory. Thus $u = +0.650c$. The speed of particle 2 in S' is $0.950c$. Also, since the two particles move in opposite directions, 2 moves in the $-x'$ direction and $v' = -0.950c$.

We want to calculate v, the speed of particle 2 in frame S.

Eq.(39-24): $v = \dfrac{v' + u}{1 + uv'/c^2} = \dfrac{-0.950c + 0.650c}{1 + (0.950c)(-0.650c)/c^2} = \dfrac{-0.300c}{1 - 0.6175} = -0.784c.$

The speed of the second particle, as masured in the laboratory, is $0.784c$.

39-23   Use the Lorentz velocity transformation equation, Eq.(39-23):

$$v' = \frac{v - u}{1 - uv/c^2}$$

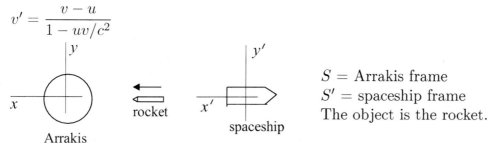

S = Arrakis frame
S' = spaceship frame
The object is the rocket.

u is the velocity of the spaceship relative to Arrakis.

$v = +0.360c; \quad v' = +0.920c$

(In each frame the rocket is moving in the positive coordinate direction.)

$$v' = \frac{v - u}{1 - uv/c^2} \text{ so } v' - u\left(\frac{vv'}{c^2}\right) = v - u \text{ and } u\left(1 - \frac{vv'}{c^2}\right) = v - v'$$

$$u = \frac{v - v'}{1 - vv'/c^2} = \frac{0.360c - 0.920c}{1 - (0.360c)(0.920c)/c^2} = -\frac{0.560c}{0.6688} = -0.837c$$

The speed of the spacecraft relative to Arrakis is $0.837c = 2.51 \times 10^8$ m/s. The minus sign in our result for U means that the spacecraft is moving in the $-x$-direction, so it is moving away from Arrakis.

39-25

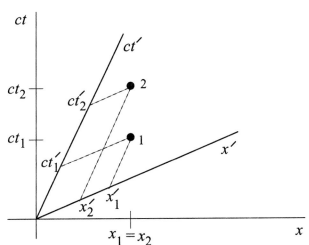

a) To find t'_2 and t'_1 on the spacetime diagram draw lines through points 2 and 1 parallel to the x' axis. The intercepts with the ct' axis give ct'_2 and ct'_1. We see that $t'_2 > t'_1$, so Mavis agrees that event 1 occurs first.

b) To find x'_2 and x'_1 on the spacetime diagram draw lines through points 2 and 1 parallel to the ct' axis. The intercepts with the x' axis give x'_2 and x'_1. We see that x'_1 is not equal to x'_2 (in fact $x'_1 > x'_2$), so according to Mavis the two events do not occur at the same position.

39-27 a) Source and observer are approaching, so use Eq.(39-26): $f = \sqrt{\dfrac{c+u}{c-u}} f_0$.

Solve for u: $f^2 = \left(\dfrac{c+u}{c-u} \right) f_0^2$

$(c-u)f^2 = (c+u)f_0^2$ and $u = \dfrac{c(f^2 - f_0^2)}{f^2 + f_0^2} = c \left(\dfrac{(f/f_0)^2 - 1}{(f/f_0)^2 + 1} \right)$

$\lambda_0 = 675$ nm, $\quad \lambda = 575$ nm

$u = \left(\dfrac{(675 \text{ nm}/575 \text{ nm})^2 - 1}{(675 \text{ nm}/575 \text{ nm})^2 + 1} \right) c = 0.159c = (0.159)(2.998 \times 10^8 \text{ m/s}) = $
4.77×10^7 m/s; definitely speeding

b) 4.77×10^7 m/s $= (4.77 \times 10^7$ m/s$)(1$ km/1000 m$)(3600$ s/1 h$) = $
1.72×10^8 km/h. Your fine would be $\$1.72 \times 10^8$ (172 million dollars).

39-29 Eq.(39-30): $F = \dfrac{dp}{dt} = \dfrac{d}{dt} \left(\dfrac{mv}{\sqrt{1 - v^2/c^2}} \right)$

If $\vec{F}$ is parallel to $\vec{v}$ then $\vec{F}$ changes the magnitude of $\vec{v}$ and not its direction.

$$F = \frac{m}{(1-v^2/c^2)^{1/2}}\left(\frac{dv}{dt}\right) + \frac{mv}{(1-v^2/c^2)^{3/2}}\left(-\frac{1}{2}\right)\left(-\frac{2v}{c^2}\right)\left(\frac{dv}{dt}\right)$$

$$F = \frac{dv}{dt}\frac{m}{(1-v^2/c^2)^{3/2}}\left(1 - \frac{v^2}{c^2} + \frac{v^2}{c^2}\right) = \frac{dv}{dt}\frac{m}{(1-v^2/c^2)^{3/2}}$$

But $\dfrac{dv}{dt} = a$, so $a = (F/m)(1-v^2/c^2)^{3/2}$.

b) $\vec{F} = \dfrac{d}{dt}\left(\dfrac{m\vec{v}}{\sqrt{1-v^2/c^2}}\right).$

If $\vec{F}$ is perpendicular to $\vec{v}$ then $\vec{F}$ changes the direction of $\vec{v}$ and not its magnitude. $\vec{a} = d\vec{v}/dt$ but the magnitude of v in the denominator of Eq.(39-30) is constant. Hence $F = \dfrac{ma}{\sqrt{1-v^2/c^2}}$ and $a = (F/m)(1-v^2/c^2)^{1/2}$.

39-35   Eq.39-38: $K = \frac{1}{2}mv^2 + \frac{3}{8}mv^4/c^2$

We want $\dfrac{K - \frac{1}{2}mv^2}{\frac{1}{2}mv^2} = 0.020$ and thus $\dfrac{\frac{3}{8}mv^4/c^2}{\frac{1}{2}mv^2} = 0.020$.

$\frac{3}{4}(v^2/c^2) = 0.020$ and $v = \sqrt{\frac{4}{3}(0.020)}c = 0.16c = 4.8 \times 10^7$ m/s

39-37   **a)** The total energy is given in terms of the momentum by Eq.(39-40):

$E = \sqrt{(mc^2)^2 + (pc)^2} =$

$\sqrt{[(6.64 \times 10^{-27})(2.998 \times 10^8)^2]^2 + [(2.10 \times 10^{-18})(2.998 \times 10^8)]^2}$ J

$E = 8.67 \times 10^{-10}$ J

b) $E = K + mc^2$ so $K = E - mc^2$

$mc^2 = (6.64 \times 10^{-27} \text{ kg})(2.998 \times 10^8 \text{ m/s})^2 = 5.97 \times 10^{-10}$ J

$K = E - mc^2 = 8.67 \times 10^{-10}$ J $- 5.97 \times 10^{-10}$ J $= 2.70 \times 10^{-10}$ J

c) $\dfrac{K}{mc^2} = \dfrac{2.70 \times 10^{-10} \text{ J}}{5.97 \times 10^{-10} \text{ J}} = 0.452$

39-41   **a)** $K = 0.420$ MeV $= 4.20 \times 10^5$ eV

b) $E = K + mc^2$ (Eq.39-39)

For an electron $mc^2 = (9.109 \times 10^{-31} \text{ kg})(2.998 \times 10^8 \text{ m/s})^2 =$

8.187×10^{-14} J$(1 \text{ eV}/1.602 \times 10^{-19} \text{ J})$

$mc^2 = 5.110 \times 10^5$ eV $= 0.511$ MeV

Then $E = K + mc^2 = 0.420$ MeV $+ 0.511$ MeV $= 0.931$ MeV $= 9.31 \times 10^5$ eV

c) $E = \dfrac{mc^2}{\sqrt{1 - v^2/c^2}}$ so $1 - \dfrac{v^2}{c^2} = \left(\dfrac{mc^2}{E}\right)^2$

$v = c\sqrt{1 - \left(\dfrac{mc^2}{E}\right)^2} = (2.998 \times 10^8 \text{ m/s})\sqrt{1 - \left(\dfrac{0.511 \text{ MeV}}{0.931 \text{ MeV}}\right)^2} = 2.51 \times 10^8 \text{ m/s}$

d) The classical relation between speed and kinetic energy is
$K = \frac{1}{2}mv^2$ so

$v = \sqrt{\dfrac{2K}{m}} = \sqrt{\dfrac{2(4.20 \times 10^5 \text{ eV})(1.602 \times 10^{-19} \text{ J/1 eV})}{9.109 \times 10^{-31} \text{ kg}}} = 3.84 \times 10^8 \text{ m/s}$

The classical result is too large by about 50%.

Problems

39-43 a) $\Delta t_0 = 2.60 \times 10^{-8}$ s

The time measured in the lab must satisfy

$d = c\,\Delta t$ so $\Delta t = \dfrac{d}{c} = \dfrac{1.20 \times 10^3 \text{ m}}{2.998 \times 10^8 \text{ m/s}} = 4.003 \times 10^{-6}$ s

$\Delta t = \dfrac{\Delta t_0}{\sqrt{1 - u^2/c^2}}$ so $(1 - u^2/c^2)^{1/2} = \dfrac{\Delta t_0}{\Delta t}$ and $(1 - u^2/c^2) = \left(\dfrac{\Delta t_0}{\Delta t}\right)^2$

Write $u = (1 - \Delta)c$ so that $(u/c)^2 = (1 - \Delta)^2 = 1 - 2\Delta + \Delta^2 \approx 1 - 2\Delta$ since Δ is small.

Using this in the above gives $1 - (1 - 2\Delta) = \left(\dfrac{\Delta t_0}{\Delta t}\right)^2$

$\Delta = \dfrac{1}{2}\left(\dfrac{\Delta t_0}{\Delta t}\right)^2 = \dfrac{1}{2}\left(\dfrac{2.60 \times 10^{-8} \text{ s}}{4.003 \times 10^{-6} \text{ s}}\right)^2 = 2.11 \times 10^{-5}$

An alternative calculation is to say that the length of the tube must contract relative to the moving pion so that the pion travels that length before decaying. The contracted length must be

$l = c\,\Delta t_0 = (2.998 \times 10^8 \text{ m/s})(2.60 \times 10^{-8} \text{ s}) = 7.79 \text{ m}.$

$l = l_0\sqrt{1 - u^2/c^2}$ so $1 - u^2/c^2 = \left(\dfrac{l}{l_0}\right)^2$

Then $u = (1 - \Delta)c$ gives $\Delta = \dfrac{1}{2}\left(\dfrac{l}{l_0}\right)^2 = \dfrac{1}{2}\left(\dfrac{7.79 \text{ m}}{1.20 \times 10^3 \text{ m}}\right)^2 = 2.11 \times 10^{-5}$, which checks.

b) $E = \gamma mc^2$

$\gamma = \dfrac{1}{\sqrt{1 - u^2/c^2}} = \dfrac{1}{\sqrt{2\Delta}} = \dfrac{1}{\sqrt{2(2.11 \times 10^{-5})}} = 154$

$$E = 154(139.6 \text{ MeV}) = 2.15 \times 10^4 \text{ MeV} = 21.5 \text{ GeV}$$

39-45 There must be a length contraction such that the length a becomes the same as b;
$l_0 = a$, $l = b$.

Eq.(39-16): $\dfrac{l}{l_0} = \sqrt{1 - u^2/c^2}$ so $\dfrac{b}{a} = \sqrt{1 - u^2/c^2}$;

$a = 1.40b$ gives $b/1.40b = \sqrt{1 - u^2/c^2}$ and thus $1 - u^2/c^2 = 1/(1.40)^2$

$u = \sqrt{1 - 1/(1.40)^2}\,c = 0.700c = 2.10 \times 10^8 \text{ m/s}$

39-49

$E = mc^2$; the mass increase is due to the heat flow into the ice to melt it.

$E = Q = mL_f = (4.00 \text{ kg})(334 \times 10^3 \text{ J/kg}) = 1.336 \times 10^6 \text{ J}$

$m = \dfrac{E}{c^2} = \dfrac{1.336 \times 10^6 \text{ J}}{(2.998 \times 10^8 \text{ m/s})^2} = 1.49 \times 10^{-11} \text{ kg}$

39-51 In crown glass the speed of light is $v = \dfrac{c}{n} = \dfrac{2.998 \times 10^8 \text{ m/s}}{1.52} = 1.972 \times 10^8 \text{ m/s}$.

Calculate the kinetic energy of an electron that has this speed:

$K = mc^2(\gamma - 1)$

$mc^2 = (9.109 \times 10^{-31} \text{ kg})(2.998 \times 10^8 \text{ m/s})^2 = 8.187 \times 10^{-14} \text{ J}(1 \text{ eV}/1.602 \times 10^{-19} \text{ J}) = 0.5111 \text{ MeV}$

$\gamma = \dfrac{1}{\sqrt{1 - v^2/c^2}} = \dfrac{1}{\sqrt{1 - ((1.972 \times 10^8 \text{ m/s})/(2.998 \times 10^8 \text{ m/s}))^2}} = 1.328$

$K = mc^2(\gamma - 1) = (0.5111 \text{ MeV})(1.328 - 1) = 0.168 \text{ MeV}$

39-55 According to Eq.(39-31), $a = \dfrac{dv}{dt} = \dfrac{F}{m}(1 - v^2/c^2)^{3/2}$.

(One-dimensional motion is assumed, and all the F, v, and a refer to x-components.)

$\dfrac{dv}{(1 - v^2/c^2)^{3/2}} = \left(\dfrac{F}{m}\right) dt$

Integrate from $t = 0$, when $v = 0$, to time t, when the velocity is v.

$\displaystyle\int_0^v \dfrac{dv}{(1 - v^2/c^2)^{3/2}} = \int_0^t \left(\dfrac{F}{m}\right) dt$

Since F is constant, $\displaystyle\int_0^t \left(\dfrac{F}{m}\right) dt = \dfrac{Ft}{m}$.

In the velocity integral make the change of variable $y = v/c$; then $dy = dv/c$.

$\displaystyle\int_0^v \dfrac{dv}{(1 - v^2/c^2)^{3/2}} = c\int_0^{v/c} \dfrac{dy}{(1 - y^2)^{3/2}} = c\left[\dfrac{y}{(1 - y^2)^{1/2}}\right]_0^{v/c} = \dfrac{v}{\sqrt{1 - v^2/c^2}}$

Thus $\dfrac{v}{\sqrt{1 - v^2/c^2}} = \dfrac{Ft}{m}$.

Solve this equation for v:

$$\frac{v^2}{1 - v^2/c^2} = \left(\frac{Ft}{m}\right)^2 \text{ and } v^2 = \left(\frac{Ft}{m}\right)^2 (1 - v^2/c^2)$$

$$v^2 \left(1 + \left(\frac{Ft}{mc}\right)^2\right) = \left(\frac{Ft}{m}\right)^2 \text{ so } v = \frac{(Ft/m)}{\sqrt{1 + (Ft/mc)^2}} = c\frac{Ft}{\sqrt{m^2c^2 + F^2t^2}}$$

As $t \to \infty$, $\dfrac{Ft}{\sqrt{m^2c^2 + F^2t^2}} \to \dfrac{Ft}{\sqrt{F^2t^2}} \to 1$, so $v \to c$.

Note also that $\dfrac{Ft}{\sqrt{m^2c^2 + F^2t^2}}$ is always less than 1, so $v < c$ always and approaches c only when $t \to \infty$.

39-57 a) Use the Lorentz coordinate transformation (Eq.39-22) for (x_1, t_1) and (x_2, t_2):

$$x_1' = \frac{x_1 - ut_1}{\sqrt{1 - u^2/c^2}}, \qquad x_2' = \frac{x_2 - ut_2}{\sqrt{1 - u^2/c^2}}$$

$$t_1' = \frac{t_1 - ux_1/c^2}{\sqrt{1 - u^2/c^2}}, \qquad t_2' = \frac{t_2 - ux_2/c^2}{\sqrt{1 - u^2/c^2}}$$

Same point in S' implies $x_1' = x_2'$. What then is $\Delta t' = t_2' - t_1'$?

$x_1' = x_2'$ implies $x_1 - ut_1 = x_2 - ut_2$

$$u(t_2 - t_1) = x_2 - x_1 \text{ and } u = \frac{x_2 - x_1}{t_2 - t_1} = \frac{\Delta x}{\Delta t}$$

From the time transformation equations,

$$\Delta t' = t_2' - t_1' = \frac{1}{\sqrt{1 - u^2/c^2}}(\Delta t - u\,\Delta x/c^2)$$

Using the result that $u = \dfrac{\Delta x}{\Delta t}$ gives

$$\Delta t' = \frac{1}{\sqrt{1 - (\Delta x)^2/((\Delta t)^2 c^2)}}(\Delta t - (\Delta x)^2/((\Delta t)c^2))$$

$$\Delta t' = \frac{\Delta t}{\sqrt{(\Delta t)^2 - (\Delta x)^2/c^2}}(\Delta t - (\Delta x)^2/((\Delta t)c^2))$$

$$\Delta t' = \frac{(\Delta t)^2 - (\Delta x)^2/c^2}{\sqrt{(\Delta t)^2 - (\Delta x)^2/c^2}} = \sqrt{(\Delta t)^2 - (\Delta x/c)^2}, \text{ as was to be shown.}$$

This equation doesn't have a physical solution (because of a negative square root) if $(\Delta x/c)^2 > (\Delta t)^2$ or $\Delta x \geq c\,\Delta t$.

b) Now require that $t'_2 = t'_1$ (the two events are simultaneous in S') and use the Lorentz coordinate transformation equations.

$t'_1 = t'_2$ implies $t_1 - ux_1/c^2 = t_2 - ux_2/c^2$

$t_2 - t_1 = \left(\dfrac{x_2 - x_1}{c^2}\right) u$ so $\Delta t = \left(\dfrac{\Delta x}{c^2}\right) u$ and $u = \dfrac{c^2 \Delta t}{\Delta x}$

From the Lorentz transformation equations,

$$\Delta x' = x'_2 - x'_1 = \left(\frac{1}{\sqrt{1 - u^2/c^2}}\right)(\Delta x - u\,\Delta t).$$

Using the result that $u = c^2 \Delta t/\Delta x$ gives

$$\Delta x' = \frac{1}{\sqrt{1 - c^2(\Delta t)^2/(\Delta x)^2}}(\Delta x - c^2(\Delta t)^2/\Delta x)$$

$$\Delta x' = \frac{\Delta x}{\sqrt{(\Delta x)^2 - c^2(\Delta t)^2}}(\Delta x - c^2(\Delta t)^2/\Delta x)$$

$$\Delta x' = \frac{(\Delta x)^2 - c^2(\Delta t)^2}{\sqrt{(\Delta x)^2 - c^2(\Delta t)^2}} = \sqrt{(\Delta x)^2 - c^2(\Delta t)^2}$$

c) $\Delta x' = \sqrt{(\Delta x)^2 - c^2(\Delta t)^2}$

Solve for Δt: $(\Delta x')^2 = (\Delta x)^2 - c^2(\Delta t)^2$

$$\Delta t = \frac{\sqrt{(\Delta x)^2 - (\Delta x')^2}}{c} = \frac{\sqrt{(5.00 \text{ m})^2 - (2.50 \text{ m})^2}}{2.998 \times 10^8 \text{ m/s}} = 1.44 \times 10^{-8} \text{ s}$$

39-59 An increase in wavelength corresponds to a decrease in frequency ($f = c/\lambda$), so the atoms are moving away from the earth.

Receding, so use Eq.(39-27): $f = \sqrt{\dfrac{c - u}{c + u}} f_0$

Solve for u: $(f/f_0)^2(c + u) = c - u$ and $u = c\left(\dfrac{1 - (f/f_0)^2}{1 + (f/f_0)^2}\right)$

$f = c/\lambda$, $f_0 = c/\lambda_0$ so $f/f_0 = \lambda_0/\lambda$

$$u = c\left(\frac{1 - (\lambda_0/\lambda)^2}{1 + (\lambda_0/\lambda)^2}\right) = c\left(\frac{1 - (656.3/953.4)^2}{1 + (656.3/953.4)^2}\right) = 0.357c = 1.07 \times 10^8 \text{ m/s}$$

CHAPTER 40
PHOTONS, ELECTRONS, AND ATOMS

Exercises 3, 5, 7, 9, 13, 15, 17, 19, 23, 25, 29, 31, 35
Problems 41, 47, 49, 51, 53, 57, 59, 61, 63

Exercises

40-3 **a)** $E = hf$ so $f = \dfrac{E}{h} = \dfrac{(2.45 \times 10^6 \text{ eV})(1.602 \times 10^{-19} \text{ J/1 eV})}{6.626 \times 10^{-34} \text{ J} \cdot \text{s}} = 5.92 \times 10^{20}$ Hz

b) $c = f\lambda$ so $\lambda = \dfrac{c}{f} = \dfrac{2.998 \times 10^8 \text{ m/s}}{5.92 \times 10^{20} \text{ Hz}} = 5.06 \times 10^{-13}$ m

c) λ is comparable to a nuclear radius

40-5 Eq.(40-3): $\frac{1}{2}mv_{\max}^2 = hf - \phi = \dfrac{hc}{\lambda} - \phi$

$\frac{1}{2}mv_{\max}^2 = \dfrac{(6.626 \times 10^{-34} \text{ J} \cdot \text{s})(2.998 \times 10^8 \text{ m/s})}{235 \times 10^{-9} \text{ m}} - (5.1 \text{ eV})(1.602 \times 10^{-19} \text{ J/1 eV})$

$\frac{1}{2}mv_{\max}^2 = 8.453 \times 10^{-19} \text{ J} - 8.170 \times 10^{-19} \text{ J} = 2.83 \times 10^{-20}$ J

$v_{\max} = \sqrt{\dfrac{2(2.83 \times 10^{-20} \text{ J})}{9.109 \times 10^{-31} \text{ kg}}} = 2.49 \times 10^5$ m/s

40-7 **a)** First find the work function ϕ:

$eV_0 = hf - \phi$ so $\phi = hf - eV_0 = \dfrac{hc}{\lambda} - eV_0$

$\phi = \dfrac{(6.626 \times 10^{-34} \text{ J} \cdot \text{s})(2.998 \times 10^8 \text{ m/s})}{254 \times 10^{-9} \text{ m}} - (1.602 \times 10^{-19})(0.181 \text{ V})$

$\phi = 7.821 \times 10^{-19} \text{ J} - 2.900 \times 10^{-20} \text{ J} = 7.531 \times 10^{-19} \text{ J}(1 \text{ eV}/1.602 \times 10^{-19} \text{ J}) = 4.70$ eV

The threshold frequency f_{th} is the smallest frequency that still produces photoelectrons. It corresponds to $K_{\max} = 0$ in Eq.(40-3), so $hf_{\text{th}} = \phi$.

$f = \dfrac{c}{\lambda}$ says $\dfrac{hc}{\lambda_{\text{th}}} = \phi$

$\lambda_{\text{th}} = \dfrac{hc}{\phi} = \dfrac{(6.626 \times 10^{-34} \text{ J} \cdot \text{s})(2.998 \times 10^8 \text{ m/s})}{7.531 \times 10^{-19} \text{ J}} = 2.64 \times 10^{-7} \text{ m} = 264$ nm

b) As calculated in part (a), $\phi = 4.70$ eV. This is the value given in Table 40-1.

40-9   **a)** Eq.(39-41): $E = pc = (8.24 \times 10^{-28} \text{ kg} \cdot \text{m/s})(2.998 \times 10^8 \text{ m/s}) = 2.47 \times 10^{-19} \text{ J}$
$E = (2.47 \times 10^{-19} \text{ J})(1 \text{ eV}/1.602 \times 10^{-19} \text{ J}) = 1.54 \text{ eV}$

b) Eq.(40-5): $p = \dfrac{h}{\lambda}$ so $\lambda = \dfrac{h}{p} = \dfrac{6.626 \times 10^{-34} \text{ J} \cdot \text{s}}{8.24 \times 10^{-28} \text{ kg} \cdot \text{m/s}} = 8.04 \times 10^{-7} \text{ m} = 804 \text{ nm}$
This wavelength is longer than visible wavelengths; it is in the infrared region of the electromagnetic spectrum.

40-13   **a)** Eq.(40-7): $\dfrac{1}{\lambda} = R\left(\dfrac{1}{2^2} - \dfrac{1}{n^2}\right)$

For the H_γ spectral line $n = 5$ and $\dfrac{1}{\lambda} = R\left(\dfrac{1}{2^2} - \dfrac{1}{5^2}\right) = R\left(\dfrac{25 - 4}{100}\right) = R\left(\dfrac{21}{100}\right)$

Thus $\lambda = \dfrac{100}{21R} = \dfrac{100}{21(1.097 \times 10^7)} \text{ m} = 4.341 \times 10^{-7} \text{ m} = 434.1 \text{ nm}$

b) $f = \dfrac{c}{\lambda} = \dfrac{2.998 \times 10^8 \text{ m/s}}{4.341 \times 10^{-7} \text{ m}} = 6.906 \times 10^{14} \text{ Hz}$

c) $E = hf = (6.626 \times 10^{-34} \text{ J} \cdot \text{s})(6.906 \times 10^{14} \text{ Hz}) = 4.576 \times 10^{-19} \text{ J} = 2.856 \text{ eV}$

40-15   Call the two lowest excited levels level 1 and level 2. If λ is the wavelength shown in the figure for the transition from one of these levels to the ground state then the transition energy is $\Delta E = hc/\lambda$.
The difference in energy between these two levels is

$$\Delta E_1 - \Delta E_2 = hc\left(\dfrac{1}{\lambda_1} - \dfrac{1}{\lambda_2}\right) = hc\left(\dfrac{\lambda_2 - \lambda_1}{\lambda_1 \lambda_2}\right) =$$

$$(4.136 \times 10^{-15} \text{ eV} \cdot \text{s})(2.998 \times 10^8 \text{ m/s})\left(\dfrac{(589.6 \times 10^{-9} \text{ m} - 589.0 \times 10^{-9} \text{ m})}{(589.6 \times 10^{-9} \text{ m})(589.0 \times 10^{-9} \text{ m})}\right) =$$

0.002 eV

40-17   **a)** If the particles are treated as point charges, $U = \dfrac{1}{4\pi\epsilon_0}\dfrac{q_1 q_2}{r}$.

$q_1 = 2e$ (alpha particle)

$q_2 = 82e$ (gold nucleus)

$U = (8.987 \times 10^9 \text{ N} \cdot \text{m}^2/\text{C}^2)\dfrac{(2)(82)(1.602 \times 10^{-19} \text{ C})^2}{6.50 \times 10^{-14} \text{ m}} = 5.82 \times 10^{-13} \text{ J}$

$U = 5.82 \times 10^{-13} \text{ J}(1 \text{ eV}/1.602 \times 10^{-19} \text{ J}) = 3.63 \times 10^6 \text{ eV} = 3.63 \text{ MeV}$

b) $K_1 + U_1 = K_2 + U_2$
Alpha particle is initially far from the lead nucleus implies $r_1 \approx \infty$ and $U_1 = 0$.
Alpha particle stops implies $K_2 = 0$.

Thus $K_1 = U_2 = 5.82 \times 10^{-13}$ J $= 3.63$ MeV

c) $K = \frac{1}{2}mv^2$ so $v = \sqrt{\dfrac{2K}{m}} = \sqrt{\dfrac{2(5.82 \times 10^{-13} \text{ J})}{6.64 \times 10^{-27} \text{ kg}}} = 1.32 \times 10^7$ m/s

40-19 The force between the electron and the nucleus in Be^{3+} is $F = \dfrac{1}{4\pi\epsilon_0}\dfrac{Ze^2}{r^2}$, where $Z = 4$ is the nuclear charge. All the equations for the hydrogen atom apply to Be^{+3} if we replace e^2 by Ze^2.

a) $E_n = -\dfrac{1}{\varepsilon_0}\dfrac{me^4}{8n^2h^2}$ (hydrogen) becomes

$E_n = -\dfrac{1}{\varepsilon_0}\dfrac{m(Ze^2)^2}{8n^2h^2} = Z^2\left(-\dfrac{1}{\varepsilon_0}\dfrac{me^4}{8n^2h^2}\right) = Z^2\left(-\dfrac{13.60 \text{ eV}}{n^2}\right)$ (for Be^{3+})

The ground-level energy of Be^{3+} is $E_1 = 16\left(-\dfrac{13.60 \text{ eV}}{1^2}\right) = -218$ eV; the ground-level energy of Be^{3+} is $Z^2 = 16$ times the ground-level energy of H.

b) The ionization energy is the energy difference between the $n \to \infty$ level energy and the $n = 1$ level energy. The $n \to \infty$ level energy is zero, so the ionization energy of Be^{3+} is 218 eV; this is 16 times the ionization energy of hydrogen.

c) $\dfrac{1}{\lambda} = R\left(\dfrac{1}{n_1^2} - \dfrac{1}{n_2^2}\right)$ just as for hydrogen but now R has a diferent value.

$R_H = \dfrac{me^4}{8\epsilon_0 h^3 c} = 1.097 \times 10^7$ m^{-1} for hydrogen becomes

$R_{Be} = Z^2\dfrac{me^4}{8\epsilon_0 h^3 c} = 16(1.097 \times 10^7 \text{ m}^{-1}) = 1.755 \times 10^8$ m^{-1} for Be^{3+}.

For $n = 2$ to $n = 1$, $\dfrac{1}{\lambda} = R_{Be}\left(\dfrac{1}{1^2} - \dfrac{1}{2^2}\right) = 3R/4$.

$\lambda = 4/(3R) = 4/(3(1.755 \times 10^8 \text{ m}^{-1})) = 7.60 \times 10^{-9}$ m $= 7.60$ nm.

This wavelength is smaller by a factor of 16 compared to the wavelength for the corresponding transition in the hydrogen atom.

d) Eq. (40-12): $r_n = \varepsilon_0\dfrac{n^2h^2}{\pi me^2}$ (hydrogen). Thus $r_n = \varepsilon_0\dfrac{n^2h^2}{\pi m(Ze^2)}$ (Be^{3+}).

For a given n the orbit radius for Be^{3+} is smaller by a factor of $Z = 4$ compared to the corresponding radius for hydrogen.

40-23 $\dfrac{n_{5s}}{n_{3p}} = e^{-(E_{5s} - E_{3p})/kT}$

From Fig.40-12a, $E_{5s} = 20.66$ eV and $E_{3p} = 18.70$ eV

$E_{5s} - E_{3p} = 20.66$ eV $- 18.70$ eV $= 1.96$ eV$(1.602 \times 10^{-19}$ J/1 eV$) =$
3.140×10^{-19} J

a) $\dfrac{n_{5s}}{n_{3p}} = e^{-(3.140 \times 10^{-19} \text{ J})/[(1.38 \times 10^{-23} \text{ J/K})(300 \text{ K})]} = e^{-75.79} = 1.2 \times 10^{-33}$

b) $\dfrac{n_{5s}}{n_{3p}} = e^{-(3.140 \times 10^{-19} \text{ J})/[(1.38 \times 10^{-23} \text{ J/K})(600 \text{ K})]} = e^{-37.90} = 3.5 \times 10^{-17}$

c) $\dfrac{n_{5s}}{n_{3p}} = e^{-(3.140 \times 10^{-19} \text{ J})/[(1.38 \times 10^{-23} \text{ J/K})(1200 \text{ K})]} = e^{-18.95} = 5.9 \times 10^{-9}$

d) At each of these temperatures the number of atoms in the $5s$ excited state, the initial state for the transition that emits 632.8 nm radiation, is quite small.

40-25 Power is energy per unit time, so in 1.00 s the energy emitted by the laser is
$(7.50 \times 10^{-3}$ W$)(1.00$ s$) = 7.50 \times 10^{-3}$ J.

The energy of each photon is

$E = \dfrac{hc}{\lambda} = \dfrac{(6.626 \times 10^{-34} \text{ J} \cdot \text{s})(2.998 \times 10^{8} \text{ m/s})}{10.6 \times 10^{-6} \text{ m}} = 1.874 \times 10^{-20}$ J.

The number of photons emited each second is the total energy emitted divided by the energy of one photon:

$\dfrac{7.50 \times 10^{-3} \text{ J/s}}{1.874 \times 10^{-20} \text{ J/photon}} = 4.00 \times 10^{17}$ photons/s

40-29 The kinetic energy K of an electron after acceleration is
$K = q\,\Delta V = (1.602 \times 10^{-19} \text{ C})(15.0 \times 10^{3} \text{ V}) = 2.403 \times 10^{-15}$ J.

The shortest wavelength x rays are those for which the kinetic energy of a photon is converted entirely to the photon energy $hf = hc/\lambda$.

$\dfrac{hc}{\lambda} = K$ gives

$\lambda = \dfrac{hc}{K} = \dfrac{(6.626 \times 10^{-34} \text{ J} \cdot \text{s})(2.998 \times 10^{8} \text{ m/s})}{2.403 \times 10^{-15} \text{ J}} = 8.27 \times 10^{-11}$ m $= 0.0827$ nm.

40-31 Eq.(40-21): $\lambda' - \lambda = \dfrac{h}{mc}(1 - \cos\phi) = \lambda_c(1 - \cos\phi)$

$\lambda' = \lambda + \lambda_c(1 - \cos\phi)$

The largest λ' corresponds to $\phi = 180°$, so $\cos\phi = -1$.

Then $\lambda' = \lambda + 2\lambda_c = 0.0665 \times 10^{-9}$ m $+ 2(2.426 \times 10^{-12}$ m$) = 7.135 \times 10^{-11}$ m $= 0.0714$ nm.

This wavelength occurs at a scattering angle of $\phi = 180°$.

40-35 The wavelength λ_m where the Planck distribution peaks is given by Eq.(40-28):
$$\lambda_m = \frac{2.90 \times 10^{-3} \text{ m} \cdot \text{K}}{2.728 \text{ K}} = 1.063 \times 10^{-3} \text{ m} = 1.063 \text{ mm}.$$
This wavelength is in the microwave portion of the electromagnetic spectrum.

Problems

40-41 a) One photon dissociates one AgBr molecule, so we need to find the energy required to dissociate a single molecule. The problem states that it requires 1.00×10^5 J to dissociate one mole of AgBr, and one mole contains Avogadro's number (6.02×10^{23}) of molecules, so the energy required to dissociate one AgBr is
$$\frac{1.00 \times 10^5 \text{ J/mol}}{6.02 \times 10^{23} \text{ molecules/mol}} = 1.66 \times 10^{-19} \text{ J/molecule}.$$
The photon is to have this energy, so
$$E = 1.66 \times 10^{-19} \text{ J}(1\text{eV}/1.602 \times 10^{-19} \text{ J}) = 1.04 \text{ eV}.$$

b) $E = \dfrac{hc}{\lambda}$ so $\lambda = \dfrac{hc}{E} = \dfrac{(6.626 \times 10^{-34} \text{ J} \cdot \text{s})(2.998 \times 10^8 \text{ m/s})}{1.66 \times 10^{-19} \text{ J}} = 1.20 \times 10^{-6} \text{ m} = 1200$ nm

c) $c = f\lambda$ so $f = \dfrac{c}{\lambda} = \dfrac{2.998 \times 10^8 \text{ m/s}}{1.20 \times 10^{-6} \text{ m}} = 2.50 \times 10^{14}$ Hz

d) $E = hf = (6.626 \times 10^{-34} \text{ J} \cdot \text{s})(100 \times 10^6 \text{ Hz}) = 6.63 \times 10^{-26}$ J
$E = 6.63 \times 10^{-26}$ J$(1 \text{ eV}/1.602 \times 10^{-19} \text{ J}) = 4.14 \times 10^{-7}$ eV

e) A photon with frequency $f = 100$ MHz has too little energy, by a large factor, to dissociate a AgBr molecule. The photons in the visible light from a firefly do individually have enough energy to dissociate AgBr. The huge number of 100 MHz photons can't compensate for the fact that individually they have too little energy.

40-47 a) The stopping potential V_0 is given by Eq.(40-4): $eV_0 = hf - \phi = (hc/\lambda) - \phi$.
Call the stopping potential V_{01} for λ_1 and V_{02} for λ_2. Thus $eV_{01} = (hc/\lambda_1) - \phi$ and $eV_{02} = (hc/\lambda_2) - \phi$. Note that the work function ϕ is a property of the material and is independent of the wavelength of the light.

Subtracting one equation from the other gives $e(V_{02} - V_{01}) = hc\left(\dfrac{\lambda_1 - \lambda_2}{\lambda_1 \lambda_2}\right)$.

b) $\Delta V_0 = \dfrac{(6.626 \times 10^{-34} \text{ J} \cdot \text{s})(2.998 \times 10^8 \text{ m/s})}{1.602 \times 10^{-19} \text{ C}} \left(\dfrac{295 \times 10^{-9} \text{ m} - 265 \times 10^{-9} \text{ m}}{(295 \times 10^{-9} \text{ m})(265 \times 10^{-9} \text{ m})}\right)$
$= 0.476$ V.

40-49 **a)** Eq.(40-18): $m_r = \dfrac{m_1 m_2}{m_1 + m_2} = \dfrac{207 m_e m_p}{207 m_e + m_p}$

$$m_r = \frac{207(9.109 \times 10^{-31} \text{ kg})(1.673 \times 10^{-27} \text{ kg})}{207(9.109 \times 10^{-31} \text{ kg}) + 1.673 \times 10^{-27} \text{ kg}} = 1.69 \times 10^{-28} \text{ kg}$$

We have used m_e to denote the electron mass.

b) In Eq.(40-16) replace $m = m_e$ by m_r: $E_n = -\dfrac{1}{\varepsilon_0^2} \dfrac{m_r e^4}{8 n^2 h^2}$.

Write as $E_n = \left(\dfrac{m_r}{m_H} \right) \left(-\dfrac{1}{\varepsilon_0^2} \dfrac{m_H e^4}{8 n^2 h^2} \right)$, since we know that $\dfrac{1}{\varepsilon_0^2} \dfrac{m_H e^4}{8 h^2} = 13.60$ eV.

Here m_H denotes the reduced mass for the hydrogen atom; $m_H = 0.99946(9.109 \times 10^{-31} \text{ kg}) = 9.104 \times 10^{-31} \text{ kg}$.

$$E_n = \left(\frac{m_r}{m_H} \right) \left(-\frac{13.60 \text{ eV}}{n^2} \right)$$

$$E_1 = \frac{1.69 \times 10^{-28} \text{ kg}}{9.104 \times 10^{-31} \text{ kg}}(-13.60 \text{ eV}) = 186(-13.60 \text{ eV}) = -2.53 \text{ keV}$$

c) From part (a), $E_n = \left(\dfrac{m_r}{m_H} \right) \left(-\dfrac{R_H c h}{n^2} \right)$, where $R_H = 1.097 \times 10^7$ m^{-1} is the Rydberg constant for the hydrogen atom.

$$\frac{hc}{\lambda} = E_i - E_f$$

The initial level for the transition is the $n_i = 2$ level and the final level is the $n_f = 1$ level.

$$\frac{hc}{\lambda} = \frac{m_r}{m_H} \left(-\frac{R_H c h}{n_i^2} - \left(-\frac{R_H c h}{n_f^2} \right) \right)$$

$$\frac{1}{\lambda} = \frac{m_r}{m_H} R_H \left(\frac{1}{n_f^2} - \frac{1}{n_i^2} \right)$$

$$\frac{1}{\lambda} = \frac{1.69 \times 10^{-28} \text{ kg}}{9.104 \times 10^{-31} \text{ kg}}(1.097 \times 10^7 \text{ m}^{-1}) \left(\frac{1}{1^2} - \frac{1}{2^2} \right) = 1.527 \times 10^9 \text{ m}^{-1}$$

$$\lambda = 0.655 \text{ nm}$$

Note: From Example 40-7 the wavelength of the radiation emitted in this transition in hydrogen is 122 nm. The wavelength for muonium is $\dfrac{m_H}{m_r} = 5.39 \times 10^{-3}$ times this.

40-51 **a)** $hf = E_f - E_i$, the energy given to the electron in the atom when a photon is absorbed.

The energy of one photon is $\dfrac{hc}{\lambda} = \dfrac{(6.626 \times 10^{-34} \text{ J} \cdot \text{s})(2.998 \times 10^8 \text{ m/s})}{85.5 \times 10^{-9} \text{ m}}$

$$\frac{hc}{\lambda} = 2.323 \times 10^{-18} \text{ J}(1 \text{ eV}/1.602 \times 10^{-19} \text{ J}) = 14.50 \text{ eV}.$$

The final energy of the electron is $E_f = E_i + hf$. In the ground state of the hydrogen atom the energy of the electron is $E_i = -13.60$ eV. Thus
$E_f = -13.60 \text{ eV} + 14.50 \text{ eV} = 0.90 \text{ eV}.$

b) At theremal equilibrium a few atoms will be in the $n = 2$ excited levels, which have an energy of $-13.6 \text{ eV}/4 = -3.40$ eV, 10.2 eV greater than the energy of the ground state.

40-53 a) Bohr's angular momentum result: $L = mvr = n(h/2\pi)$, so $v = \dfrac{nh}{2\pi mr}$.

$\sum \vec{F} = m\vec{a}$ gives $Dr = m(v^2/r)$ and $r = \sqrt{(m/D)}v$.

Use the Bohr equation to replace v: $r = \sqrt{\dfrac{m}{D}}\dfrac{nh}{2\pi mr}$.

$$r^2 = \frac{nh}{2\pi D^{1/2}m^{1/2}} \text{ and } r_n = \left(\frac{n^2h^2}{4\pi^2 mD}\right)^{1/4}$$

b) The force has the same dependence on position as $F = -kx$ for a one-dimensional simple harmonic oscillator. For $F = -kx$ the potential energy is $U = \frac{1}{2}kx^2$, so for $F = -Dr$ it is $U = \frac{1}{2}Dr^2$. (This does give $U = 0$ when $r = 0$.)
$E = K + U = \frac{1}{2}mv^2 + \frac{1}{2}Dr^2$

$v = \sqrt{\dfrac{D}{m}}r$, so $E = \frac{1}{2}m\left(\dfrac{D}{m}r^2\right) + \frac{1}{2}Dr^2 = Dr^2$

Use the result for r obtained in part (a):

$$E_n = D\frac{nh}{2\pi\sqrt{m}\sqrt{D}} = \sqrt{\frac{D}{m}}\frac{nh}{2\pi}$$

c) $\Delta E = E_i - E_f = (n_i - n_f)\sqrt{\dfrac{D}{m}}\dfrac{h}{2\pi}$

$n_i - n_f$ can be any positive integer, so can write the possible photon energies as

$n\sqrt{\dfrac{D}{m}}\dfrac{h}{2\pi}$, where $n = 1, 2, \ldots$

d) This could describe a charged mass attached to a spring and traveling in a circle. Such spring type forces describe the vibrational motion of atoms in molecules.

40-57 a) Let E_{tr} be the transition energy, E_{ph} be the energy of the photon with wavelength λ', and E_{r} be the kinetic energy of the recoiling atom.
$E_{\text{ph}} + E_{\text{r}} = E_{\text{tr}}$

$$E_{\text{ph}} = \frac{hc}{\lambda'} \text{ so } \frac{hc}{\lambda'} = E_{\text{tr}} - E_{\text{r}} \text{ and } \lambda' = \frac{hc}{E_{\text{tr}} - E_{\text{r}}}.$$

If the recoil energy is neglected then the photon wavelength is $\lambda = hc/E_{\text{tr}}$.

$$\Delta\lambda = \lambda' - \lambda = hc\left(\frac{1}{E_{\text{tr}} - E_{\text{r}}} - \frac{1}{E_{\text{tr}}}\right) = \left(\frac{hc}{E_{\text{tr}}}\right)\left(\frac{1}{1 - E_{\text{r}}/E_{\text{tr}}} - 1\right)$$

$$\frac{1}{1 - E_{\text{r}}/E_{\text{tr}}} = \left(1 - \frac{E_{\text{r}}}{E_{\text{tr}}}\right)^{-1} \approx 1 + \frac{E_{\text{r}}}{E_{\text{tr}}} \text{ since } \frac{E_{\text{r}}}{E_{\text{tr}}} << 1$$

(We have ued the binomial theorem, Appendix B.)

Thus $\Delta\lambda = \dfrac{hc}{E_{\text{tr}}}\left(\dfrac{E_{\text{r}}}{E_{\text{tr}}}\right)$, or since $E_{\text{tr}} = hc/\lambda$, $\Delta\lambda = \left(\dfrac{E_{\text{r}}}{hc}\right)\lambda^2$.

Use conservation of linear momentum to find E_{r}:

Assuming that the atom is initially at rest, the momentum p_{r} of the recoiling atom must be equal in magnitude and opposite in direction to the momentum $p_{\text{ph}} = h/\lambda$ of the emitted photon: $h/\lambda = p_{\text{r}}$.

$$E_{\text{r}} = \frac{p_{\text{r}}^2}{2m}, \text{ where } m \text{ is the mass of the atom, so } E_{\text{r}} = \frac{h^2}{2m\lambda^2}.$$

Use this result in the above equation:

$$\Delta\lambda = \left(\frac{E_{\text{r}}}{hc}\right)\lambda^2 = \left(\frac{h^2}{2m\lambda^2}\right)\left(\frac{\lambda^2}{hc}\right) = \frac{h}{2mc};$$

note that this result for $\Delta\lambda$ is independent of the atomic transition energy.

b) For a hydrogen atom $m = m_{\text{p}}$ and

$$\Delta\lambda = \frac{h}{2m_{\text{p}}c} = \frac{6.626 \times 10^{-34} \text{ J} \cdot \text{s}}{2(1.673 \times 10^{-27} \text{ kg})(2.998 \times 10^8 \text{ m/s})} = 6.61 \times 10^{-16} \text{ m}$$

The correction is independent of n.

40-59 **a)** $\Delta\lambda = \dfrac{h}{mc}(1 - \cos\phi) = \lambda_{\text{c}}(1 - \cos\phi)$

Largest $\Delta\lambda$ is for $\phi = 180°$: $\Delta\lambda = 2\lambda_{\text{c}} = 2(2.426 \text{ pm}) = 4.85 \text{ pm}$

b) $\lambda' - \lambda = \lambda_{\text{c}}(1 - \cos\phi)$

Wavelength doubles implies $\lambda' = 2\lambda$ so $\lambda' - \lambda = \lambda$. Thus $\lambda = \lambda_{\text{c}}(1 - \cos\phi)$.

$E = hc/\lambda$, so smallest energy photon means largest wavelength photon, so $\phi = 180°$ and $\lambda = 2\lambda_{\text{c}} = 4.85 \text{ pm}$.

Then $E = \dfrac{hc}{\lambda} = \dfrac{(6.626 \times 10^{-34} \text{ J} \cdot \text{s})(2.998 \times 10^8 \text{ m/s})}{4.85 \times 10^{-12} \text{ m}} =$

$4.096 \times 10^{-14} \text{ J}(1 \text{ eV}/1.602 \times 10^{-19} \text{ J}) = 0.256 \text{ MeV}$.

40-61 **a)** Conservation of energy applied to the collision gives $E_\lambda = E_{\lambda'} + E_{\text{e}}$, where E_e is the kinetic energy of the electron after the collision and E_λ and $E_{\lambda'}$ are the energies of the photon before and after the collision.

$$E_e = hc\left(\frac{1}{\lambda} - \frac{1}{\lambda'}\right) = hc\left(\frac{\lambda' - \lambda}{\lambda\lambda'}\right)$$

$$E_e = (6.626 \times 10^{-34} \text{ J} \cdot \text{s})(2.998 \times 10^8 \text{ m/s})\left(\frac{0.0032 \times 10^{-9} \text{ m}}{(0.1100 \times 10^{-9} \text{ m})(0.1132 \times 10^{-9} \text{ m})}\right)$$

$$E_e = 5.105 \times 10^{-17} \text{ J} = 319 \text{ eV}$$

$$E_e = \tfrac{1}{2}mv^2 \text{ so } v = \sqrt{\frac{2E_e}{m}} = \sqrt{\frac{2(5.105 \times 10^{-17} \text{ J})}{9.109 \times 10^{-31} \text{ kg}}} = 1.06 \times 10^7 \text{ m/s}$$

b) The wavelength λ of a photon with energy E_e is given by $E_e = hc/\lambda$ so

$$\lambda = \frac{hc}{E_e} = \frac{(6.626 \times 10^{-34} \text{ J} \cdot \text{s})(2.998 \times 10^8 \text{ m/s})}{5.105 \times 10^{-17} \text{ J}} = 3.89 \text{ nm}$$

40-63 a) The wavelength of a 1 MeV photon is

$$\lambda = \frac{hc}{E} = \frac{(4.136 \times 10^{-15} \text{ eV} \cdot \text{s})(2.998 \times 10^8 \text{ m/s})}{1 \times 10^6 \text{ eV}} = 1 \times 10^{-12} \text{ m}$$

The total change in wavelength therefore is

500×10^{-9} m $- 1 \times 10^{-12}$ m $= 500 \times 10^{-9}$ m.

If this shift is produced in 10^{26} Compton scattering events, the wavelength shift in each scattering event is $\Delta\lambda = \dfrac{500 \times 10^{-9} \text{ m}}{1 \times 10^{26}} = 5 \times 10^{-33}$ m.

b) Use this $\Delta\lambda$ in $\Delta\lambda = \dfrac{h}{mc}(1 - \cos\phi)$ and solve for ϕ. We anticipate that ϕ will be very small, since $\Delta\lambda$ is much less than h/mc, so can use $\cos\phi \approx 1 - \phi^2/2$.

$$\Delta\lambda = \frac{h}{mc}\left(1 - \left(1 - \phi^2/2\right)\right) = \frac{h}{2mc}\phi^2$$

$$\phi = \sqrt{\frac{2\,\Delta\lambda}{(h/mc)}} = \sqrt{\frac{2(5 \times 10^{-33} \text{ m})}{2.426 \times 10^{-12} \text{ m}}} = 6.4 \times 10^{-11} \text{ rad} = (4 \times 10^{-9})^\circ$$

ϕ in radians is much less than 1 so the approximation we used is valid.

c) The total time to travel from the core to the surface is $(10^6 \text{ y})(3.156 \times 10^7 \text{ s/y}) = 3.2 \times 10^{13}$ s. There are 10^{26} scatterings during this time, so the average time between scaterings is $t = \dfrac{3.2 \times 10^{13} \text{ s}}{10^{26}} = 3.2 \times 10^{-13}$ s.

The distance light travels in this time is

$d = ct = (3.0 \times 10^8 \text{ m/s})(3.2 \times 10^{-13} \text{ s}) = 0.1$ mm

CHAPTER 41
THE WAVE NATURE OF PARTICLES

Exercises 5, 7, 11, 13, 17, 21, 23, 25, 29
Problems 35, 37, 39, 43, 45, 49, 51, 53

Exercises

41-5 **a)** The de Broglie wavelength is $\lambda = \dfrac{h}{p} = \dfrac{h}{mv}$.

In the Bohr model, $mvr_n = n(h/2\pi)$, so $mv = nh/(2\pi r_n)$.

Then $\lambda = h\left(\dfrac{2\pi r_n}{nh}\right) = \dfrac{2\pi r_n}{n}$.

For $n = 1$, $\lambda = 2\pi r_1$ with $r_1 = a_0 = 0.529 \times 10^{-10}$ m, so

$\lambda = 2\pi(0.529 \times 10^{-10} \text{ m}) = 3.32 \times 10^{-10}$ m

$\lambda = 2\pi r_1$; the de Broglie wavelength equals the circumference of the orbit.

b) For $n = 4$, $\lambda = 2\pi r_4/4$.

$r_n = n^2 a_0$ so $r_4 = 16a_0$.

$\lambda = 2\pi(16a_0)/4 = 4(2\pi a_0) = 4(3.32 \times 10^{-10} \text{ m}) = 1.33 \times 10^{-9}$ m

$\lambda = 2\pi r_4/4$; the de Broglie wavelength is $\dfrac{1}{n} = \dfrac{1}{4}$ times the circumference of the orbit.

41-7 $\lambda = \dfrac{h}{p} = \dfrac{h}{mv} = \dfrac{6.626 \times 10^{-34} \text{ J} \cdot \text{s}}{(5.00 \times 10^{-3} \text{ kg})(340 \text{ m/s})} = 3.90 \times 10^{-34}$ m

This wavelength is extremely short; the bullet will not exhibit wavelike properties.

41-11 Since the alpha particles are scattered from the surface plane of the crystal
Eq.(41-3) applies:

$d \sin \theta = m\lambda$.

$m = 1$ implies $\sin \theta = \lambda/d$

Use the de Broglie relation to calculate the wavelength of the particles: $\lambda = h/p$.
$E = \frac{1}{2}mv^2 = p^2/2m$ so $p = \sqrt{2mE}$ and

$\lambda = \dfrac{h}{\sqrt{2mE}} = \dfrac{6.626 \times 10^{-34} \text{ J} \cdot \text{s}}{\sqrt{2(6.64 \times 10^{-27} \text{ kg})(840 \text{ eV})(1.602 \times 10^{-19} \text{ eV}/1 \text{ eV})}} =$

4.957×10^{-13} m

$$\sin\theta = \frac{\lambda}{d} = \frac{4.957 \times 10^{-13} \text{ m}}{0.0834 \times 10^{-9} \text{ m}} = 5.94 \times 10^{-3} \text{ and } \theta = 0.341°$$

41-13 a) $\Delta x \Delta p_x \geq h/2\pi$

$$\Delta p_x \geq \frac{h}{2\pi \Delta x} = \frac{6.626 \times 10^{-34} \text{ J} \cdot \text{s}}{2\pi(1.00 \times 10^{-6} \text{ m})} = 1.055 \times 10^{-28} \text{ kg} \cdot \text{m/s}$$

$$\Delta v_x = \frac{\Delta p_x}{m} = \frac{1.055 \times 10^{-28} \text{ kg} \cdot \text{m/s}}{1200 \text{ kg}} = 8.79 \times 10^{-32} \text{ m/s}$$

b) Even for this very small Δx the minimum Δv_x required by the Heisenberg uncertainty principle is very small. The uncertainty principle does not impose any practical limit on the simultaneous measurements of the positions and velocities of ordinary objects.

41-17 $\Delta x \Delta p_x \geq h/2\pi$

$$\Delta x \approx (0.215 \text{ nm})/2 = 0.1075 \times 10^{-9} \text{ m}$$

$$\Delta p_x \geq \frac{h}{2\pi \Delta x} = \frac{6.626 \times 10^{-34} \text{ J} \cdot \text{s}}{2\pi(0.1075 \times 10^{-9} \text{ m})} = 9.8 \times 10^{-25} \text{ kg} \cdot \text{m/s}$$

b) $K = \dfrac{p^2}{2m} = \dfrac{(9.8 \times 10^{-25} \text{ kg} \cdot \text{m/s})^2}{2(9.75 \times 10^{-26} \text{ kg})} = 4.9 \times 10^{-24} \text{ J} = 3.1 \times 10^{-5} \text{ eV}$

c) The number of atoms in 1.00 kg is $N = \dfrac{1.00 \text{ kg}}{9.75 \times 10^{-26} \text{ kg}} = 1.03 \times 10^{25}$.

$$K_{\text{tot}} = NK = (1.03 \times 10^{25})(4.9 \times 10^{-24} \text{ J}) = 50 \text{ J}$$

d) $K_{\text{tot}} = mgy$

$$y = \frac{K_{\text{tot}}}{mg} = \frac{50 \text{ J}}{(1.00 \text{ kg})(9.80 \text{ m/s}^2)} = 5.1 \text{ m}$$

e) The kinetic energy cannot become zero because then the x-component of the momentum and the uncertainty in the x-component of the momentum of the atoms in the crystal would be zero and this would violate the Heisneberg uncertainty principle.

41-21 Find the uncertainty ΔE in the particle's energy: $E = mc^2$ so $\Delta E = (\Delta m)c^2$.

$m = 4.50m_{\text{p}}$; $\Delta m = (0.145)m = 0.145(4.50m_{\text{p}}) = (0.6525)m_{\text{p}} = 0.6525(1.673 \times 10^{-27} \text{ kg}) = 1.092 \times 10^{-27} \text{ kg}$.

Then $\Delta E = (\Delta m)c^2 = (1.092 \times 10^{-27} \text{ kg})(2.998 \times 10^8 \text{ m/s})^2 = 9.815 \times 10^{-11} \text{ J}$.

Then use the energy uncertainty principle $\Delta E \Delta t \geq h/2\pi$ to estimate the lifetime:

$$\Delta t \approx \frac{h}{2\pi \Delta E} = \frac{6.626 \times 10^{-34} \text{ J} \cdot \text{s}}{2\pi(9.815 \times 10^{-11} \text{ J})} = 1.1 \times 10^{-24} \text{ s}$$

41-23 a) The relation between the kinetic energy K of the particle of charge q and the potential difference through which it has been accelerated is $K = |q|\,\Delta V$. And $K = p^2/2m$, so $p = \sqrt{2mK} = \sqrt{2m|q|\,\Delta V}$.

The de Broglie wavelength is $\lambda = \dfrac{h}{p} = \dfrac{h}{\sqrt{2m|q|\,\Delta V}} =$

$$\frac{6.626 \times 0^{-34} \text{ J} \cdot \text{s}}{\sqrt{2(9.109 \times 10^{-31} \text{ kg})(1.602 \times 10^{-19} \text{ C})(800 \text{ V})}} = 4.34 \times 10^{-11} \text{ m} = 43.4 \text{ pm}$$

b) The only change is the mass m of the particle:

$$\lambda = \frac{h}{\sqrt{2m|q|\,\Delta V}} = \frac{6.626 \times 10^{-34} \text{ J} \cdot \text{s}}{\sqrt{2(1.673 \times 10^{-27} \text{ kg})(1.602 \times 10^{-19} \text{ C})(800 \text{ V})}} =$$

$1.01 \times 10^{-12} \text{ m} = 1.01 \text{ pm}$

The proton wavelength is smaller than the electron wavelength by a factor of $\sqrt{m_e/m_p} = 1/\sqrt{1836}$.

41-25 a) $\psi(x) = A \sin kx$

The position probability density is given by $|\psi(x)|^2 = A^2 \sin^2 kx$.

The probability is highest where $\sin kx = 1$ so $kx = 2\pi x/\lambda = n\pi/2$, $n = 1, 3, 5, \ldots$.

$x = n\lambda/4$, $n = 1, 3, 5, \ldots$ so $x = \lambda/4,\ 3\lambda/4,\ 5\lambda/4, \ldots$

b) The probability of finding the particle is zero where $|\psi|^2 = 0$, which occurs where $\sin kx = 0$ and $kx = 2\pi x/\lambda = n\pi$, $n = 0,\ 1,\ 2, \ldots$

$x = n\lambda/2$, $n = 0, 1, 2, \ldots$ so $x = 0,\ \lambda/2,\ 3\lambda/2, \ldots$

41-29 a) At the point $(L/4,\ L/4,\ L/4)$ the wavefunction has the value $\psi = (2/L)^{3/2} \sin(\pi/4)^3$

$P = |\psi|^2\,\Delta V = (2/L)^3 \sin(\pi/4)^6 (0.01L)^3 = 2^3 (0.01)^3 (\sin \pi/4)^6 = 1.0 \times 10^{-6}$

b) At the point $(L/2,\ L/2,\ L/2)$ the wavefunction has the value

$\psi = (2/L)^{3/2} \sin(\pi/2)^3 = (2/L)^{3/2}$

$P = |\psi|^2\,\Delta V = (2/L)^3 (0.01L)^3 = 2^3 (0.01)^3 = 8.0 \times 10^{-6}$

Problems

41-35 a) The expression in Problem 41-34 says $\lambda = \dfrac{hc}{\sqrt{K(K + 2mc^2)}}$.

With $K = 3mc^2$ this becomes $\lambda = \dfrac{hc}{\sqrt{3mc^2(3mc^2 + 2mc^2)}} = \dfrac{h}{\sqrt{15}\,mc}$.

b) (i) $K = 3mc^2 = 3(9.109 \times 10^{-31} \text{ kg})(2.998 \times 10^8 \text{ m/s})^2 = 2.456 \times 10^{-13} \text{ J} =$

1.53 MeV

$$\lambda = \frac{h}{\sqrt{15}mc} = \frac{6.626 \times 10^{-34} \text{ J} \cdot \text{s}}{\sqrt{15}(9.109 \times 10^{-31} \text{ kg})(2.998 \times 10^8 \text{ m/s})} = 6.26 \times 10^{-13} \text{ m}$$

(ii) K is proportional to m, so for a proton

$K = (m_e/m_p)(1.53 \text{ MeV}) = 1836(1.53 \text{ MeV}) = 2810 \text{ MeV}$

λ is proportional to $1/m$, so for a proton

$\lambda = (m_e/m_p)(6.26 \times 10^{-13} \text{ m}) = (1/1836)(6.26 \times 10^{-13} \text{ m}) = 3.41 \times 10^{-16} \text{ m}$

41-37 $\Delta x = 0.40\lambda = 0.40(h/p)$

$\Delta x \Delta p_x \geq h/2\pi$. So $(\Delta p_x)_{min} = h/(2\pi \Delta x)$ (as in Example 41-3).

$$(\Delta p_x)_{min} = \frac{h}{2\pi(0.40 h/p)} = \frac{p}{2\pi(0.40)} = 0.40p$$

41-39 **a)** $\Delta x \Delta p_x \geq h/2\pi$

Estimate Δx as $\Delta x \approx 5.0 \times 10^{-15}$ m. Then the minimum allowed Δp_x is

$$\Delta p_x \approx \frac{h}{2\pi \Delta x} = \frac{6.626 \times 10^{-34} \text{ J} \cdot \text{s}}{2\pi(5.0 \times 10^{-15} \text{ m})} = 2.1 \times 10^{-20} \text{ kg} \cdot \text{m/s}$$

b) Assume $p \approx 2.1 \times 10^{-20}$ kg $\cdot$ m/s

Eq.(39-40): $E = \sqrt{(mc^2)^2 + (pc)^2}$

$mc^2 = (9.109 \times 10^{-31} \text{ kg})(2.998 \times 10^8 \text{ m/s})^2 = 8.187 \times 10^{-14}$ J

$pc = (2.1 \times 10^{-20} \text{ kg} \cdot \text{m/s})(2.998 \times 10^8 \text{ m/s}) = 6.296 \times 10^{-12}$ J

$E = \sqrt{(8.187 \times 10^{-14} \text{ J})^2 + (6.296 \times 10^{-12} \text{ J})^2} = 6.297 \times 10^{-12}$ J

$K = E - mc^2 = 6.297 \times 10^{-12} \text{ J} - 8.187 \times 10^{-14} \text{ J} =$
$6.215 \times 10^{-12} \text{ J}(1 \text{ eV}/1.602 \times 10^{-19} \text{ J}) = 39 \text{ MeV}$

c) The Coulomb potential energy is

$$U = -\frac{ke^2}{r} = -\frac{(8.988 \times 10^9 \text{ N} \cdot \text{m}^2/\text{C}^2)(1.602 \times 10^{-19} \text{ C})}{5.0 \times 10^{-15} \text{ m}} = -4.6 \times 10^{-14} \text{ J} =$$
-0.29 MeV

The kinetic energy of the electron required by the uncertainty principle would be much larger than the magnitude of the negative Coulomb potential energy. The total energy of the electron would be large and positive and the electron could not be bound within the nucleus.

41-43 **a)** $\lambda = \dfrac{h}{mv}$, so

$$v = \frac{h}{m\lambda} = \frac{6.626 \times 10^{-34} \text{ J} \cdot \text{s}}{(60.0 \text{ kg})(1.0 \text{ m})} = 1.1 \times 10^{-35} \text{ m/s}$$

b) $t = \dfrac{\text{distance}}{\text{velocity}} = \dfrac{0.80 \text{ m}}{1.1 \times 10^{-35} \text{ m/s}} = 7.3 \times 10^{34} \text{ s}(1 \text{ y}/3.156 \times 10^{7} \text{ s}) =$

2.3×10^{27} y

Since you walk through doorways much more quickly than this, you will not experience diffraction effects.

41-45 $h = 6.63 \times 10^{-22}$ J $\cdot$ s

$$\Delta E \Delta t \geq h/2\pi \text{ so } \Delta E \approx \frac{h}{2\pi \, \Delta t} = \frac{6.63 \times 10^{-22} \text{ J} \cdot \text{s}}{2\pi (2.24 \times 10^{-3} \text{ s})} =$$

4.71×10^{-20} J$(1 \text{ eV}/1.602 \times 10^{-19}$ J$) = 0.294$ eV

41-49 **a)** $U = A|x|$

$$F = -\frac{dU}{dx} \quad \text{(Eq.7-17)}$$

For $x > 0$, $|x| = x$ so $U = Ax$ and $F = -\dfrac{d(Ax)}{dx} = -A$

For $x < 0$, $|x| = -x$ so $U = -Ax$ and $F = -\dfrac{d(-Ax)}{dx} = +A$

We can write this result as $F = -A|x|/x$, valid for all x except for $x = 0$.

b) $E = K + U = \dfrac{p^2}{2m} + A|x|$

$px \approx h$, so $p \approx h/x$

Then $E \approx \dfrac{h^2}{2mx^2} + A|x|$.

For $x > 0$, $E = \dfrac{h^2}{2mx^2} + Ax$.

To find the value of x that gives minimum E set $\dfrac{dE}{dx} = 0$.

$$0 = \frac{-2h^2}{2mx^3} + A$$

$$x^3 = \frac{h^2}{mA} \text{ and } x = \left(\frac{h^2}{mA}\right)^{1/3}$$

With this x the minimum E is

$$E = \frac{h^2}{2m}\left(\frac{mA}{h^2}\right)^{2/3} + A\left(\frac{h^2}{mA}\right)^{1/3} = \frac{1}{2}h^{2/3}m^{-1/3}A^{2/3} + h^{2/3}m^{-1/3}A^{2/3}$$

$$E = \tfrac{3}{2}\left(\frac{h^2 A^2}{m}\right)^{1/3}$$

41-51 a) Let the y-direction be from the thrower to the catcher, and let the x-direction be horizontal and perpendicular to the y-direction.

A cube with volume $V = 125$ cm$^3 = 0.125 \times 10^{-3}$ m^3 has side length $l = V^{1/3} = (0.125 \times 10^{-3}$ m$^3)^{1/3} = 0.050$ m. Thus estimate Δx as $\Delta x \approx 0.050$ m.

$\Delta x \Delta p_x \geq h/2\pi$ then gives $\Delta p_x \approx \dfrac{h}{2\pi\,\Delta x} = \dfrac{0.0663 \text{ J} \cdot \text{s}}{2\pi(0.050 \text{ m})} = 0.21$ kg $\cdot$ m/s

(The value of h in this other universe has been used.)

b) The uncertainty in the ball's horizontal velocity is

$\Delta v_x = \dfrac{\Delta p_x}{m} = \dfrac{0.21 \text{ kg} \cdot \text{m/s}}{0.25 \text{ kg}} = 0.84$ m/s

The time it takes the ball to travel to the second student is $t = \dfrac{12 \text{ m}}{6.0 \text{ m/s}} = 2.0$ s.

The uncertainty in the x-coordinate of the ball when it reaches the second student that is introduced by Δv_x is $\Delta x = (\Delta v_x)t = (0.84 \text{ m/s})(2.0 \text{ s}) = 1.7$ m.

The ball could miss the second student by about 1.7 m.

41-53 a) The probability is $P = |\psi|^2\, dV$ with $dV = 4\pi r^2\, dr$

$|\psi|^2 = A^2 e^{-2\alpha r^2}$ so $P = 4\pi A^2 r^2 e^{-2\alpha r^2}\, dr$

b) P is maximum where $\dfrac{dP}{dr} = 0$

$\dfrac{d}{dr}\left(r^2 e^{-2\alpha r^2} \right) = 0$

$2r e^{-2\alpha r^2} - 4\alpha r^3 e^{-2\alpha r^2} = 0$ and this reduces to $2r - 4\alpha r^3 = 0$

$r = 0$ is a solution of the equation but corresponds to a minimum not a maximum. Seek r not equal to 0 so divide by r and get $2 - 4\alpha r^2 = 0$

This gives $r = \dfrac{1}{\sqrt{2\alpha}}$ (We took the positive square root since r must be positive.)

This is different from the value of r, $r = 0$, where $|\psi|^2$ is a maximum. At $r = 0$ $|\psi|^2$ has a maximum but the volume element $dV = 4\pi r^2 dr$ is zero here so P does not have a maximum at $r = 0$.

CHAPTER 42
QUANTUM MECHANICS

Exercises 1, 3, 5, 11, 15, 21, 23, 25, 27
Problems 29, 31, 33, 35, 43, 45, 49

Exercises

42-1 **a)** The energy levels for a particle in a box are given by $E_n = \dfrac{n^2 h^2}{8mL^2}$.

The lowest level is for $n = 1$, and $E_1 = \dfrac{(1)(6.626 \times 10^{-34} \text{ J} \cdot \text{s})^2}{8(0.20 \text{ kg})(1.5 \text{ m})^2} = 1.2 \times 10^{-67}$ J.

b) $E = \frac{1}{2}mv^2$ so $v = \sqrt{\dfrac{2E}{m}} = \sqrt{\dfrac{2(1.2 \times 10^{-67} \text{ J})}{0.20 \text{ kg}}} = 1.1 \times 10^{-33}$ m/s.

If the ball has this speed the time it would take it to travel from one side of the table to the other is $t = \dfrac{1.5 \text{ m}}{1.1 \times 10^{-33} \text{ m/s}} = 1.4 \times 10^{33}$ s.

c) $E_1 = \dfrac{h^2}{8mL^2}$, $E_2 = 4E_1$,

so $\Delta E = E_2 - E_1 = 3E_1 = 3(1.2 \times 10^{-67} \text{ J}) = 3.6 \times 10^{-67}$ J

d) No, quantum mechanical effects are not important for the game of billiards. The discrete, quantized nature of the energy levels is completely unobservable.

42-3 Eq.(42-5): $E_n = \dfrac{n^2 h^2}{8mL^2}$

Ground state energy is $E_1 = \dfrac{h^2}{8mL^2}$; first excited state energy is $E_2 = \dfrac{4h^2}{8mL^2}$.

The energy separation between these two levels is $\Delta E = E_2 - E_1 = \dfrac{3h^2}{8mL^2}$.

This gives $L = h\sqrt{\dfrac{3}{8m \, \Delta E}} =$

$L = 6.626 \times 10^{-34} \text{ J} \cdot \text{s} \sqrt{\dfrac{3}{8(9.109 \times 10^{-31} \text{ kg})(3.0 \text{ eV})(1.602 \times 10^{-19} \text{ J/1 eV})}} =$

6.1×10^{-10} m $= 0.61$ nm.

42-5 For the $n = 2$ first excited state the normalized wave function is given by Eq.(42-9):

$$\psi_2(x) = \sqrt{\dfrac{2}{L}} \sin\left(\dfrac{2\pi x}{L}\right).$$

$$|\psi_2(x)|^2\, dx = \frac{2}{L} \sin^2\left(\frac{2\pi x}{L}\right) dx$$

a) $|\psi_2|^2\, dx = 0$ implies $\sin\left(\dfrac{2\pi x}{L}\right) = 0$

$$\frac{2\pi x}{L} = m\pi, \quad m = 0,\ 1,\ 2,\ \ldots; \quad x = m(L/2)$$

For $m = 0$, $x = 0$; for $m = 1$, $x = L/2$; for $m = 2$, $x = L$

The probability of finding the particle is zero at $x = 0$, $L/2$, and L.

b) $|\psi_2|^2\, dx$ is maximum when $\sin\left(\dfrac{2\pi x}{L}\right) = \pm 1$

$$\frac{2\pi x}{L} = m(\pi/2), \quad m = 1,\ 3,\ 5,\ \ldots; \quad x = m(L/4)$$

For $m = 1$, $x = L/4$; for $m = 3$, $x = 3L/4$

The probability of finding the particle is largest at $x = L/4$ and $3L/4$.

c) The answers to part (a) corespond to the zeros of $|\psi|^2$ shown in Fig.42-4 and the answers to part (b) correspond to the two values of x where $|\psi|^2$ in the figure is maximum.

42-11 a) $\psi = A\cos kx$

Eq.(42-13): $\quad -\dfrac{h^2}{8\pi^2 m}\dfrac{d^2\psi}{dx^2} = E\psi$

$$\frac{d\psi}{dx} = A(-k\sin kx) = -Ak\sin kx$$

$$\frac{d^2\psi}{dx^2} = -Ak(k\cos kx) = -Ak^2\cos kx$$

Thus Eq.(42-13) requires $-\dfrac{h^2}{8\pi^2 m}(-Ak^2\cos kx) = E(A\cos kx)$.

This says $\dfrac{h^2 k^2}{8\pi^2 m} = E$; $\ k = \dfrac{\sqrt{2mE}}{(h/2\pi)} = \dfrac{\sqrt{2mE}}{\hbar}$

$\psi = A\cos kx$ is a solution to Eq.(42-13) if $k = \dfrac{\sqrt{2mE}}{\hbar}$.

b) The wave function for a particle in a box with rigid walls at $x = 0$ and $x = L$ must satisfy the boundry conditions $\psi = 0$ at $x = 0$ and $\psi = 0$ at $x = L$.

$\psi(0) = A\cos 0 = A$, since $\cos 0 = 1$.

Thus ψ is not 0 at $x = 0$ and this wave function isn't acceptable because it doesn't satisfy the required boundary condition.

42-15 $U_0 = 6E_\infty$, as in Fig.42-8, so $E_1 = 0.625E_\infty$ and $E_3 = 5.09E_\infty$ with

$E_\infty = \dfrac{\pi^2\hbar^2}{2mL^2}$.   In this problem the particle bound in the well is a proton, so $m = 1.673 \times 10^{-27}$ kg. Then

$$E_\infty = \frac{\pi^2\hbar^2}{2mL^2} = \frac{\pi^2(1.055 \times 10^{-34}\ \text{J} \cdot \text{s})^2}{2(1.673 \times 10^{-27}\ \text{kg})(4.0 \times 10^{-15}\ \text{m})^2} = 2.052 \times 10^{-12}\ \text{J}.$$

The transition energy is $\Delta E = E_3 - E_1 = (5.09 - 0.625)E_\infty = 4.465E_\infty$.
$\Delta E = 4.465(2.052 \times 10^{-12}\ \text{J}) = 9.162 \times 10^{-12}\ \text{J}$

The wavelength of the photon that is absorbed is related to the transition energy by $\Delta E = hc/\lambda$, so

$$\lambda = \frac{hc}{\Delta E} = \frac{(6.626 \times 10^{-34}\ \text{J} \cdot \text{s})(2.998 \times 10^8\ \text{m/s})}{9.162 \times 10^{-12}\ \text{J}} = 2.2 \times 10^{-14}\ \text{m} = 22\ \text{fm}.$$

42-21 The probability is $T = Ae^{-2\kappa L}$, with $A = 16\dfrac{E}{U_0}\left(1 - \dfrac{E}{U_0}\right)$

and $\kappa = \dfrac{\sqrt{2m(U_0 - E)}}{\hbar}$.

$E = 32$ eV, $U_0 = 41$ eV, $L = 0.25 \times 10^{-9}$ m

a) $A = 16\dfrac{E}{U_0}\left(1 - \dfrac{E}{U_0}\right) = 16\dfrac{32}{41}\left(1 - \dfrac{32}{41}\right) = 2.741.$

$\kappa = \dfrac{\sqrt{2m(U_0 - E)}}{\hbar}$

$\kappa = \dfrac{\sqrt{2(9.109 \times 10^{-31}\ \text{kg})(41\ \text{eV} - 32\ \text{eV})(1.602 \times 10^{-19}\ \text{J/eV})}}{1.055 \times 10^{-34}\ \text{J} \cdot \text{s}} =$

$1.536 \times 10^{10}\ \text{m}^{-1}$

$T = Ae^{-2\kappa L} = (2.741)e^{-2(1.536\times10^{10}\ \text{m}^{-1})(0.25\times10^{-9}\ \text{m})} = 2.741e^{-7.68} = 0.0013$

b) The only change is the mass m, which appears in κ.

$\kappa = \dfrac{\sqrt{2m(U_0 - E)}}{\hbar}$

$\kappa = \dfrac{\sqrt{2(1.673 \times 10^{-27}\ \text{kg})(41\ \text{eV} - 32\ \text{eV})(1.602 \times 10^{-19}\ \text{J/eV})}}{1.055 \times 10^{-34}\ \text{J} \cdot \text{s}} =$

$6.584 \times 10^{11}\ \text{m}^{-1}$

Then
$T = Ae^{-2\kappa L} = (2.741)e^{-2(6.584\times10^{11}\ \text{m}^{-1})(0.25\times10^{-9}\ \text{m})} = 2.741e^{-392.2} = 10^{-143}$

42-23 $\omega = \sqrt{\dfrac{k'}{m}} = \sqrt{\dfrac{110\ \text{N/m}}{0.250\ \text{kg}}} = 21.0\ \text{rad/s}$

The ground state energy is given by Eq.(42-29):

$E_0 = \frac{1}{2}\hbar\omega = \frac{1}{2}(1.055 \times 10^{-34} \text{ J} \cdot \text{s})(21.0 \text{ rad/s}) =$

$1.11 \times 10^{-33} \text{ J}(1 \text{ eV}/1.602 \times 10^{-19} \text{ J}) = 6.93 \times 10^{-15} \text{ eV}$

$E_n = (n + \frac{1}{2})\hbar\omega; \; E_{(n+1)} = (n + 1 + \frac{1}{2})\hbar\omega$

The energy separation between these adjacent levels is

$\Delta E = E_{n+1} - E_n = \hbar\omega = 2E_0 = 2(1.11 \times 10^{-33} \text{ J}) = 2.22 \times 10^{-33} \text{ J} =$

$1.39 \times 10^{-14} \text{ eV}$

These energies are extremely small; quantum effects are not important for this oscillator.

42-25 The photon wavelength λ is related to the transition energy by $\Delta E = hc/\lambda$.

Ground state energy is $E_0 = \frac{1}{2}\hbar\omega$; first excited state energy is $E_1 = \frac{3}{2}\hbar\omega$, so the

transition energy is $\Delta E = (\frac{3}{2} - \frac{1}{2})\hbar\omega = \hbar\omega = \dfrac{h}{2\pi}\sqrt{k'/m}$.

Equating these two expressions for ΔE gives $\dfrac{hc}{\lambda} = \dfrac{h}{2\pi}\sqrt{k'/m}$.

Thus $k' = m\left(\dfrac{2\pi c}{\lambda}\right)^2 = (9.4 \times 10^{-26} \text{ kg})\left(\dfrac{2\pi(2.998 \times 10^8 \text{ m/s})}{525 \times 10^{-6} \text{ m}}\right)^2 = 1.2 \text{ N/m}$

42-27 The total energy of a Newtonian oscillator is given by $E = \frac{1}{2}k'A^2$ where k' is the force constant and A is the amplitude of the oscillator. Set this equal to the energy $E = (n + \frac{1}{2})\hbar\omega$ of an excited level that has quantum number n, where

$\omega = \sqrt{\dfrac{k'}{m}}$, and solve for A:

$\frac{1}{2}k'A^2 = (n + \frac{1}{2})\hbar\omega$

$A = \sqrt{\dfrac{(2n + 1)\hbar\omega}{k'}}$

The total energy of the Newtonian oscillator can also be written as $E = \frac{1}{2}mv_{\text{max}}^2$. Set this equal to $E = (n + \frac{1}{2})\hbar\omega$ and solve for v_{max}:

$\frac{1}{2}mv_{\text{max}}^2 = (n + \frac{1}{2})\hbar\omega$

$v_{\text{max}} = \sqrt{\dfrac{(2n + 1)\hbar\omega}{m}}$

Thus the maximum linear momentum of the oscillator is

$p_{\text{max}} = mv_{\text{max}} = \sqrt{(2n + 1)\hbar m\omega}$.

Assume that A represents the uncertainty Δx in position and that p_{max} is the corresponding uncertainty Δp_x in momentum. Then the uncertainty product is

$$\Delta x \Delta p_x = \sqrt{\frac{(2n+1)\hbar\omega}{k'}} \sqrt{(2n+1)\hbar m\omega} = (2n+1)\hbar\omega\sqrt{\frac{m}{k'}} = (2n+1)\hbar\omega\left(\frac{1}{\omega}\right) =$$
$$(2n+1)\hbar.$$

For $n = 1$ this gives $\Delta x \Delta p_x = 3\hbar$, in agreement with the result derived in Section 42-6. The uncertainty product $\Delta x \Delta p_x$ increases with n.

Problems

42-29 a) The normalized wave function for the ground state is $\psi_1 = \sqrt{\frac{2}{L}} \sin\left(\frac{\pi x}{L}\right)$.

The probability P of the particle being between $x = L/4$ and $x = 3L/4$ is

$$P = \int_{L/4}^{3L/4} |\psi_1|^2 \, dx = \frac{2}{L} \int_{L/4}^{3L/4} \sin^2\left(\frac{\pi x}{L}\right) dx.$$

Let $y = \pi x/L$; $dx = (L/\pi)\, dy$ and the integration limits become $\pi/4$ and $3\pi/4$.

$$P = \frac{2}{L}\left(\frac{L}{\pi}\right) \int_{\pi/4}^{3\pi/4} \sin^2 y \, dy = \frac{2}{\pi}\left[\frac{1}{2}y - \frac{1}{4}\sin 2y\right]_{\pi/4}^{3\pi/4}$$

$$P = \frac{2}{\pi}\left[\frac{3\pi}{8} - \frac{\pi}{8} - \frac{1}{4}\sin\left(\frac{3\pi}{2}\right) + \frac{1}{4}\sin\left(\frac{\pi}{2}\right)\right]$$

$$P = \frac{2}{\pi}\left(\frac{\pi}{4} - \frac{1}{4}(-1) + \frac{1}{4}(1)\right) = \frac{1}{2} + \frac{1}{\pi} = 0.818.$$

(Note: The integral formula $\int \sin^2 y \, dy = \frac{1}{2}y - \frac{1}{4}\sin 2y$ was used.)

b) The normalized wave function for the first excited state is $\psi_2 = \sqrt{\frac{2}{L}} \sin\left(\frac{2\pi x}{L}\right)$

$$P = \int_{L/4}^{3L/4} |\psi_2|^2 \, dx = \frac{2}{L} \int_{L/4}^{3L/4} \sin^2\left(\frac{2\pi x}{L}\right) dx.$$

Let $y = 2\pi x/L$; $dx = (L/2\pi)\, dy$ and the integration limits become $\pi/2$ and $3\pi/2$.

$$P = \frac{2}{L}\left(\frac{L}{2\pi}\right) \int_{\pi/2}^{3\pi/2} \sin^2 y \, dy = \frac{1}{\pi}\left[\frac{1}{2}y - \frac{1}{4}\sin 2y\right]_{\pi/2}^{3\pi/2} = \frac{1}{\pi}\left(\frac{3\pi}{4} - \frac{\pi}{4}\right) = 0.500$$

c) These results are consistent with Fig.42-4b. That figure shows that $|\psi|^2$ is more concentrated near the center of the box for the ground state than for the first excited state; this is consistent with the answer to part (a) being larger than the answer to part (b). Also, this figure shows that for the first excited state half the area under the $|\psi|^2$ curve lies between $L/4$ and $3L/4$, consistent with our answer to part (b).

42-31 The normalized wave function for the $n = 2$ first excited level is

$$\psi_2 = \sqrt{\frac{2}{L}} \sin\left(\frac{2\pi x}{L}\right).$$

$P = |\psi(x)|^2 \, dx$ is the probability that the particle will be found in the interval x to $x + dx$.

a) $x = L/4$

$$\psi(x) = \sqrt{\frac{2}{L}} \sin\left(\left(\frac{2\pi}{L}\right)\left(\frac{L}{4}\right)\right) = \sqrt{\frac{2}{L}} \sin\left(\frac{\pi}{2}\right) = \sqrt{\frac{2}{L}}.$$

$P = (2/L)\, dx$

b) $x = L/2$

$$\psi(x) = \sqrt{\frac{2}{L}} \sin\left(\left(\frac{2\pi}{L}\right)\left(\frac{L}{2}\right)\right) = \sqrt{\frac{2}{L}} \sin(\pi) = 0$$

$P = 0$

c) $x = 3L/4$

$$\psi(x) = \sqrt{\frac{2}{L}} \sin\left(\left(\frac{2\pi}{L}\right)\left(\frac{3L}{4}\right)\right) = \sqrt{\frac{2}{L}} \sin\left(\frac{3\pi}{2}\right) = -\sqrt{\frac{2}{L}}.$$

$P = (2/L)\, dx$

42-33 a) The energy levels are given by Eq.(42-5): $E_n = \dfrac{n^2 h^2}{8mL^2}$

Ground level, $n = 1$, $E_1 = \dfrac{h^2}{8mL^2}$

First excited level, $n = 2$, $E_2 = \dfrac{4h^2}{8mL^2}$

The transition energy is $\Delta E = E_2 - E_1 = \dfrac{3h^2}{8mL^2}$.

The transition energy is related to the wavelength of the emitted photon by $\Delta E = hc/\lambda$. Combining these two equations for ΔE gives $\dfrac{hc}{\lambda} = \dfrac{3h^2}{8mL^2}$.

$$\lambda = \frac{8mcL^2}{3h} = \frac{8(9.109 \times 10^{-31}\ \text{kg})(2.998 \times 10^8\ \text{m/s})(4.18 \times 10^{-9}\ \text{m})^2}{3(6.626 \times 10^{-34}\ \text{J} \cdot \text{s})}$$

$\lambda = 1.92 \times 10^{-5}\ \text{m} = 19.2\ \mu\text{m}.$

b) Second excited level has $n = 3$ and $E_3 = \dfrac{9h^2}{8mL^2}$. The transition energy is

$$\Delta E = E_3 - E_2 = \frac{9h^2}{8mL^2} - \frac{4h^2}{8mL^2} = \frac{5h^2}{8mL^2}.$$

$\dfrac{hc}{\lambda} = \dfrac{5h^2}{8mL^2}$ so $\lambda = \dfrac{8mcL^2}{5h} = \dfrac{3}{5}(19.2\ \mu\text{m}) = 11.5\ \mu\text{m}.$

42-35 Eq.(42-9): $\psi_n = \sqrt{\dfrac{2}{L}}\sin\left(\dfrac{n\pi x}{L}\right)$ for a particle in a box with walls at $x = 0$ and $x = L$. (This is the wave function for $0 \le x \le L$. For $x < 0$ or $x > L$, $\psi_n = 0$.)

a) For $n = 1$, $\psi_1 = \sqrt{\dfrac{2}{L}}\sin\left(\dfrac{\pi x}{L}\right)$

$\dfrac{d\psi_1}{dx} = \sqrt{\dfrac{2}{L}}\dfrac{\pi}{L}\cos\left(\dfrac{\pi x}{L}\right)$

$\cos\theta = 1 - \theta^2/2 + \ldots$

So for x slightly greater than 0, $\cos\left(\dfrac{\pi x}{L}\right) = 1$ and

$\dfrac{d\psi_1}{dx} = \sqrt{\dfrac{2}{L}}\dfrac{\pi}{L} = \sqrt{\dfrac{2\pi^2}{L^3}}.$

b) For $n = 2$, $\psi_2 = \sqrt{\dfrac{2}{L}}\sin\left(\dfrac{2\pi x}{L}\right)$

$\dfrac{d\psi_2}{dx} = \sqrt{\dfrac{2}{L}}\dfrac{2\pi}{L}\cos\left(\dfrac{2\pi x}{L}\right)$

For x slightly greater than 0, $\cos\left(\dfrac{2\pi x}{L}\right) = 1$ and

$\dfrac{d\psi_2}{dx} = \sqrt{\dfrac{2}{L}}\dfrac{2\pi}{L} = \sqrt{\dfrac{8\pi^2}{L^3}}.$

c) For x very close to L, $\cos\left(\dfrac{\pi x}{L}\right) \approx \cos\pi = -1$ and, from part (a), $\dfrac{d\psi_1}{dx} = -\sqrt{\dfrac{2\pi^2}{L^3}}.$

d) For x very close to L, $\cos\left(\dfrac{2\pi x}{L}\right) \approx \cos 2\pi = 1$ and, from part (b), $\dfrac{d\psi_2}{dx} = \sqrt{\dfrac{8\pi^2}{L^3}}.$

e) The magnitude of the slope of the wavefunction near the walls is greater for the $n = 2$ level.

These results all agree with Fig.42-4a. For $n = 2$ the slope near both walls is positive, for $n = 1$ the slope is positive near $x = 0$ and negative near $x = L$, and the magnitude of the slope near the walls is larger for $n = 2$ than for $n = 1$.

42-43 **a)** $T = \left[1 + \dfrac{(U_0\sinh\kappa L)^2}{4E(U_0 - E)}\right]^{-1}$

$\sinh\kappa L = \dfrac{e^{\kappa L} - e^{-\kappa L}}{2}$

For $\kappa L \gg 1$, $\sinh \kappa L \to \dfrac{e^{\kappa L}}{2}$ and

$$T \to \left[1 + \frac{U_0^2 e^{2\kappa L}}{16E(U_0 - E)}\right]^{-1} = \frac{16E(U_0 - E)}{16E(U_0 - E) + U_0^2 e^{2\kappa L}}$$

For $\kappa L \gg 1$, $16E(U_0 - E) + U_0^2 e^{2\kappa L} \to U_0^2 e^{2\kappa L}$

$$T \to \frac{16E(U_0 - E)}{U_0^2 e^{2\kappa L}} = 16\left(\frac{E}{U_0}\right)\left(1 - \frac{E}{U_0}\right)e^{-2\kappa L}, \text{ which is Eq.(42-24)}.$$

b) $\kappa L = \dfrac{L\sqrt{2m(U_0 - E)}}{\hbar}$. So $\kappa L \gg 1$ when L is large (barrier is wide) or $U_0 - E$ is large. (E is small compared to U_0.)

c) $\kappa = \dfrac{\sqrt{2m(U_0 - E)}}{\hbar}$; κ becomes small as E approaches U_0.

For κ small, $\sinh \kappa L \to \kappa L$ and

$$T \to \left[1 + \frac{U_0^2 \kappa^2 L^2}{4E(U_0 - E)}\right]^{-1} = \left[1 + \frac{U_0^2 2m(U_0 - E)L^2}{\hbar^2 4E(U_0 - E)}\right]^{-1} \quad \text{(using the definiton of } \kappa\text{)}$$

Thus $T \to \left[1 + \dfrac{2U_0^2 L^2 m}{4E\hbar^2}\right]^{-1}$

$U_0 \to E$ so $\dfrac{U_0^2}{E} \to E$ and $T \to \left[1 + \dfrac{2EL^2 m}{4\hbar^2}\right]^{-1}$

But $k^2 = \dfrac{2mE}{\hbar^2}$, so $T \to \left[1 + \left(\dfrac{kL}{2}\right)^2\right]^{-1}$, as was to be shown.

42-45 Calculate the angular frequency ω of the pendulum:

$$\omega = \frac{2\pi}{T} = \frac{2\pi}{0.500 \text{ s}} = 4\pi \text{ s}^{-1}$$

The ground-state energy is

$E_0 = \frac{1}{2}\hbar\omega = \frac{1}{2}(1.055 \times 10^{-34} \text{ J} \cdot \text{s})(4\pi \text{ s}^{-1}) = 6.63 \times 10^{-34} \text{ J}.$

$E_0 = 6.63 \times 10^{-34} \text{ J}(1 \text{ eV}/1.602 \times 10^{-19} \text{ J}) = 4.14 \times 10^{-15} \text{ eV}$

$E_n = (n + \frac{1}{2})\hbar\omega$

$E_{n+1} = (n + 1 + \frac{1}{2})\hbar\omega$

The energy difference between the adjacent energy levels is

$\Delta E = E_{n+1} - E_n = \hbar\omega = 2E_0 = 1.33 \times 10^{-33} \text{ J} = 8.30 \times 10^{-15} \text{ eV}$

These energies are much too small to detect.

42-49 a) Eq.(42-32): $-\dfrac{\hbar^2}{2m}\left(\dfrac{\partial^2\psi}{\partial x^2}+\dfrac{\partial^2\psi}{\partial y^2}+\dfrac{\partial^2\psi}{\partial z^2}\right)+U\psi = E\psi$

ψ_{n_x}, ψ_{n_y}, ψ_{n_z} are each solutions of Eq.(42-25), so

$$-\frac{\hbar^2}{2m}\frac{d^2\psi_{n_x}}{dx^2}+\frac{1}{2}k'x^2\psi_{n_x}=E_{n_x}\psi_{n_x}$$

$$-\frac{\hbar^2}{2m}\frac{d^2\psi_{n_y}}{dy^2}+\frac{1}{2}k'y^2\psi_{n_y}=E_{n_y}\psi_{n_y}$$

$$-\frac{\hbar^2}{2m}\frac{d^2\psi_{n_z}}{dz^2}+\frac{1}{2}k'z^2\psi_{n_z}=E_{n_z}\psi_{n_z}$$

$\psi = \psi_{n_x}(x)\psi_{n_y}(y)\psi_{n_z}(z)$, $U = \frac{1}{2}k'x^2+\frac{1}{2}k'y^2+\frac{1}{2}k'z^2$

$$\frac{\partial^2\psi}{\partial x^2}=\left(\frac{d^2\psi_{n_x}}{dx^2}\right)\psi_{n_y}\psi_{n_z},\quad \frac{\partial^2\psi}{\partial y^2}=\left(\frac{d^2\psi_{n_y}}{dy^2}\right)\psi_{n_x}\psi_{n_z},\quad \frac{\partial^2\psi}{\partial z^2}=\left(\frac{d^2\psi_{n_z}}{dz^2}\right)\psi_{n_x}\psi_{n_y}.$$

So $-\dfrac{\hbar^2}{2m}\left(\dfrac{\partial^2\psi}{\partial x^2}+\dfrac{\partial^2\psi}{\partial y^2}+\dfrac{\partial^2\psi}{\partial z^2}\right)+U\psi = \left(-\dfrac{\hbar^2}{2m}\dfrac{d^2\psi_{n_x}}{dx^2}+\dfrac{1}{2}k'x^2\psi_{n_x}\right)\psi_{n_y}\psi_{n_z}$

$$+\left(-\frac{\hbar^2}{2m}\frac{d^2\psi_{n_y}}{dy^2}+\frac{1}{2}k'y^2\psi_{n_y}\right)\psi_{n_x}\psi_{n_z}+\left(-\frac{\hbar^2}{2m}\frac{d^2\psi_{n_z}}{dz^2}+\frac{1}{2}k'z^2\psi_{n_z}\right)\psi_{n_x}\psi_{n_y}$$

$$-\frac{\hbar^2}{2m}\left(\frac{\partial^2\psi}{\partial x^2}+\frac{\partial^2\psi}{\partial y^2}+\frac{\partial^2\psi}{\partial z^2}\right)+U\psi = \left(E_{n_x}+E_{n_y}+E_{n_z}\right)\psi$$

Therefore, we have shown that this ψ is a solution to Eq.(42-32), with energy
$E_{n_x n_y n_z}=E_{n_x}+E_{n_y}+E_{n_z}=(n_x+n_y+n_z+\frac{3}{2})\hbar\omega$

b) and **c)** The ground state has $n_x=n_y=n_z=0$, so the energy is $E_{000}=\frac{3}{2}\hbar\omega$.
There is only one set of n_x, n_y and n_z that give this energy.

First-excited state: $n_x=1$, $n_y=n_z=0$ or $n_y=1$, $n_x=n_z=0$ or $n_z=1$,
$n_x=n_y=0$
$E_{100}=E_{010}=E_{001}=\frac{5}{2}\hbar\omega$

There are three different sets of n_x, n_y, n_z quantum numbers that give this energy,
so there are three different quantum states that have this same energy.

CHAPTER 43
ATOMIC STRUCTURE

Exercises 1, 3, 9, 13, 17, 19, 21, 23, 27, 29
Problems 33, 35, 37, 41, 43, 45

Exercises

43-1 Eq.(43-4): $L = \sqrt{l(l+1)}\hbar$, $l = 0, 1, 2, \ldots$

$$l(l+1) = \left(\frac{L}{\hbar}\right)^2 = \left(\frac{4.716 \times 10^{-34} \text{ kg} \cdot \text{m}^2/\text{s}}{1.055 \times 10^{-34} \text{ J} \cdot \text{s}}\right)^2 = 20$$

And then $l(l+1) = 20$ gives that $l = 4$.

43-3 $L = \sqrt{l(l+1)}\hbar$

The maximum l, $l_{\max}$, for a given n is $l_{\max} = n - 1$.

For $n = 2$, $l_{\max} = 1$ and $L = \sqrt{2}\hbar = 1.414\hbar$.

For $n = 20$, $l_{\max} = 19$ and $L = \sqrt{(19)(20)}\hbar = 19.49\hbar$.

For $n = 200$, $l_{\max} = 199$ and $L = \sqrt{(199)(200)}\hbar = 199.5\hbar$.

As n increases, the maximum L gets closer to the value $n\hbar$ postulated in the Bohr model.

43-9 Eq.(43-8): $a = \dfrac{4\pi\epsilon_0\hbar^2}{m_r e^2} = \dfrac{\epsilon_0 h^2}{\pi m_r e^2}$

a) $m_r = m$

$$a = \frac{\epsilon_0 h^2}{\pi m_r e^2} = \frac{(8.854 \times 10^{-12} \text{ C}^2/\text{N} \cdot \text{m}^2)(6.626 \times 10^{-34} \text{ J} \cdot \text{s})^2}{\pi(9.109 \times 10^{-31} \text{ kg})(1.602 \times 10^{-19} \text{ C})^2} = 0.5293 \times 10^{-10} \text{ m}$$

b) $m_r = m/2$

$$a = 2\left(\frac{\epsilon_0 h^2}{\pi m_r e^2}\right) = 1.059 \times 10^{-11} \text{ m}$$

c) $m_r = 185.8m$

$$a = \frac{1}{185.8}\left(\frac{\epsilon_0 h^2}{\pi m_r e^2}\right) = 2.849 \times 10^{-13} \text{ m}$$

43-13 a) For the $5g$ level, $l = 4$ and there are $2l + 1 = 9$ different m_l states. The $5g$ level is split into 9 levels by the magnetic field.

b) Each m_l level is shifted in energy an amount given by Eq.(43-18): $U = m_l \mu_{\rm B} B$. Adjacent levels differ in m_l by one, so $\Delta U = \mu_{\rm B} B$.

$$\mu_{\rm B} = \frac{e\hbar}{2m} = \frac{(1.602 \times 10^{-19} \text{ C})(1.055 \times 10^{-34} \text{ J} \cdot \text{s})}{2(9.109 \times 10^{-31} \text{ kg})} = 9.277 \times 10^{-24} \text{ A} \cdot \text{m}^2$$

$\Delta U = \mu_{\rm B} B = (9.277 \times 10^{-24} \text{ A/m}^2)(0.600 \text{ T}) =$
$5.566 \times 10^{-24} \text{ J}(1 \text{ eV}/1.602 \times 10^{-19} \text{ J}) = 3.47 \times 10^{-5} \text{ eV}$

c) The level of highest energy is for the largest m_l, which is $m_l = l = 4$; $U_4 = 4\mu_{\rm B} B$. The level of lowest energy is for the smallest m_l, which is $m_l = -l = -4$; $U_{-4} = -4\mu_{\rm B} B$.

The separation between these two levels is $U_4 - U_{-4} = 8\mu_{\rm B} B = 8(3.47 \times 10^{-5} \text{ eV}) = 2.78 \times 10^{-4} \text{ eV}$.

43-17 $U = -\vec{\mu} \cdot \vec{B} = +\mu_z B$, since the magnetic field is in the negative z-direction.

$\mu_z = -(2.00232)\left(\dfrac{e}{2m}\right) S_z$, so $U = -(2.00232)\left(\dfrac{e}{2m}\right) S_z B$

$S_z = m_s \hbar$, so $U = -2.00232 \left(\dfrac{e\hbar}{2m}\right) m_s B$

$\dfrac{e\hbar}{2m} = \mu_B = 5.788 \times 10^{-5} \text{ eV/T}$
$U = -2.00232 \mu_B m_s B$

The $m_s = +\frac{1}{2}$ level has lower energy.
$\Delta U = U(m_s = -\frac{1}{2}) - U(m_s = +\frac{1}{2}) = -2.00232\mu_B B(-\frac{1}{2} - (+\frac{1}{2})) = +2.00232\mu_B B$
$\Delta U = +2.00232(5.788 \times 10^{-5} \text{ eV/T})(1.45 \text{ T}) = 1.68 \times 10^{-4} \text{ eV}$

43-19 j can have the values $l + 1/2$ and $l - 1/2$. Thus if j takes the values $7/2$ and $9/2$ it must be that $l - 1/2 = 7/2$ and $l = 8/2 = 4$. The letter that labels this l is g.

43-21 **a)** For a classical particle $L = I\omega$.

For a uniform sphere with mass m and radius R, $I = \dfrac{2}{5}mR^2$, so $L = \left(\dfrac{2}{5}mR^2\right)\omega$.

$L = \sqrt{\dfrac{3}{4}}\hbar$ so $\dfrac{2}{5}mR^2\omega = \sqrt{\dfrac{3}{4}}\hbar$

$\omega = \dfrac{5\sqrt{3/4}\hbar}{2mR^2} = \dfrac{5\sqrt{3/4}(1.055 \times 10^{-34} \text{ J} \cdot \text{s})}{2(9.109 \times 10^{-31} \text{ kg})(1.0 \times 10^{-17} \text{ m})^2} = 2.5 \times 10^{30} \text{ rad/s}$

b) $v = r\omega = (1.0 \times 10^{-17} \text{ m})(2.5 \times 10^{30} \text{ rad/s}) = 2.5 \times 10^{13} \text{ m/s}$. This is much greater than the speed of light c, so the model cannot be valid.

43-23 Eq.(43-27): $E_n = -\left(\dfrac{Z_{\text{eff}}^2}{n^2}\right)(13.6 \text{ eV})$

$n = 5$ and $Z_{\text{eff}} = 2.771$ gives $E_5 = -\dfrac{(2.771)^2}{5^2}(13.6 \text{ eV}) = -4.18 \text{ eV}$

The ionization energy is 4.18 eV.

43-27 **a)** $Z = 7$ for nitrogen so a nitrogen atom has 7 electrons.
N^{2+} has 5 electrons: $1s^2 2s^2 2p$.

b) $Z_{\text{eff}} = 7 - 4 = 3$ for the $2p$ level.

$E_n = -\left(\dfrac{Z_{\text{eff}}^2}{n^2}\right)(13.6 \text{ eV}) = -\dfrac{3^2}{2^2}(13.6 \text{ eV}) = -30.6 \text{ eV}$

c) $Z = 15$ for phosphorus so a phosphorus atom has 15 electrons.
P^{2+} has 13 electrons: $1s^2 2s^2 2p^6 3s^2 3p$

d) $Z_{\text{eff}} = 15 - 12 = 3$ for the $3p$ level.

$E_n = -\left(\dfrac{Z_{\text{eff}}^2}{n^2}\right)(13.6 \text{ eV}) = -\dfrac{3^2}{3^2}(13.6 \text{ eV}) = -13.6 \text{ eV}$

43-29 Eq.(43-27): $E_n = -\left(\dfrac{Z_{\text{eff}}^2}{n^2}\right)(13.6 \text{ eV})$

a) The element Be has nuclear charge $Z = 4$. The ion Be^+ has 3 electrons. The outermost electron sees the nuclear charge screened by the other two electrons so $Z_{\text{eff}} = 4 - 2 = 2$.

$E_2 = -\dfrac{2^2}{2^2}(13.6 \text{ eV}) = -13.6 \text{ eV}$

b) The outermost electron in Ca^+ sees a $Z_{\text{eff}} = 2$.

$E_4 = -\dfrac{2^2}{4^2}(13.6 \text{ eV}) = -3.4 \text{ eV}$

Problems

43-33 **a)** $E_{1s} = -\dfrac{1}{(4\pi\epsilon_0)^2}\dfrac{me^4}{2\hbar^2}$; $U(r) = -\dfrac{1}{4\pi\epsilon_0}\dfrac{e^2}{r}$

$E_{1s} = U(r)$ gives $-\dfrac{1}{(4\pi\epsilon_0)^2}\dfrac{me^4}{2\hbar^2} = -\dfrac{1}{4\pi\epsilon_0}\dfrac{e^2}{r}$

$r = \dfrac{(4\pi\epsilon_0)2\hbar^2}{me^2} = 2a$

b) For the $1s$ state the probability that the electron is in the classically forbidden region is $P(r > 2a) = \int_{2a}^{\infty} |\psi_{1s}|^2 \, dV = 4\pi \int_{2a}^{\infty} |\psi_{1s}|^2 r^2 \, dr$.

The normalized wave function of the $1s$ state of hydrogen is given in Example 43-3:

$$\psi_{1s}(r) = \frac{1}{\sqrt{\pi a^3}} e^{-r/a}$$

$$P(r > 2a) = 4\pi \left(\frac{1}{\pi a^3} \right) \int_{2a}^{\infty} r^2 e^{-2r/a} \, dr$$

Use the integral fomula $\int r^2 e^{-\alpha r} \, dr = -e^{-\alpha r} \left(\dfrac{r^2}{\alpha} + \dfrac{2r}{\alpha^2} + \dfrac{2}{\alpha^3} \right)$, with $\alpha = 2/a$.

$$P(r > 2a) = -\frac{4}{a^3} \left[e^{-2r/a} \left(\frac{ar^2}{2} + \frac{a^2 r}{2} + \frac{a^3}{4} \right) \right]_{2a}^{\infty} = +\frac{4}{a^3} e^{-4}(2a^3 + a^3 + a^3/4)$$

$$P(r > 2a) = 4e^{-4}(13/4) = 13e^{-4} = 0.238.$$

43-35 $\psi_{2s}(r) = \dfrac{1}{\sqrt{32\pi a^3}} \left(2 - \dfrac{r}{a} \right) e^{-r/2a}$

a) Let $I = \displaystyle\int_0^{\infty} |\psi_{2s}|^2 \, dV = 4\pi \int_0^{\infty} |\psi_{2s}|^2 r^2 \, dr$. If ψ_{2s} is normalized then we will find that $I = 1$.

$$I = 4\pi \left(\frac{1}{32\pi a^3} \right) \int_0^{\infty} \left(2 - \frac{r}{a} \right)^2 e^{-r/a} r^2 \, dr = \frac{1}{8a^3} \int_0^{\infty} \left(4r^2 - \frac{4r^3}{a} + \frac{r^4}{a^2} \right) e^{-r/a} \, dr$$

Use the integral formula $\int_0^{\infty} x^n e^{-\alpha x} \, dx = \dfrac{n!}{\alpha^{n+1}}$, with $\alpha = 1/a$

$$I = \frac{1}{8a^3} \left(4(2!)(a^3) - \frac{4}{a}(3!)(a)^4 + \frac{1}{a^2}(4!)(a)^5 \right) = \frac{1}{8}(8 - 24 + 24) = 1; \text{ this } \psi_{2s} \text{ is}$$

normalized.

b) $P(r < 4a) = \displaystyle\int_0^{4a} |\psi_{2s}|^2 \, dV = 4\pi \int_0^{4a} |\psi_{2s}|^2 r^2 \, dr =$

$$\frac{1}{8a^3} \int_0^{4a} \left(4r^2 - \frac{4r^3}{a} + \frac{r^4}{a^2} \right) e^{-r/a} \, dr$$

Let $P(r < 4a) = \dfrac{1}{8a^3}(I_1 + I_2 + I_3)$.

$$I_1 = 4 \int_0^{4a} r^2 e^{-r/a} \, dr$$

Use the integral formula $\int r^2 e^{-\alpha r} \, dr = -e^{-\alpha r} \left(\dfrac{r^2}{\alpha} + \dfrac{2r}{\alpha^2} + \dfrac{2}{\alpha^3} \right)$ with $\alpha = 1/a$.

$$I_1 = -4 \left[e^{-r/a}(r^2 a + 2ra^2 + 2a^3) \right]_0^{4a} = (-104e^{-4} + 8)a^3.$$

$$I_2 = -\frac{4}{a} \int_0^{4a} r^3 e^{-r/a} \, dr$$

Use the integral formula

$$\int r^3 e^{-\alpha r}\, dr = -e^{-\alpha r}\left(\frac{r^3}{\alpha} + \frac{3r^2}{\alpha^2} + \frac{6r}{\alpha^3} + \frac{6}{\alpha^4}\right) \text{ with } \alpha = 1/a.$$

$$I_2 = \frac{4}{a}\left[e^{-r/a}(r^3 a + 3r^2 a^2 + 6ra^3 + 6a^4)\right]_0^{4a} = (568e^{-4} - 24)a^3.$$

$$I_3 = \frac{1}{a^2}\int_0^{4a} r^4 e^{-r/a}\, dr$$

Use the integral formula $\int r^4 e^{-\alpha r}\, dr = -e^{-\alpha r}\left(\frac{r^4}{\alpha} + \frac{4r^3}{\alpha^2} + \frac{12r^2}{\alpha^3} + \frac{24r}{\alpha^4} + \frac{24}{\alpha^5}\right)$

with $\alpha = 1/a.$

$$I_3 = -\frac{1}{a^2}\left[e^{-r/a}(r^4 a + 4r^3 a^2 + 12r^2 a^3 + 24ra^4 + 24a^5)\right]_0^{4a} = (-824e^{-4} + 24)a^3.$$

Thus $P(r < 4a) = \dfrac{1}{8a^3}(I_1 + I_2 + I_3) = \dfrac{1}{8a^3}a^3([8 - 24 + 24] + e^{-4}[-104 + 568 - 824])$

$$P(r < 4a) = \frac{1}{8}(8 - 360e^{-4}) = 1 - 45e^{-4} = 0.176.$$

43-37 a) $\cos\theta_L = \dfrac{L_z}{L}$ so $\theta_L = \arccos\left(\dfrac{L_z}{L}\right)$

The smallest angle $(\theta_L)_{min}$ is for the state with the largest L and the largest L_z. This is the state with $l = n - 1$ and $m_l = l = n - 1$.

$L_z = m_l \hbar = (n - 1)\hbar$

$L = \sqrt{l(l+1)}\hbar = \sqrt{(n-1)n}\hbar$

$(\theta_L)_{min} = \arccos\left(\dfrac{(n-1)\hbar}{\sqrt{(n-1)n}\hbar}\right) = \arccos\left(\dfrac{(n-1)}{\sqrt{(n-1)n}}\right) = \arccos\left(\sqrt{\dfrac{n-1}{n}}\right) =$

$\arccos(\sqrt{1 - 1/n}).$

Note that $(\theta_L)_{min}$ approaches $0°$ as $n \to \infty.$

b) The largest angle $(\theta_L)_{max}$ is for $l = n - 1$ and $m_l = -l = -(n - 1)$. A similar calculation to part (a) yields $(\theta_L)_{max} = \arccos(-\sqrt{1 - 1/n})$

Note that $(\theta_L)_{max}$ approaches $180°$ as $n \to \infty.$

43-41 Eq.(40-19): $\dfrac{n_1}{n_0} = e^{-(E_1 - E_0)/kT}$, where 1 is the higher energy state and 0

is the lower energy state.

The interaction energy with the magnetic field is

$$U = -\mu_z B = 2.00232\left(\frac{e\hbar}{2m}\right)m_s B \text{ (Example 43-6).}$$

The energy of the $m_s = +\frac{1}{2}$ level is increased and the energy of the $m_s = -\frac{1}{2}$ level is decreased.

$$\frac{n_{1/2}}{n_{-1/2}} = e^{-(U_{1/2} - U_{-1/2})/kT}$$

$$U_{1/2} - U_{-1/2} = 2.00232 \left(\frac{e\hbar}{2m}\right) B(\tfrac{1}{2} - (-\tfrac{1}{2})) = 2.00232 \left(\frac{e\hbar}{2m}\right) B = 2.00232 \mu_{\text{B}} B$$

$$\frac{n_{1/2}}{n_{-1/2}} = e^{-(2.00232)\mu_{\text{B}} B/kT}$$

a) $B = 5.00 \times 10^{-5} T$

$$\frac{n_{1/2}}{n_{-1/2}} = e^{-2.00232(9.274 \times 10^{-24} \text{ A/m}^2)(5.00 \times 10^{-5} \text{ T})/([1.381 \times 10^{-23} \text{ J/K}][300K])}$$

$$\frac{n_{1/2}}{n_{-1/2}} = e^{-2.24 \times 10^{-7}} = 0.99999978 = 1 - 2.2 \times 10^{-7}$$

b) $B = 0.500$ T, $\dfrac{n_{1/2}}{n_{-1/2}} = e^{-2.24 \times 10^{-3}} = 0.9978$

c) $B = 5.00$ T, $\dfrac{n_{1/2}}{n_{-1/2}} = e^{-2.24 \times 10^{-2}} = 0.978$

43-43 **a)** The energy of the photon equals the transition energy of the atom: $\Delta E = hc/\lambda$

$$E_n = -\frac{13.60 \text{ eV}}{n^2} \text{ so } E_2 = -\frac{13.60 \text{ eV}}{4} \text{ and } E_1 = -\frac{13.60 \text{ eV}}{1}$$

$\Delta E = E_2 - E_1 = 13.60 \text{ eV}(-\tfrac{1}{4} + 1) = \tfrac{3}{4}(13.60 \text{ eV}) = 10.20 \text{ eV} =$
$(10.20 \text{ eV})(1.602 \times 10^{-19} \text{ J/eV}) = 1.634 \times 10^{-18} \text{ J}$

$$\lambda = \frac{hc}{\Delta E} = \frac{(6.626 \times 10^{-34} \text{ J} \cdot \text{s})(2.998 \times 10^8 \text{ m/s})}{1.634 \times 10^{-18} \text{ J}} = 1.22 \times 10^{-7} \text{ m} = 122 \text{ nm}$$

b) The shift of a level due to the energy of interaction with the magnetic field in the z-direction is $U = m_l \mu_{\text{B}} B$.

The ground state has $m_l = 0$ so is unaffected by the magnetic field.

The $n = 2$ initial state has $m_l = -1$ so its energy is shifted downward an amount
$U = m_l \mu_{\text{B}} B = (-1)(9.274 \times 10^{-24} \text{ A/m}^2)(2.20 \text{ T}) =$
$(-2.040 \times 10^{-23} \text{ J})(1 \text{ eV}/1.602 \times 10^{-19} \text{ J}) = 1.273 \times 10^{-4} \text{ eV}$

Note that the shift in energy due to the magnetic field is a very small fraction of the 10.2 eV transition energy. Problem 41-44(c) shows that in this situation $|\Delta\lambda/\lambda| = |\Delta E/E|$.

This gives $|\Delta\lambda| = \lambda|\Delta E/E| = 122 \text{ nm} \left(\dfrac{1.273 \times 10^{-4} \text{ eV}}{10.2 \text{ eV}}\right) = 1.52 \times 10^{-3} \text{ nm} = 1.52 \text{ pm}$.

The upper level in the transition is lowered in energy so the transition energy is decreased. A smaller ΔE means a larger λ; the magnetic field increases the wavelength.

43-45 **a)** vanadium, $Z = 23$

minimum wavelength; corresponds to largest transition energy

The highest occupied shell is the N shell ($n = 4$). The highest energy transition is $N \rightarrow K$, with transition energy $\Delta E = E_N - E_K$. Since the shell energies scale like $1/n^2$ neglect E_N relative to E_K, so $\Delta E = E_K = (Z-1)^2(13.6 \text{ eV}) = (23-1)^2(13.6 \text{ eV}) = 6.582 \times 10^3 \text{ eV} = 1.055 \times 10^{-15} \text{ J}$.

The energy of the emitted photon equals this transition energy, so the photon's wavelength is given by $\Delta E = hc/\lambda$ so $\lambda = hc/\Delta E$.

$$\lambda = \frac{(6.626 \times 10^{-34} \text{ J} \cdot \text{s})(2.998 \times 10^8 \text{ m/s})}{1.055 \times 10^{-15} \text{ J}} = 1.88 \times 10^{-10} \text{m} = 0.188 \text{ nm}.$$

maximum wavelength; coresponds to smallest transition energy, so for the K_α transition

The frequency of the photon emitted in this transition is given by Moseley's law (Eq.43-29):

$$f = (2.48 \times 10^{15} \text{ Hz})(Z-1)^2 = (2.48 \times 10^{15} \text{ Hz})(23-1)^2 = 1.200 \times 10^{18} \text{ Hz}$$

$$\lambda = \frac{c}{f} = \frac{2.998 \times 10^8 \text{ m/s}}{1.200 \times 10^{18} \text{ Hz}} = 2.50 \times 10^{-10} \text{ m} = 0.250 \text{ nm}$$

b) rhenium, $Z = 45$

Apply the analysis of part (a), just with this different value of Z.

minimum wavelength

$\Delta E = E_K = (Z-1)^2(13.6 \text{ eV}) = (45-1)^2(13.6 \text{ eV}) = 2.633 \times 10^4 \text{ eV} = 4.218 \times 10^{-15} \text{ J}$.

$$\lambda = hc/\Delta E = \frac{(6.626 \times 10^{-34} \text{ J} \cdot \text{s})(2.998 \times 10^8 \text{ m/s})}{4.218 \times 10^{-15} \text{ J}} = 4.71 \times 10^{-11} \text{m} = 0.0471 \text{ nm}.$$

maximum wavelength

$$f = (2.48 \times 10^{15} \text{ Hz})(Z-1)^2 = (2.48 \times 10^{15} \text{ Hz})(45-1)^2 = 4.801 \times 10^{18} \text{ Hz}$$

$$\lambda = \frac{c}{f} = \frac{2.998 \times 10^8 \text{ m/s}}{4.801 \times 10^{18} \text{ Hz}} = 6.24 \times 10^{-11} \text{ m} = 0.0624 \text{ nm}$$

CHAPTER 44
MOLECULES AND CONDENSED MATTER

Exercises 3, 7, 9, 11, 15, 17, 23, 25, 27
Problems 29, 31, 33, 37, 41, 45

Exercises

44-3 a) The energy of a rotational level with quantum number l is

$E_l = l(l+1)\hbar^2/2I$ (Eq.(44-3)).

$I = m_{\mathrm{r}} r^2$

$$m_{\mathrm{r}} = \frac{m_1 m_2}{m_1 + m_2} = \frac{m_{\mathrm{Li}} m_{\mathrm{H}}}{m_{\mathrm{Li}} + m_{\mathrm{H}}} = \frac{(1.17 \times 10^{-26}\ \mathrm{kg})(1.67 \times 10^{-27}\ \mathrm{kg})}{1.17 \times 10^{-26}\ \mathrm{kg} + 1.67 \times 10^{-27}\ \mathrm{kg}} =$$

1.461×10^{-27} kg

$I = m_{\mathrm{r}} r^2 = (1.461 \times 10^{-27}\ \mathrm{kg})(0.159 \times 10^{-9}\ \mathrm{m})^2 = 3.694 \times 10^{-47}\ \mathrm{kg \cdot m^2}$

$$l = 3:\ E = 3(4)\left(\frac{\hbar^2}{2I}\right) = 6\left(\frac{\hbar^2}{I}\right)$$

$$l = 4:\ E = 4(5)\left(\frac{\hbar^2}{2I}\right) = 10\left(\frac{\hbar^2}{I}\right)$$

$$\Delta E = E_3 - E_2 = 4\left(\frac{\hbar^2}{I}\right) = 4\left(\frac{(1.055 \times 10^{-34}\ \mathrm{J \cdot s})^2}{3.694 \times 10^{-47}\ \mathrm{kg \cdot m^2}}\right) = 1.20 \times 10^{-21}\ \mathrm{J} =$$

7.49×10^{-3} eV

b) $\Delta E = hc/\lambda$ so $\lambda = \dfrac{hc}{\Delta E} = \dfrac{(4.136 \times 10^{-15}\ \mathrm{eV})(2.998 \times 10^8\ \mathrm{m/s})}{7.49 \times 10^{-3}\ \mathrm{eV}} = 166\ \mu\mathrm{m}$

44-7 From Example 44-2, $E_l = l(l+1)\dfrac{\hbar^2}{2I}$, with $\dfrac{\hbar^2}{2I} = 0.2395 \times 10^{-3}$ eV.

From Example 44-3, $\Delta E = 0.2690$ eV is the spacing between vibratonal levels. Thus $E_n = (n + \frac{1}{2})\hbar\omega$, with $\hbar\omega = 0.2690$ eV.

By Eq.(44-9), $E = E_n + E_l = (n + \frac{1}{2})\hbar\omega + l(l+1)\dfrac{\hbar^2}{2I}$.

a) $n = 0 \to n = 1$ and $l = 1 \to l = 2$

For $n = 0$, $l = 1$, $E_i = \frac{1}{2}\hbar\omega + 2\left(\dfrac{\hbar^2}{2I}\right)$.

For $n = 1$, $l = 2$, $E_f = \frac{3}{2}\hbar\omega + 6\left(\dfrac{\hbar^2}{2I}\right)$.

$$\Delta E = E_f - E_i = \hbar\omega + 4\left(\frac{\hbar^2}{2I}\right) = 0.2690 \text{ eV} + 4(0.2395 \times 10^{-3} \text{ eV}) = 0.2700 \text{ eV}$$

$$\frac{hc}{\lambda} = \Delta E \text{ so}$$

$$\lambda = \frac{hc}{\Delta E} = \frac{(4.136 \times 10^{-15} \text{ eV} \cdot \text{s})(2.998 \times 10^8 \text{ m/s})}{0.2700 \text{ eV}} = 4.592 \times 10^{-6} \text{ m} = 4.592 \text{ } \mu\text{m}$$

b) $n = 0 \rightarrow n = 1$ and $l = 2 \rightarrow l = 1$

For $n = 0$, $l = 2$, $E_i = \frac{1}{2}\hbar\omega + 6\left(\frac{\hbar^2}{2I}\right)$.

For $n = 1$, $l = 1$, $E_f = \frac{3}{2}\hbar\omega + 2\left(\frac{\hbar^2}{2I}\right)$.

$$\Delta E = E_f - E_i = \hbar\omega - 4\left(\frac{\hbar^2}{2I}\right) = 0.2690 \text{ eV} - 4(0.2395 \times 10^{-3} \text{ eV}) = 0.2680 \text{ eV}$$

$$\lambda = \frac{hc}{\Delta E} = \frac{(4.136 \times 10^{-15} \text{ eV} \cdot \text{s})(2.998 \times 10^8 \text{ m/s})}{0.2680 \text{ eV}} = 4.627 \times 10^{-6} \text{ m} = 4.627 \text{ } \mu\text{m}$$

c) $n = 0 \rightarrow n = 1$ and $l = 3 \rightarrow l = 2$

For $n = 0$, $l = 3$, $E_i = \frac{1}{2}\hbar\omega + 12\left(\frac{\hbar^2}{2I}\right)$.

For $n = 1$, $l = 2$, $E_f = \frac{3}{2}\hbar\omega + 6\left(\frac{\hbar^2}{2I}\right)$.

$$\Delta E = E_f - E_i = \hbar\omega - 6\left(\frac{\hbar^2}{2I}\right) = 0.2690 \text{ eV} - 6(0.2395 \times 10^{-3} \text{ eV}) = 0.2676 \text{ eV}$$

$$\lambda = \frac{hc}{\Delta E} = \frac{(4.136 \times 10^{-15} \text{ eV} \cdot \text{s})(2.998 \times 10^8 \text{ m/s})}{0.2676 \text{ eV}} = 4.634 \times 10^{-6} \text{ m} = 4.634 \text{ } \mu\text{m}$$

44-9 **a)** $f = \frac{\omega}{2\pi} = \frac{1}{2\pi}\sqrt{\frac{k'}{m_r}}$, so $k' = m_r(2\pi f)^2$

$$m_r = \frac{m_1 m_2}{m_1 + m_2} = \frac{m_H m_F}{m_H + m_F} = \frac{(1.67 \times 10^{-27} \text{ kg})(3.15 \times 10^{-26} \text{ kg})}{1.67 \times 10^{-27} \text{ kg} + 3.15 \times 10^{-26} \text{ kg}} =$$

$1.586 \times 10^{-27} \text{ kg}$

$k' = m_r(2\pi f)^2 = (1.586 \times 10^{-27} \text{ kg})(2\pi[1.24 \times 10^{14} \text{ Hz}])^2 = 963 \text{ N/m}$

b) $E_n = (n + \frac{1}{2})\hbar\omega = (n + \frac{1}{2})hf$, since $\hbar\omega = (h/2\pi)\omega$ and $(\omega/2\pi) = f$.
The energy spacing between adjacent levels is
$\Delta E = E_{n+1} - E_n = (n + 1 + \frac{1}{2} - n - \frac{1}{2})hf = hf$, independent of n.
$\Delta E = hf = (6.626 \times 10^{-34} \text{ J} \cdot \text{s})(1.24 \times 10^{14} \text{ Hz}) = 8.22 \times 10^{-20} \text{ J} = 0.513 \text{ eV}$

c) The photon energy equals the transition energy so $\Delta E = hc/\lambda$.

$$hf = hc/\lambda \text{ so } \lambda = \frac{c}{f} = \frac{2.998 \times 10^8 \text{ m/s}}{1.24 \times 10^{14} \text{ Hz}} = 2.42 \times 10^{-6} \text{ m} = 2.42 \text{ } \mu\text{m}$$

This photon is in the infrared.

44-11 Each atom occupies a cube with side length 0.282 nm. Therefore, the volume occupied by each atom is $V = (0.282 \times 10^{-9} \text{ m})^3 = 2.24 \times 10^{-29} \text{ m}^3$.

In NaCl there are equal numbers of Na and Cl atoms, so the average mass of the atoms in the crystal is

$$m = \tfrac{1}{2}(m_{\text{Na}} + m_{\text{Cl}}) = \tfrac{1}{2}(3.82 \times 10^{-26} \text{ kg} + 5.89 \times 10^{-26} \text{ kg}) = 4.855 \times 10^{-26} \text{ kg}$$

The density then is $\rho = \dfrac{m}{V} = \dfrac{4.855 \times 10^{-26} \text{ kg}}{2.24 \times 10^{-29} \text{ m}^3} = 2.17 \times 10^3 \text{ kg/m}^3$.

44-15 a) The photon energy must be at least $E = 1.12 \text{ eV} = 1.794 \times 10^{-19} \text{ J}$.

The wavelength is given by $E = hc/\lambda$, so

$$\lambda = \frac{hc}{E} = \frac{(6.626 \times 10^{-34} \text{ J} \cdot \text{s})(2.998 \times 10^8 \text{ m/s})}{1.794 \times 10^{-19} \text{ J}} = 1.11 \times 10^{-6} \text{ m} = 1.11 \text{ } \mu\text{m; this}$$

photon is in the infrared.

b) Photons of wavelength less than 1.11 μm have energy greater than 1.12 eV so can be absorbed in transitions from the valence band into the conduction band. This includes the entire visible range of wavelengths so silicon absorbs all visible wavelengths and hence is opaque to visible light.

44-17 a) The three-dimensional Schrodinger equation is

$$-\frac{\hbar^2}{2m}\left(\frac{\partial^2 \psi}{\partial x^2} + \frac{\partial^2 \psi}{\partial y^2} + \frac{\partial^2 \psi}{\partial z^2}\right) + U\psi = E\psi \text{ (Eq.42-32).}$$

(Eq.44-10): $\psi = A \sin\left(\dfrac{n_x \pi x}{L}\right) \sin\left(\dfrac{n_y \pi y}{L}\right) \sin\left(\dfrac{n_z \pi z}{L}\right)$

For free electrons, $U = 0$.

$$\frac{\partial \psi}{\partial x} = \frac{n_x \pi}{L} A \cos\left(\frac{n_x \pi x}{L}\right) \sin\left(\frac{n_y \pi y}{L}\right) \sin\left(\frac{n_z \pi z}{L}\right)$$

$$\frac{\partial^2 \psi}{\partial x^2} = -\left(\frac{n_x \pi}{L}\right)^2 A \sin\left(\frac{n_x \pi x}{L}\right) \sin\left(\frac{n_y \pi y}{L}\right) \sin\left(\frac{n_z \pi z}{L}\right) = -\left(\frac{n_x \pi}{L}\right)^2 \psi$$

Similarly $\dfrac{\partial^2 \psi}{\partial y^2} = -\left(\dfrac{n_y \pi}{L}\right)^2 \psi$ and $\dfrac{\partial^2 \psi}{\partial z^2} = -\left(\dfrac{n_z \pi}{L}\right)^2 \psi$.

Therefore, $-\dfrac{\hbar^2}{2m}\left(\dfrac{\partial^2 \psi}{\partial x^2} + \dfrac{\partial^2 \psi}{\partial y^2} + \dfrac{\partial^2 \psi}{\partial z^2}\right) = \dfrac{\hbar^2}{2m}\left(-\dfrac{\pi^2}{L^2}\right)(n_x^2 + n_y^2 + n_z^2)\psi = $

$$\frac{(n_x^2 + n_y^2 + n_z^2)\pi^2 \hbar^2}{2mL^2}\psi$$

This equals $E\psi$, with $E = \dfrac{(n_x^2 + n_y^2 + n_z^2)\pi^2\hbar^2}{2mL^2}$, which is Eq.(44-11).

b) Ground level: lowest E so $n_x = n_y = n_z = 1$ and $E = \dfrac{3\pi^2\hbar^2}{2mL^2}$.

No other combination of n_x, n_y, and n_z gives this same E, so the only degeneracy is the degeneracy of two due to spin.

First excited level: next lower E so one n equals 2 and the others equal 1.

$$E = (2^2 + 1^2 + 1^2)\frac{\pi^2\hbar^2}{2mL^2} = \frac{6\pi^2\hbar^2}{2mL^2}$$

There are three different sets of n_x, n_y, n_z values that give this E:

$n_x = 2,\, n_y = 1,\, n_z = 1$; $n_x = 1,\, n_y = 2,\, n_z = 1$; $n_x = 1,\, n_y = 1,\, n_z = 2$

This gives a degeneracy of 3 so the total degeneracy, with the factor of 2 from spin, is 6.

Second excited level: next lower E so two of n_x, n_y, n_z equal 2 and the other equals 1.

$$E = (2^2 + 2^2 + 1^2)\frac{\pi^2\hbar^2}{2mL^2} = \frac{9\pi^2\hbar^2}{2mL^2}$$

There are three different sets of n_x, n_y, n_z values that give this E:

$n_x = 2,\, n_y = 2,\, n_z = 1$; $n_x = 2,\, n_y = 1,\, n_z = 2$; $n_x = 1,\, n_y = 2,\, n_z = 2$.

Thus, as for the first excited level, the total degeneracy, including spin, is 6.

44-23 a) The electron contribution to the molar heat capacity at constant volume of a metal is $C_V = \left(\dfrac{\pi^2 kT}{2E_F}\right) R$.

$$C_V = \frac{\pi^2(1.381 \times 10^{-23}\ \text{J/K})(300\ \text{K})}{2(5.48\ \text{eV})(1.602 \times 10^{-19}\ \text{J/eV})} R = 0.0233R.$$

b) The electron contribution found in part (a) is $0.0233R = 0.194\ \text{J/mol} \cdot \text{K}$. This is $0.194/25.3 = 7.67 \times 10^{-3} = 0.767\%$ of the total C_V.

c) Only a small fraction of C_V is due to the electrons. Most of C_V is due to the vibrational motion of the ions.

44-25 Eq.(44-17): $f(E) = \dfrac{1}{e^{(E-E_F)/kT} + 1}$

$$e^{(E-E_F)/kT} = \frac{1}{f(E)} - 1$$

The problem states that $f(E) = 4.4 \times 10^{-4}$ for E at the bottom of the conduction band.

$$e^{(E-E_F)/kT} = \frac{1}{4.4 \times 10^{-4}} - 1 = 2.272 \times 10^3.$$

$E - E_F = kT \ln(2.272 \times 10^3) = (1.3807 \times 10^{-23} \text{ J/T})(300 \text{ K}) \ln(2.272 \times 10^3) = 3.201 \times 10^{-20} \text{ J} = 0.20 \text{ eV}$

$E_F = E - 0.20$ eV; the Fermi level is 0.20 eV below the bottom of the conduction band.

44-27 The voltage-current relation is given by Eq.(44-23): $I = I_s \left(e^{eV/kT} - 1 \right)$.

a) Find I_s: $V = +15.0 \times 10^{-3}$ V gives $I = 9.25 \times 10^{-3}$ A

$$\frac{eV}{kT} = \frac{(1.602 \times 10^{-19} \text{ C})(15.0 \times 10^{-3} \text{ V})}{(1.381 \times 10^{-23} \text{ J/K})(300 \text{ K})} = 0.5800$$

$$I_s = \frac{I}{e^{eV/kT} - 1} = \frac{9.25 \times 10^{-3} \text{ A}}{e^{0.5800} - 1} = 1.177 \times 10^{-2} = 11.77 \text{ mA}$$

Then can calculate I for $V = 10.0$ mV:

$$\frac{eV}{kT} = \frac{(1.602 \times 10^{-19} \text{ C})(10.0 \times 10^{-3} \text{ V})}{(1.381 \times 10^{-23} \text{ J/K})(300 \text{ K})} = 0.3867$$

$$I = I_s \left(e^{eV/kT} - 1 \right) = (11.77 \text{ mA}) \left(e^{0.3867} - 1 \right) = 5.56 \text{ mA}$$

b) $\frac{eV}{kT}$ has the same magnitude as in part (a) but now V is negative so $\frac{eV}{kT}$ is negative.

$\underline{V = -15.0 \text{ mV}}$: $\frac{eV}{kT} = -0.5800$ and

$$I = I_s \left(e^{eV/kT} - 1 \right) = (11.77 \text{ mA}) \left(e^{-0.5800} - 1 \right) = -5.18 \text{ mA}$$

$\underline{V = -10.0 \text{ mV}}$: $\frac{eV}{kT} = -0.3867$ and

$$I = I_s \left(e^{eV/kT} - 1 \right) = (11.77 \text{ mA}) \left(e^{-0.3867} - 1 \right) = -3.77 \text{ mA}$$

Problems

44-29 **a)** Eq.(22-14): electric dipole moment $p = qd$

Point charges $+e$ and $-e$ separated by distance d, so

$p = ed = (1.602 \times 10^{-19} \text{ C})(0.24 \times 10^{-9} \text{ m}) = 3.8 \times 10^{-29}$ C·m

b) $p = qd$ so $q = \frac{p}{d} = \frac{3.0 \times 10^{-29} \text{ C/m}}{0.24 \times 10^{-9} \text{ m}} = 1.3 \times 10^{-19}$ C

c) $\frac{q}{e} = \frac{1.3 \times 10^{-19} \text{ C}}{1.602 \times 10^{-19} \text{ C}} = 0.81$

d) $q = \dfrac{p}{d} = \dfrac{1.5 \times 10^{-30} \text{ C/m}}{0.16 \times 10^{-9} \text{ m}} = 9.37 \times 10^{-21}$ C

$\dfrac{q}{e} = \dfrac{9.37 \times 10^{-21} \text{ C}}{1.602 \times 10^{-19} \text{ C}} = 0.058$

The fractional ionic character for the bond in HI is much less than the fractional ionic character for the bond in NaCl. The bond in HI is mostly covalent and not very ionic.

44-31 a) U must make up for the difference between the ionization energy of Na (5.1 eV) and the electron affinity of Cl (3.6 eV).

Thus $U = -(5.1 \text{ eV} - 3.6 \text{ eV}) = -1.5 \text{ eV} = -2.4 \times 10^{-19}$ J, and

$-\dfrac{1}{4\pi\epsilon_0}\dfrac{e^2}{r} = -2.4 \times 10^{-19}$ J

$r = \left(\dfrac{1}{4\pi\epsilon_0}\right)\dfrac{e^2}{2.4 \times 10^{-19} \text{ J}} = (8.988 \times 10^9 \text{ N}\cdot\text{m}^2/\text{C}^2)\dfrac{(1.602 \times 10^{-19} \text{ C})^2}{2.4 \times 10^{-19} \text{ J}}$

$r = 9.6 \times 10^{-10}$ m $= 0.96$ nm

b) ionization energy of K $= 4.3$ eV; electron affinity of Br $= 3.5$ eV

Thus $U = -(4.3 \text{ eV} - 3.5 \text{ eV}) = -0.8 \text{ eV} = -1.28 \times 10^{-19}$ J, and

$-\dfrac{1}{4\pi\epsilon_0}\dfrac{e^2}{r} = -1.28 \times 10^{-19}$ J

$r = \left(\dfrac{1}{4\pi\epsilon_0}\right)\dfrac{e^2}{1.28 \times 10^{-19} \text{ J}} = (8.988 \times 10^9 \text{ N}\cdot\text{m}^2/\text{C}^2)\dfrac{(1.602 \times 10^{-19} \text{ C})^2}{1.28 \times 10^{-19} \text{ J}}$

$r = 1.8 \times 10^{-9}$ m $= 1.8$ nm

44-33 a) Problem 44-32 gives $I = 2.71 \times 10^{-47}$ kg$\cdot$m^2.

$I = m_r r^2$

$m_r = \dfrac{m_H m_{Cl}}{m_H + m_{Cl}} = \dfrac{(1.67 \times 10^{-27} \text{ kg})(5.81 \times 10^{-26} \text{ kg})}{1.67 \times 10^{-27} \text{ kg} + 5.81 \times 10^{-26} \text{ kg}} = 1.623 \times 10^{-27}$ kg

$r = \sqrt{\dfrac{I}{m_r}} = \sqrt{\dfrac{2.71 \times 10^{-47} \text{ kg}\cdot\text{m}^2}{1.623 \times 10^{-27} \text{ kg}}} = 1.29 \times 10^{-10}$ m $= 0.129$ nm

b) Each transition is from the level l to the level $l-1$.

$E_l = l(l+1)\hbar^2/2I$, so $\Delta E = E_l - E_{l-1} = [l(l+1) - l(l-1)]\left(\dfrac{\hbar^2}{2I}\right) = l\left(\dfrac{\hbar^2}{I}\right).$

The transition energy is related to the photon wavelength by $\Delta E = hc/\lambda$.

Combining these two equations for ΔE gives $l\left(\dfrac{\hbar^2}{I}\right) = \dfrac{hc}{\lambda}$

$$l = \frac{2\pi c I}{\hbar \lambda} = \frac{2\pi (2.998 \times 10^8 \text{ m/s})(2.71 \times 10^{-47} \text{ kg} \cdot \text{m}^2)}{(1.055 \times 10^{-34} \text{ J} \cdot \text{s})\lambda} = \frac{4.843 \times 10^{-4} \text{ m}}{\lambda}$$

For $\lambda = 60.4 \ \mu$m, $l = \dfrac{4.843 \times 10^{-4} \text{ m}}{60.4 \times 10^{-6} \text{ m}} = 8.$

For $\lambda = 69.0 \ \mu$m, $l = \dfrac{4.843 \times 10^{-4} \text{ m}}{69.0 \times 10^{-6} \text{ m}} = 7.$

For $\lambda = 80.4 \ \mu$m, $l = \dfrac{4.843 \times 10^{-4} \text{ m}}{80.4 \times 10^{-6} \text{ m}} = 6.$

For $\lambda = 96.4 \ \mu$m, $l = \dfrac{4.843 \times 10^{-4} \text{ m}}{96.4 \times 10^{-6} \text{ m}} = 5.$

For $\lambda = 120.4 \ \mu$m, $l = \dfrac{4.843 \times 10^{-4} \text{ m}}{120.4 \times 10^{-6} \text{ m}} = 4.$

c) Longest λ implies smallest ΔE, and this is for the transition from $l = 1$ to $l = 0$.

$$\Delta E = l \left(\frac{\hbar^2}{I} \right) = (1) \frac{(1.055 \times 10^{-34} \text{ J} \cdot \text{s})^2}{2.71 \times 10^{-47} \text{ kg} \cdot \text{m}^2} = 4.099 \times 10^{-22} \text{ J}$$

$$\lambda = \frac{hc}{\Delta E} = \frac{(6.626 \times 10^{-34} \text{ J} \cdot \text{s})(2.998 \times 10^8 \text{ m/s})}{4.099 \times 10^{-22} \text{ J}} = 4.85 \times 10^{-4} \text{ m} = 485 \ \mu\text{m}.$$

d) What changes is m_r, the reduced mass of the molecule. The transition energy is $\Delta E = l \left(\dfrac{\hbar^2}{I} \right)$ and $\Delta E = \dfrac{hc}{\lambda}$, so $\lambda = \dfrac{2\pi c I}{l \hbar}$ (part (b)).

$I = m_\text{r} r^2$, so λ is directly proportional to m_r.

$$\frac{\lambda(\text{HCl})}{m_\text{r}(\text{HCl})} = \frac{\lambda(\text{DCl})}{m_\text{r}(\text{DCl})} \text{ so } \lambda(\text{DCl}) = \lambda(\text{HCl}) \frac{m_\text{r}(\text{DCl})}{m_\text{r}(\text{HCl})}$$

The mass of a deuterium atom is approximatey twice the mass of a hydrogen atom, so $m_\text{D} = 3.34 \times 10^{-27}$ kg.

$$m_\text{r}(\text{DCl}) = \frac{m_\text{D} m_\text{Cl}}{m_\text{D} + m_\text{Cl}} = \frac{(3.34 \times 10^{-27} \text{ kg})(5.81 \times 10^{-26} \text{ kg})}{3.34 \times 10^{-27} \text{ kg} + 5.81 \times 10^{-26} \text{ kg}} = 3.158 \times 10^{-27} \text{ kg}$$

$$\lambda(\text{DCl}) = \lambda(\text{HCl}) \left(\frac{3.158 \times 10^{-27} \text{ kg}}{1.623 \times 10^{-27} \text{ kg}} \right) = (1.946)\lambda(\text{HCl})$$

$l = 8 \rightarrow l = 7$; $\lambda = (60.4 \ \mu\text{m})(1.946) = 118 \ \mu\text{m}$

$l = 7 \rightarrow l = 6$; $\lambda = (69.0 \ \mu\text{m})(1.946) = 134 \ \mu\text{m}$

$l = 6 \rightarrow l = 5$; $\lambda = (80.4 \ \mu\text{m})(1.946) = 156 \ \mu\text{m}$

$l = 5 \rightarrow l = 4$; $\lambda = (96.4 \ \mu\text{m})(1.946) = 188 \ \mu\text{m}$

$l = 4 \rightarrow l = 3$; $\lambda = (120.4 \ \mu\text{m})(1.946) = 234 \ \mu\text{m}$

44-37 a) $E_l = l(l+1)\hbar^2 / 2I$

Calculate I for Na^{35}Cl:

$$m_r = \frac{m_{Na}m_{Cl}}{m_{Na} + m_{Cl}} = \frac{(3.8176 \times 10^{-26} \text{ kg})(5.8068 \times 10^{-26} \text{ kg})}{3.8176 \times 10^{-26} \text{ kg} + 5.8068 \times 10^{-26} \text{ kg}} = 2.303 \times 10^{-26} \text{ kg}$$

$$I = m_r r^2 = (2.303 \times 10^{-26} \text{ kg})(0.2361 \times 10^{-9} \text{ m})^2 = 1.284 \times 10^{-45} \text{ kg} \cdot \text{m}^2$$

$l = 2 \rightarrow l = 1$ transition

$$\Delta E = E_2 - E_1 = (6 - 2)\left(\frac{\hbar^2}{2I}\right) = \frac{2\hbar^2}{I} = \frac{2(1.055 \times 10^{-34} \text{ J} \cdot \text{s})^2}{1.284 \times 10^{-45} \text{ kg} \cdot \text{m}^2} = 1.734 \times 10^{-23} \text{ J}$$

$$\Delta E = \frac{hc}{\lambda} \text{ so}$$

$$\lambda = \frac{hc}{\Delta E} = \frac{(6.626 \times 10^{-34} \text{ J} \cdot \text{s})(2.998 \times 10^8 \text{ m/s})}{1.734 \times 10^{-23} \text{ J}} = 1.146 \times 10^{-2} \text{ m} = 1.146 \text{ cm}$$

$l = 1 \rightarrow l = 0$ transition

$$\Delta E = E_1 - E_0 = (2 - 0)\left(\frac{\hbar^2}{2I}\right) = \frac{\hbar^2}{I} = \frac{1}{2}(1.734 \times 10^{-23} \text{ J}) = 8.67 \times 10^{-24} \text{ J}$$

$$\lambda = \frac{hc}{\Delta E} = \frac{(6.626 \times 10^{-34} \text{ J} \cdot \text{s})(2.998 \times 10^8 \text{ m/s})}{8.67 \times 10^{-24} \text{ J}} = 2.291 \text{ cm}$$

b) Calculate I for Na^{37}Cl:

$$m_r = \frac{m_{Na}m_{Cl}}{m_{Na} + m_{Cl}} = \frac{(3.8176 \times 10^{-26} \text{ kg})(6.1384 \times 10^{-26} \text{ kg})}{3.8176 \times 10^{-26} \text{ kg} + 6.1384 \times 10^{-26} \text{ kg}} = 2.354 \times 10^{-26} \text{ kg}$$

$$I = m_r r^2 = (2.354 \times 10^{-26} \text{ kg})(0.2361 \times 10^{-9} \text{ m})^2 = 1.312 \times 10^{-45} \text{ kg} \cdot \text{m}^2$$

$l = 2 \rightarrow l = 1$ transition

$$\Delta E = \frac{2\hbar^2}{I} = \frac{2(1.055 \times 10^{-34} \text{ J} \cdot \text{s})^2}{1.312 \times 10^{-45} \text{ kg} \cdot \text{m}^2} = 1.697 \times 10^{-23} \text{ J}$$

$$\lambda = \frac{hc}{\Delta E} = \frac{(6.626 \times 10^{-34} \text{ J} \cdot \text{s})(2.998 \times 10^8 \text{ m/s})}{1.697 \times 10^{-23} \text{ J}} = 1.171 \times 10^{-2} \text{ m} = 1.171 \text{ cm}$$

$l = 1 \rightarrow l = 0$ transition

$$\Delta E = \frac{\hbar^2}{I} = \frac{1}{2}(1.697 \times 10^{-23} \text{ J}) = 8.485 \times 10^{-24} \text{ J}$$

$$\lambda = \frac{hc}{\Delta E} = \frac{(6.626 \times 10^{-34} \text{ J} \cdot \text{s})(2.998 \times 10^8 \text{ m/s})}{8.485 \times 10^{-24} \text{ J}} = 2.341 \text{ cm}$$

The differences in the wavelengths for the two isotopes are:

$l = 2 \rightarrow l = 1$ transition: $1.171 \text{ cm} - 1.146 \text{ cm} = 0.025 \text{ cm}$

$l = 1 \rightarrow l = 0$ transition: $2.341 \text{ cm} - 2.291 \text{ cm} = 0.050 \text{ cm}$

44-41 a) $I = m_r r^2$

$$m_r = \frac{m_H m_I}{m_H + m_I} = \frac{(1.67 \times 10^{-27} \text{ kg})(2.11 \times 10^{-25} \text{ kg})}{1.67 \times 10^{-27} \text{ kg} + 2.11 \times 10^{-25} \text{ kg}} = 1.657 \times 10^{-27} \text{ kg}$$

$$I = m_r r^2 = (1.657 \times 10^{-27} \text{ kg})(0.160 \times 10^{-9} \text{ m})^2 = 4.24 \times 10^{-47} \text{ kg} \cdot \text{m}^2$$

b) The energy levels are $E_{nl} = l(l+1)\left(\dfrac{\hbar^2}{2I}\right) + (n + \frac{1}{2})\hbar\sqrt{\dfrac{k'}{m_r}}$ (Eq.(44-9))

$$\sqrt{\frac{k'}{m}} = \omega = 2\pi f \text{ so } E_{nl} = l(l+1)\left(\frac{\hbar^2}{2I}\right) + (n + \tfrac{1}{2})hf$$

(i) transition $n = 1 \to n = 0$, $l = 1 \to l = 0$

$$\Delta E = (2 - 0)\left(\frac{\hbar^2}{2I}\right) + (1 + \tfrac{1}{2} - \tfrac{1}{2})hf = \frac{\hbar^2}{I} + hf$$

$$\Delta E = \frac{hc}{\lambda} \text{ so } \lambda = \frac{hc}{\Delta E} = \frac{hc}{(\hbar^2/I) + hf} = \frac{c}{(\hbar/2\pi I) + f}$$

$$\frac{\hbar}{2\pi I} = \frac{1.055 \times 10^{-34} \text{ J} \cdot \text{s}}{2\pi(4.24 \times 10^{-47} \text{ kg} \cdot \text{m}^2)} = 3.960 \times 10^{11} \text{ Hz}$$

$$\lambda = \frac{c}{(\hbar/2\pi I) + f} = \frac{2.998 \times 10^8 \text{ m/s}}{3.960 \times 10^{11} \text{ Hz} + 6.93 \times 10^{13} \text{ Hz}} = 4.30 \ \mu\text{m}$$

(ii) transition $n = 1 \to n = 0$, $l = 2 \to l = 1$

$$\Delta E = (6 - 2)\left(\frac{\hbar^2}{2I}\right) + hf = \frac{2\hbar^2}{I} + hf$$

$$\lambda = \frac{c}{2(\hbar/2\pi I) + f} = \frac{2.998 \times 10^8 \text{ m/s}}{2(3.960 \times 10^{11} \text{ Hz}) + 6.93 \times 10^{13} \text{ Hz}} = 4.28 \ \mu\text{m}$$

(iii) transition $n = 2 \to n = 1$, $l = 2 \to l = 3$

$$\Delta E = (6 - 12)\left(\frac{\hbar^2}{2I}\right) + hf = -\frac{3\hbar^2}{I} + hf$$

$$\lambda = \frac{c}{-3(\hbar/2\pi I) + f} = \frac{2.998 \times 10^8 \text{ m/s}}{-3(3.960 \times 10^{11} \text{ Hz}) + 6.93 \times 10^{13} \text{ Hz}} = 4.40 \ \mu\text{m}$$

44-45 **a)** $p = -\dfrac{dE_{\text{tot}}}{dV}$

$$E_{\text{tot}} = N E_{\text{av}} = \frac{3N}{5} E_{F0} = \frac{3}{5}\left(\frac{3^{2/3}\pi^{4/3}\hbar^2}{2m}\right) N^{5/3} V^{-2/3}$$

$$\frac{dE_{\text{tot}}}{dV} = \frac{3}{5}\left(\frac{3^{2/3}\pi^{4/3}\hbar^2}{2m}\right) N^{5/3}\left(-\frac{2}{3}V^{-5/3}\right) \text{ so}$$

$$p = \left(\frac{3^{2/3}\pi^{4/3}\hbar^2}{5m}\right)\left(\frac{N}{V}\right)^{5/3}, \text{ as was to be shown.}$$

b) $N/V = 8.45 \times 10^{28} \text{ m}^{-3}$

$$p = \left(\frac{3^{2/3} \pi^{4/3} (1.055 \times 10^{-34} \text{ J} \cdot \text{s})^2}{5(9.109 \times 10^{-31} \text{ kg})} \right) \left(8.45 \times 10^{28} \text{ m}^{-3} \right)^{5/3} = 3.81 \times 10^{10} \text{ Pa} = 3.76 \times 10^5 \text{ atm}.$$

c) The electrons are held in the metal by the attractive force exerted on them by the copper ions.

CHAPTER 45

NUCLEAR PHYSICS

Exercises 3, 5, 7, 9, 13, 19, 23, 25, 29, 31, 33, 35
Problems 43, 47, 49, 53, 55, 57, 59, 61

Exercises

45-3 When the spin component is parallel to the field the interaction energy is
$U = -\mu_z B$. When the spin component is antiparallel to the field the interaction energy is $U = +\mu_z B$. The transition energy for a transition between these two states is $\Delta E = 2\mu_z B$, where $\mu_z = 2.7928\mu_n$.

The transition energy is related to the photon frequency by $\Delta E = hf$, so
$2\mu_z B = hf$.
$$B = \frac{hf}{2\mu_z} = \frac{(6.626 \times 10^{-34} \text{ J} \cdot \text{s})(22.7 \times 10^6 \text{ Hz})}{2(2.7928)(5.051 \times 10^{-27} \text{ J/T})} = 0.533 \text{ T}$$

45-5 a) From Example 45-2, when the z-component of $\vec{S}$ (and $\vec{\mu}$) is parallel to $\vec{B}$,
$U = -|\mu_z|B = -2.7928\mu_n B$.

When the z-component of $\vec{S}$ (and $\vec{\mu}$) is antiparallel to $\vec{B}$,
$U = +|\mu_z|B = +2.7928\mu_n B$.

The state with the proton spin component parallel to the field lies lower in energy.
The energy difference between these two states is $\Delta E = 2(2.7928\mu_n B)$.
$$\Delta E = hf \text{ so } f = \frac{\Delta E}{h} = \frac{2(2.7928\mu_n B)}{h} = \frac{2(2.7928)(5.051 \times 10^{-27} \text{ J/T})(1.65 \text{ T})}{6.626 \times 10^{-34} \text{ J}}$$
$f = 7.03 \times 10^7 \text{ Hz} = 70.3 \text{ MHz}$

And then $\lambda = \dfrac{c}{f} = \dfrac{2.998 \times 10^8 \text{ m/s}}{7.03 \times 10^7 \text{ Hz}} = 4.26 \text{ m}$

From Fig.33-15, these are radio waves.

b) From Eqs.(28-27) and (43-22) and Fig.43-12, the state with the z-component of $\vec{\mu}$ parallel to $\vec{B}$ has lower energy. But, since the charge of the electron is negative, this is the state with the electron spin component antiparallel to $\vec{B}$.

That is, the $m_s = -\frac{1}{2}$ state lies lower in energy.

For the $m_s = +\frac{1}{2}$ state,

$$U = +(2.00232)\left(\frac{e}{2m}\right)\left(+\frac{\hbar}{2}\right)B = +\frac{1}{2}(2.00232)\left(\frac{e\hbar}{2m}\right)B = +\frac{1}{2}(2.00232)\mu_B B.$$

For the $m_s = -\frac{1}{2}$ state, $U = -\frac{1}{2}(2.00232)\mu_B B$.

The energy difference between these two states is $\Delta E = (2.00232)\mu_B B$.

$\Delta E = hf$ so $f = \dfrac{\Delta E}{h} = \dfrac{2.00232\mu_B B}{h} = \dfrac{(2.00232)(9.274 \times 10^{-24} \text{ J/T})(1.65 \text{ T})}{6.626 \times 10^{-34} \text{ J}} = 4.62 \times 10^{10} \text{ Hz} = 46.2 \text{ GHz}.$

And $\lambda = \dfrac{c}{f} = \dfrac{2.998 \times 10^8 \text{ m/s}}{4.62 \times 10^{10} \text{ Hz}} = 6.49 \times 10^{-3} \text{ m} = 6.49 \text{ mm}.$

From Fig.33-15, these are microwaves.

45-7 The text calculates that the binding energy of the deuteron is 2.224 MeV.

A photon that breaks the deuteron up into a proton and a neutron must have at least this much energy.

$E = \dfrac{hc}{\lambda}$ so $\lambda = \dfrac{hc}{E}$

$\lambda = \dfrac{(4.136 \times 10^{-15} \text{ eV} \cdot \text{s})(2.998 \times 10^8 \text{ m/s})}{2.224 \times 10^6 \text{ eV}} = 5.575 \times 10^{-13} \text{ m} = 0.5575 \text{ pm}.$

45-9 a) A $^{11}_{5}\text{B}$ atom has 5 protons, $11 - 5 = 6$ neutrons, and 5 electrons. The mass defect therefore is $\Delta M = 5m_p + 6m_n + 5m_e - M(^{11}_{5}\text{B})$.

$\Delta M = 5(1.0072765 \text{ u}) + 6(1.0086649 \text{ u}) + 5(0.0005485799 \text{ u}) - 11.009305 \text{ u} = 0.08181 \text{ u}.$

The energy equivalent is $E_B = (0.08181 \text{ u})(931.5 \text{ MeV/u}) = 76.21 \text{ MeV}.$

b) Eq.(45-9): $E_B = C_1 A - C_2 A^{2/3} - C_3 Z(Z-1)/A^{1/3} - C_4 (A - 2Z)^2/A$

The fifth term is zero since Z is odd but N is even.

$E_B = (15.75 \text{ MeV})(11) - (17.80 \text{ MeV})(11)^{2/3} - (0.7100 \text{ MeV})5(4)/11^{1/3} - (23.69 \text{ MeV})(11 - 10)^2/11.$

$E_B = +173.25 \text{ MeV} - 88.04 \text{ MeV} - 6.38 \text{ MeV} - 2.15 \text{ MeV} = 76.68 \text{ MeV}$

The percentage difference between the calculated and measured E_B is

$\dfrac{76.68 \text{ MeV} - 76.21 \text{ MeV}}{76.21 \text{ MeV}} = 0.6\%.$

Eq.(45-9) has a greater percentage accuracy for ^{62}Ni.

45-13 a) α-decay: Z decreases by 2, $A = N+Z$ decreases by 4 (an α particle is a ^4_2He nucleus)

$^{239}_{94}\text{Pu} \rightarrow {}^4_2\text{He} + {}^{235}_{92}\text{U}$

b) β^- decay: Z increases by 1, $A = N + Z$ remains the same (a β^- particle is an electron, $_{-1}^{0}e$)

$^{24}_{11}\text{Na} \rightarrow {}_{-1}^{0}e + {}^{24}_{12}\text{Mg}$

c) β^+ decay: Z decreases by 1, $A = N + Z$ remains the same (a β^+ particle is a positron, $_{+1}^{0}e$)

$$^{15}_{8}O \rightarrow \, _{+1}^{0}e + \, ^{15}_{7}N$$

Note: In each case the total charge and total number of nucleons for the decay products equals the charge and number of nucleons for the parent nucleus; these two quantities are conserved in the decay.

45-19 a) The β^- decay reaction is $^{3}_{1}H \rightarrow \, _{-1}^{0}e + \, ^{3}_{2}He$.

$^{3}_{1}H$ is unstable with respect to β^- decay since the mass of the $^{3}_{2}He$ nucleus plus the emitted electron is less than the mass of the $^{3}_{1}H$ nucleus.

The masses in Table 45-2 are for the neutral atoms, so the mass of the $^{3}_{1}H$ nucleus is $3.016049 \text{ u} - 0.0005486 \text{ u} = 3.0155004 \text{ u}$ and the mass of the $^{3}_{2}He$ nucleus is $3.016029 \text{ u} - 2(0.0005486 \text{ u}) = 3.0149318 \text{ u}$.

The mass decrease in the decay is $3.0155004 \text{ u} - 3.0149318 \text{ u} - 0.0005486 \text{ u} = 0.000020 \text{ u}$.

b) Find the energy equivalent of the mass decrease found in part (a):

$0.000020 \text{ u}(931.5 \text{ MeV/u}) = 0.019 \text{ MeV} = 19 \text{ keV}$.

45-23 a) $|dN/dt| = \lambda N$

$$\lambda = \frac{0.693}{T_{1/2}} = \frac{0.693}{(1.28 \times 10^9 \text{ y})(3.156 \times 10^7 \text{ s/1 y})} = 1.715 \times 10^{-17} \text{ s}^{-1}$$

The mass of one ^{40}K atom is approximately 40 u, so the number of ^{40}K nuclei in th sample is

$$N = \frac{1.63 \times 10^{-9} \text{ kg}}{40 \text{ u}} = \frac{1.63 \times 10^{-9} \text{ kg}}{40(1.66054 \times 10^{-27} \text{ kg})} = 2.454 \times 10^{16}.$$

Then $|dN/dt| = \lambda N = (1.715 \times 10^{-17} \text{ s}^{-1})(2.454 \times 10^{16}) = 0.421$ decays/s

b) $|dN/dt| = (0.421 \text{ decays/s})(1 \text{ Ci}/(3.70 \times 10^{10} \text{ decays/s})) = 1.14 \times 10^{-11} \text{ Ci}$

45-25 a) Eq.(45-17): $\lambda = \frac{0.693}{T_{1/2}} = \frac{0.693}{(4.47 \times 10^9 \text{ y})(3.156 \times 10^7 \text{ s/1 y})} = 4.91 \times 10^{-18} \text{ s}^{-1}$.

b) Activity of 12.0×10^{-6} Ci implies that $dN/dt = 4.44 \times 10^5$ decays/s

Eq.(45-14) says $|dN/dt| = \lambda N$ so

$$N = \frac{|dN/dt|}{\lambda} = \frac{4.44 \times 10^5 \text{ decays/s}}{4.91 \times 10^{-18} \text{ s}^{-1}} = 9.04 \times 10^{22} \text{ nuclei}$$

The sample must contain 9.04×10^{22} uranium nuclei.

The mass of one uranium ^{238}U atom is approximatey 238 u, so the mass of the sample is

$$m = N(238 \text{ u}) = (9.04 \times 10^{22})(238)(1.66054 \times 10^{-27} \text{ kg}) = 35.7 \text{ g}.$$

c) One α particle is emitted in each decay, so the problem is asking for the activity, in decays/s, of this sample.

Compute N, the number of nuclei in the sample. The mass of one ^{238}U atom is approximately 238 u, so

$$N = \frac{60.0 \times 10^{-3} \text{ kg}}{238 \text{ u}} = \frac{60.0 \times 10^{-3} \text{ kg}}{238(1.66054 \times 10^{-27} \text{ kg})} = 1.518 \times 10^{23} \text{ nuclei.}$$

$$|dN/dt| = \lambda N = (4.91 \times 10^{-18} \text{ s}^{-1})(1.518 \times 10^{23}) = 7.45 \times 10^5 \text{ decays/s.}$$

45-29 a) The energy E of each photon is

$$E = \frac{hc}{\lambda} = \frac{(6.626 \times 10^{-34} \text{ J} \cdot \text{s})(2.998 \times 10^8 \text{ m/s})}{0.0200 \times 10^{-9} \text{ m}} = 9.932 \times 10^{-15} \text{ J.}$$

The total energy absorbed is the number of photons absorbed times the energy of each photon,

$$(6.50 \times 10^{10} \text{ photons})(9.932 \times 10^{-15} \text{ J/photon}) = 6.46 \times 10^{-4} \text{ J.}$$

b) The absorbed dose is the energy absorbed divided by the mass of the absorbing tissue:

$$\text{absorbed dose} = \frac{6.46 \times 10^{-4} \text{ J}}{0.600 \text{ kg}} = (1.08 \times 10^{-3} \text{ J/kg})(1 \text{ rad}/(0.01 \text{ J/kg}))$$

$$\text{absorbed dose} = 0.108 \text{ rad}$$

The equivalent dose in rem equals the RBE times the absorbed dose in rad. For x rays, RBE $=1$, so the equivalent dose is 0.108 rem.

45-31 a) X must have $A = +2 + 9 - 4 = 7$ and $Z = +1 + 4 - 2 = 3$.

Thus X is $^{7}_{3}$Li and the reaction is $^{2}_{1}$H $+$ $^{9}_{4}$Be $=$ $^{7}_{3}$Li $+$ $^{4}_{2}$He

b) If we use the neutral atom masses then there are the same number of electrons (five) in the reactants as in the products. Their masses cancel, so we get the same mass defect whether we use nuclear masses or neutral atom masses.

The neutral atom masses are given in Table 45-2:

$^{2}_{1}$H $+$ $^{9}_{4}$Be has mass 2.014102 u $+$ 9.012182 u $=$ 11.026284 u

$^{7}_{3}$Li $+$ $^{4}_{2}$He has mass 7.016003 u $+$ 4.002603 u $=$ 11.018606 u

The mass decrease is 11.026284 u $-$ 11.018606 u $=$ 0.007678 u.

This corresponds to an energy release of 0.007678 u(931.5 MeV/1 u) $=$ 7.152 MeV.

c) Estimate the threshold energy by calculating the Coulomb potential energy when the $^{2}_{1}$H and $^{9}_{4}$Be nuclei just touch.

The radius R_{Be} of the $^{9}_{4}$Be nucleus is $R_{Be} = (1.2 \times 10^{-15} \text{ m})(9)^{1/3} = 2.5 \times 10^{-15} \text{ m.}$

The radius R_H of the $^{2}_{1}$H nucleus is $R_H = (1.2 \times 10^{-15} \text{ m})(2)^{1/3} = 1.5 \times 10^{-15} \text{ m.}$

The nuclei touch when their center-to-center separation is

$R = R_{\text{Be}} + R_{\text{H}} = 4.0 \times 10^{-15}$ m.

The Coulomb potential energy of the two reactant nuclei at this separation is

$$U = \frac{1}{4\pi\epsilon_0}\frac{q_1 q_2}{r} = \frac{1}{4\pi\epsilon_0}\frac{e(4e)}{r}$$

$$U = (8.988 \times 10^9 \text{ N} \cdot \text{m}^2/\text{C}^2)\frac{4(1.602 \times 10^{-19} \text{ C})^2}{(4.0 \times 10^{-15} \text{ m})(1.602 \times 10^{-19} \text{ J/eV})} = 1.4 \text{ MeV}$$

This is an estimate of the threshold energy for this reaction.

45-33 a) X must have $A = 2 + 14 - 10 = 6$ and $Z = 1 + 7 - 5 = 3$.

Thus X is ^6_3Li and the reaction is $^2_1\text{H} + {}^{14}_7\text{N} \rightarrow {}^6_3\text{Li} + {}^{10}_5\text{B}$

b) The neutral atoms on each side of the reaction equation have a total of 8 electrons, so the electron masses cancel when neutral atom masses are used. The neutral atom masses are found in Table 45-2.

mass of $^2_1\text{H} + {}^{14}_7\text{N}$ is 2.014102 u + 14.003074 u = 16.017176 u

mass of $^6_3\text{Li} + {}^{10}_5\text{B}$ is 6.015121 u + 10.012937 u = 16.028058 u

The mass increases, so energy is absorbed by the reaction. The Q value is

(16.017176 u − 16.028058 u)(931.5 Mev/u) = −10.14 MeV

c) The kinetic energy that must be available to cause the reaction is 10.14 MeV. Thus in Eq.(45-22) $K_{\text{cm}} = 10.14$ MeV. The mass M of the stationary target ($^{14}_7\text{N}$) is $M = 14$ u. The mass m of the colliding particle (^2_1H) is 2 u. Then by Eq.(45-22) the minumum kinetic energy K that the ^2_1H must have is

$$K = \left(\frac{M+m}{M}\right)K_{\text{cm}} = \left(\frac{14 \text{ u} + 2 \text{ u}}{14 \text{ u}}\right)(10.14 \text{ MeV}) = 11.59 \text{ MeV}$$

45-35 $^{235}_{92}\text{U} + {}^1_0\text{n} \rightarrow {}^{236}_{92}\text{U}$

The mass defect is $\Delta M = M\left({}^{235}_{92}\text{U}\right) + M\left({}^1_0\text{n}\right) - M\left({}^{236}_{92}\text{U}\right)$.

$\Delta M = 235.043923 \text{ u} + 1.008665 \text{ u} - 236.045562 \text{ u} = +0.007026 \text{ u}$

$Q = (\Delta M)c^2 = (0.007026 \text{ u})(931.5 \text{ MeV}/1 \text{ u}) = 6.545 \text{ MeV}$

This much energy is liberated in the reaction. By conservaton of linear momentum, since there is essentially zero momentum before the reaction, the product nucleus must be essentially at rest and therefore have zero kinetic energy. All the energy released must go into the internal excitation of the $^{236}_{92}\text{U}$.

Problems

45-43 a) The heavier nucleus will decay into the lighter one; $^{25}_{13}\text{Al}$ will decay into $^{25}_{12}\text{Mg}$.

b) Balance A and Z in the decay reaction: $^{25}_{13}\text{Al} \rightarrow {}^{25}_{12}\text{Mg} + {}^0_{+1}\text{e}$. The emitted particle must have charge $+e$ and its nucleon number must be zero. Therefore, it is a β^+ particle, a positron.

c) Calculate the energy defect ΔM for the reaction. Use the nuclear masses for $^{25}_{13}\text{Al}$ and $^{25}_{12}\text{Mg}$, to avoid confusion in including the correct number of electrons if neutral atom masses are used.

The nuclear mass for $^{25}_{13}\text{Al}$ is $M_{\text{nuc}}\left(^{25}_{13}\text{Al}\right) = 24.990429 \text{ u} - 13(0.000548580 \text{ u}) = 24.983297 \text{ u}$.

The nuclear mass for $^{25}_{12}\text{Mg}$ is $M_{\text{nuc}}\left(^{25}_{12}\text{Mg}\right) = 24.985837 \text{ u} - 12(0.000548580 \text{ u}) = 24.979254 \text{ u}$.

The mass defect for the reaction is

$\Delta M = M_{\text{nuc}}\left(^{25}_{13}\text{Al}\right) - M_{\text{nuc}}\left(^{25}_{12}\text{Mg}\right) - M\left(^{0}_{+1}e\right) = 24.983297 \text{ u} - 24.979254 \text{ u} - 0.00054858 \text{ u} = 0.003494 \text{ u}$

$Q = (\Delta M)c^2 = 0.003494 \text{ u}(931.5 \text{ MeV}/1 \text{ u}) = 3.255 \text{ MeV}$

The mass decreases in the decay and energy is released.

Note: $^{25}_{13}\text{Al}$ can also decay into $^{25}_{12}\text{Mg}$ by electron capture.

$^{25}_{13}\text{Al} + {}^{0}_{-1}e \rightarrow {}^{25}_{12}\text{Mg}$

The ${}^{0}_{-1}e$ electron in the reaction is an orbital electron in the neutral $^{25}_{13}\text{Al}$ atom. The mass defect can be calculated using the nuclear masses:

$\Delta M = M_{\text{nuc}}\left(^{25}_{13}\text{Al}\right) + M\left(^{0}_{-1}e\right) - M_{\text{nuc}}\left(^{25}_{12}\text{Mg}\right) = 24.983297 \text{ u} + 0.00054858 \text{ u} - 24.979254 \text{ u} = 0.004592 \text{ u}$.

$Q = (\Delta M)c^2 = (0.004592 \text{ u})(931.5 \text{ MeV}/1 \text{ u}) = 4.277 \text{ MeV}$

The mass decreases in the decay and energy is released.

45-47 $^{198}_{79}\text{Au} \rightarrow {}^{198}_{80}\text{Hg} + {}^{0}_{-1}e$

The mass change is $197.968225 \text{ u} - 197.966752 \text{ u} = 1.473 \times 10^{-3} \text{ u}$

(The neutral atom masses include 79 electrons before the decay and 80 electrons after the decay. This one additional electron in the products accounts correctly for the electron emitted by the nucleus.)

The total energy released in the decay is $(1.473 \times 10^{-3} \text{ u})(931.5 \text{ MeV}/\text{u}) = 1.372$ MeV. This energy is divided between the energy of the emitted photon and the kinetic energy of the β^- particle. Thus the β^- particle has kinetic energy equal to $1.372 \text{ MeV} - 0.412 \text{ MeV} = 0.960 \text{ MeV}$.

45-49 $^{13}_{7}\text{N} \rightarrow {}^{0}_{+1}e + {}^{13}_{6}\text{C}$

To avoid confusion in including the correct number of electrons with neutral atom masses, use nuclear masses, obtained by subtracting the mass of the atomic electrons from the neutral atom masses.

The nuclear mass for $^{13}_{7}\text{N}$ is $M_{\text{nuc}}\left(^{13}_{7}\text{N}\right) = 13.005739 \text{ u} - 7(0.00054858 \text{ u}) = 13.001899 \text{ u}$.

The nuclear mass for $^{13}_{6}\text{C}$ is $M_{\text{nuc}}\left(^{13}_{6}\text{C}\right) = 13.003355\ \text{u} - 6(0.00054858\ \text{u}) = 13.000064\ \text{u}$.

The mass defect for the reaction is $\Delta M = M_{\text{nuc}}\left(^{13}_{7}\text{N}\right) - M_{\text{nuc}}\left(^{13}_{6}\text{C}\right) - M\left(^{0}_{+1}\text{e}\right)$
$\Delta M = 13.001899\ \text{u} - 13.000064\ \text{u} - 0.00054858\ \text{u} = 0.001286\ \text{u}$.

The mass decreases in the decay, so energy is released. This decay is energetically possible.

45-53 We have to be careful; after ^{87}Rb has undergone radioactive decay it is no longer a rubidium atom. Let N_{85} be the number of ^{85}Rb atoms; this number doesn't change. Let N_0 be the number of ^{87}Rb atoms on earth when the solar system was formed. Let N be the present number of ^{87}Rb atoms.

The present measurements say that $0.2783 = N/(N + N_{85})$.

$(N + N_{85})(0.2783) = N$, so $N = 0.3856 N_{85}$.

The percentage we are asked to calculate is $N_0/(N_0 + N_{85})$.

N and N_0 are related by $N = N_0 e^{-\lambda t}$ so $N_0 = e^{+\lambda t} N$.

Thus $\dfrac{N_0}{N_0 + N_{85}} = \dfrac{N e^{\lambda t}}{N e^{\lambda t} + N_{85}} = \dfrac{(0.3856 e^{\lambda t}) N_{85}}{(0.3856 e^{\lambda t}) N_{85} + N_{85}} = \dfrac{0.3856 e^{\lambda t}}{0.3856 e^{\lambda t} + 1}$.

$t = 4.6 \times 10^9\ \text{y}$; $\lambda = \dfrac{0.693}{T_{1/2}} = \dfrac{0.693}{4.75 \times 10^{10}\ \text{y}} = 1.459 \times 10^{-11}\ \text{y}^{-1}$

$e^{\lambda t} = e^{(1.459 \times 10^{-11}\ \text{y}^{-1})(4.6 \times 10^9\ \text{y})} = e^{0.06711} = 1.0694$

Thus $\dfrac{N_0}{N_0 + N_{85}} = \dfrac{(0.3856)(1.0694)}{(0.3856)(1.0694) + 1} = 29.2\%$.

45-55 a) First find the number of decays each second:

$2.6 \times 10^{-4}\ \text{Ci} \left(\dfrac{3.70 \times 10^{10}\ \text{decays/s}}{1\ \text{Ci}} \right) = 9.6 \times 10^6\ \text{decays/s}$

The average energy per decay is 1.25 MeV, and one-half of this energy is deposited in the tumor. The energy delivered to the tumor per second then is
$\frac{1}{2}(9.6 \times 10^6\ \text{decays/s})(1.25 \times 10^6\ \text{eV/decay})(1.602 \times 10^{-19}\ \text{J/eV}) = 9.6 \times 10^{-7}\ \text{J/s}$.

b) The absorbed dose is the energy absorbed divided by the mass of the tissue:
$\dfrac{9.6 \times 10^{-7}\ \text{J/s}}{0.500\ \text{kg}} = (1.9 \times 10^{-6}\ \text{J/kg}\cdot\text{s})(1\ \text{rad}/(0.01\ \text{J/kg})) = 1.9 \times 10^{-4}\ \text{rad/s}$

c) equivalent dose (REM) = RBE $\times$ absorbed dose (rad)

In one second the equivalent dose is $0.70(1.9 \times 10^{-4}\ \text{rad}) = 1.3 \times 10^{-4}\ \text{rem}$.

d) $(200\ \text{rem}/1.3 \times 10^{-4}\ \text{rem/s}) = 1.5 \times 10^6\ \text{s}(1\ \text{h}/3600\ \text{s}) = 420\ \text{h} = 17\ \text{days}$.

45-57 $N = N_0 e^{-\lambda t}$

The problem says $N/N_0 = 0.21$; solve for t.

$0.21 = e^{-\lambda t}$ so $\ln(0.21) = -\lambda t$ and $t = -\ln(0.21)/\lambda$

Example 45-9 gives $\lambda = 1.209 \times 10^{-4}$ y^{-1} for ^{14}C.

Thus $t = \dfrac{-\ln(0.21)}{1.209 \times 10^{-4} \text{ y}} = 1.3 \times 10^4$ y.

45-59 a) The radius of $_1^2$H is $R = (1.2 \times 10^{-15} \text{ m})(2)^{1/3} = 1.51 \times 10^{-15}$ m.

The barrier energy is the Coulomb potential energy of two $_1^2$H nuclei with their centers separated by twice this distance:

$$U = \frac{1}{4\pi\epsilon_0}\frac{e^2}{r} = (8.988 \times 10^9 \text{ N}\cdot\text{m}^2/\text{C}^2)\frac{(1.602 \times 10^{-19} \text{ C})^2}{2(1.51 \times 10^{-15} \text{ m})} = 7.64 \times 10^{-14} \text{ J} =$$

0.48 MeV

b) $_1^2$H $+ {}_1^2$H $\rightarrow {}_2^3$He $+ {}_0^1$n

If we use neutral atom masses there are two electrons on each side of the reaction equation, so their masses cancel. The neutral atom masses are given in Table 45-2.

$_1^2$H $+ {}_1^2$H has mass $2(2.014102$ u$) = 4.028204$ u

$_2^3$He $+ {}_0^1$n has mass 3.016029 u $+ 1.008665$ u $= 4.024694$ u

The mass decrease is 4.028204 u $- 4.024694$ u $= 3.510 \times 10^{-3}$ u. This corrresponds to a liberated energy of $(3.510 \times 10^{-3}$ u$)(931.5$ MeV/u$) = 3.270$ MeV, or

$(3.270 \times 10^6$ eV$)(1.602 \times 10^{-19}$ J/eV$) = 5.239 \times 10^{-13}$ J.

c) Each reaction takes two $_1^2$H nuclei. Each mole of D$_2$ has 6.022×10^{23} molecules, so 6.022×10^{23} pairs of atoms. The energy liberated when one more of deuterium undergoes fusion is $(6.022 \times 10^{23})(5.239 \times 10^{-13}$ J$) = 3.155 \times 10^{11}$ J/mol.

45-61 Consider 1.00 kg of body tissue. The mass of ^{40}K in 1.00 kg of tissue is

$(0.21 \times 10^{-2})(0.012 \times 10^{-2})(1.00 \text{ kg}) = 2.52 \times 10^{-7}$ kg.

The mass of a ^{40}K atom is approximately 40 u, so the number of ^{40}K nuclei in 1.00 kg of tissue is

$$\frac{2.52 \times 10^{-7} \text{ kg}}{40 \text{ u}} = \frac{2.52 \times 10^{-7} \text{ kg}}{40 \text{ u}(1.66054 \times 10^{-27} \text{ kg/u})} = 3.794 \times 10^{18}.$$

The activity is $|dN/dt| = \lambda N$. $\lambda = \dfrac{0.693}{T_{1/2}} = \dfrac{0.693}{1.28 \times 10^9 \text{ y}} = 5.414 \times 10^{-10}$ y^{-1}.

Thus $|dN/dt| = (5.414 \times 10^{-10}$ y$^{-1})(3.794 \times 10^{18}) = 2.054 \times 10^9$ decays/y.

In 50 y there are $(50 \text{ y})(2.054 \times 10^9 \text{ decays/y}) = 1.027 \times 10^{11}$ decays.

For each decay an average of 0.50 MeV of energy is absorbed. The energy absorbed by 1.00 kg of tissue is $(1.027 \times 10^{11}$ decays$)(0.50$ MeV/decay$) = 5.14 \times 10^{10}$ MeV $=$ 5.14×10^{16} eV$(1.602 \times 10^{-19}$ J/eV$) = 0.0082$ J.

The absobed dose is $(0.0082 \text{ J/kg})(1 \text{ rad}/ 0.01 \text{ J/kg})=0.82 \text{ rad}$. RBE = 1.0, so the equivalent dose is 0.82 rem.

CHAPTER 46
PARTICLE PHYSICS AND COSMOLOGY

Exercises 1, 7, 9, 11, 13, 17, 21, 25, 27, 29, 33
Problems 35, 37, 39, 43, 47

Exercises

46-1 By momentum conservation the two photons must have equal and opposite momenta.

Then $E = pc$ says the photons must have equal energies.

The mass of the pion is $270m_e$, so the rest energy of the pion is $270(0.511 \text{ MeV}) = 138 \text{ MeV}$. Each photon has half this energy, or 69 MeV.

$$E = hf \text{ so } f = \frac{E}{h} = \frac{(69 \times 10^6 \text{ eV})(1.602 \times 10^{-19} \text{ J/eV})}{6.626 \times 10^{-34} \text{ J} \cdot \text{s}} = 1.7 \times 10^{22} \text{ Hz}$$

$$\lambda = \frac{c}{f} = \frac{2.998 \times 10^8 \text{ m/s}}{1.7 \times 10^{22} \text{ Hz}} = 1.8 \times 10^{-14} \text{ m} = 18 \text{ fm}.$$

These photons are in the gamma ray part of the electromagnetic spectrum.

46-7 **a)** Eq.(46-7): $\omega = |q|B/m$ so $B = m\omega/|q|$.

And since $\omega = 2\pi f$, this becomes $B = 2\pi m f/|q|$.

A deuteron is a deuterium nucleus $\left(^2_1\text{H}\right)$. Its charge is $q = +e$. Its mass is the mass of the neutral ^2_1H atom (Table 45-2) minus the mass of the one atomic electron:

$m = 2.014102 \text{ u} - 0.0005486 \text{ u} = 2.013553 \text{ u}(1.66054 \times 10^{-27} \text{ kg/1 u}) = 3.344 \times 10^{-27} \text{ kg}$

$$B = \frac{2\pi m f}{|q|} = \frac{2\pi (3.344 \times 10^{-27} \text{ kg})(9.00 \times 10^6 \text{ Hz})}{1.602 \times 10^{-19} \text{ C}} = 1.18 \text{ T}$$

b) Eq.(46-8): $K = \dfrac{q^2 B^2 R^2}{2m} = \dfrac{[(1.602 \times 10^{-19} \text{ C})(1.18 \text{ T})(0.320 \text{ m})]^2}{2(3.344 \times 10^{-27} \text{ kg})}$.

$K = 5.471 \times 10^{-13} \text{ J} = (5.471 \times 10^{-13} \text{ J})(1 \text{ eV}/1.602 \times 10^{-19} \text{ J}) = 3.42 \text{ MeV}$

$$K = \tfrac{1}{2}mv^2 \text{ so } v = \sqrt{\frac{2K}{m}} = \sqrt{\frac{2(5.471 \times 10^{-13} \text{ J})}{3.344 \times 10^{-27} \text{ kg}}} = 1.81 \times 10^7 \text{ m/s}$$

Note: $v/c = 0.06$, so it is ok to use the nonrelativistic expression for kinetic energy.

46-9 **a)** The masses of the target and projectile particles are equal, so Eq.(46-10) can be used:

$$E_a^2 = 2mc^2(E_m + mc^2)$$

$$E_m = \frac{E_a^2}{2mc^2} - mc^2$$

The mass of the alpha particle can be calculated by subtracting two electron masses from the $_2^4$He atomic mass:

$$m = m_\alpha = 4.002603 \text{ u} - 2(0.0005486 \text{ u}) = 4.001506 \text{ u}$$

Then $mc^2 = (4.001506 \text{ u})(931.5 \text{ MeV/u}) = 3.727 \text{ GeV}$.

$$E_m = \frac{E_a^2}{2mc^2} - mc^2 = \frac{(16.0 \text{ GeV})^2}{2(3.727 \text{ GeV})} - 3.727 \text{ GeV} = 30.6 \text{ GeV}.$$

b) Each beam must have $\frac{1}{2}E_a = 8.0$ GeV.

46-11 **a)** For a proton beam on a stationary proton target and since E_a is much larger than the proton rest energy we can use Eq.(46-11): $E_a^2 = 2mc^2 E_m$.

$$E_m = \frac{E_a^2}{2mc^2} = \frac{(77.4 \text{ GeV})^2}{2(0.938 \text{ GeV})} = 3200 \text{ GeV}$$

b) For colliding beams the total momentum is zero and the available energy E_a is the total energy of the two colliding particles. For proton-proton collisions the colliding beams each have the same energy, so the total energy of each beam is $\frac{1}{2}E_a = 38.7$ GeV.

46-13 The mass decrease is $m(\Sigma^+) - m(\text{p}) - m(\pi^0)$ and the energy released is

$$mc^2(\Sigma^+) - mc^2(\text{p}) - mc^2(\pi^0) = 1189 \text{ MeV} - 938.3 \text{ MeV} - 135.0 \text{ MeV} = 116 \text{ MeV}.$$

(The mc^2 values for each particle were taken from Table 46-3.)

46-17 **a)** $\text{K}^+ \rightarrow \mu^+ + \nu_\mu$; $S_{\text{K}^+} = +1, S_{\mu^+} = 0, S_{\nu_\mu} = 0$

$S = 1$ initially; $S = 0$ for the products; S is <u>not conserved</u>

b) $\text{n} + \text{K}^+ \rightarrow \text{p} + \pi^0$; $S_\text{n} = 0, S_{\text{K}^+} = +1, S_\text{p} = 0, S_{\pi^0} = 0$

$S = 1$ initially; $S = 0$ for the products; S is <u>not conserved</u>

c) $\text{K}^+ + \text{K}^- \rightarrow \pi^0 + \pi^0$; $S_{\text{K}^+} = +1$; $S_{\text{K}^-} = -1$; $S_{\pi^0} = 0$

$S = +1 - 1 = 0$ initially; $S = 0$ for the products; S <u>is conserved</u>

d) $\text{p} + \text{K}^- \rightarrow \Lambda^0 + \pi^0$; $S_\text{p} = 0, S_{\text{K}^-} = -1, S_{\Lambda^0} = -1, S_{\pi^0} = 0.$

$S = -1$ initially; $S = -1$ for the products; S <u>is conserved</u>

46-21 Each value for the combination is the sum of the values for each quark.

a) uds

$$Q = \tfrac{2}{3}e - \tfrac{1}{3}e - \tfrac{1}{3}e = 0$$
$$B = \tfrac{1}{3} + \tfrac{1}{3} + \tfrac{1}{3} = 1$$

$S = 0 + 0 - 1 = -1$

$C = 0 + 0 + 0 = 0$

b) c$\bar{\text{u}}$

The values for $\bar{\text{u}}$ are the negative of those for **u**.

$Q = \frac{2}{3}e - \frac{2}{3}e = 0$

$B = \frac{1}{3} - \frac{1}{3} = 0$

$S = 0 + 0 = 0$

$C = +1 + 0 = +1$

c) ddd

$Q = -\frac{1}{3}e - \frac{1}{3}e - \frac{1}{3}e = -e$

$B = \frac{1}{3} + \frac{1}{3} + \frac{1}{3} = +1$

$S = 0 + 0 + 0 = 0$

$C = 0 + 0 + 0 = 0$

d) d$\bar{\text{c}}$

$Q = -\frac{1}{3}e - \frac{2}{3}e = -e$

$B = \frac{1}{3} - \frac{1}{3} = 0$

$S = 0 + 0 = 0$

$C = 0 - 1 = -1$

46-25 a) $r = 5210$ Mly

Eq.(46-15): $v = H_0 r = ((20 \text{ km/s})/\text{Mly})(5210 \text{ Mly}) = 1.04 \times 10^5$ km/s

b) $\dfrac{\lambda_0}{\lambda_S} = \sqrt{\dfrac{c+v}{c-v}} = \sqrt{\dfrac{1+v/c}{1-v/c}}$

$\dfrac{v}{c} = \dfrac{1.04 \times 10^8 \text{ m/s}}{2.9980 \times 10^8 \text{ m/s}} = 0.3469$

$\dfrac{\lambda_0}{\lambda_S} = \sqrt{\dfrac{1+0.3469}{1-0.3469}} = 1.44$

46-27 a) Eq.(46-14): $v = \left[\dfrac{(\lambda_0/\lambda_S)^2 - 1}{(\lambda_0/\lambda_s)^2 + 1}\right]c$

$v = \left[\dfrac{(658.5 \text{ nm}/590 \text{ nm})^2 - 1}{(658.5 \text{ nm}/590 \text{ nm})^2 + 1}\right]c = 0.1094c$

$v = (0.1094)(2.998 \times 10^8 \text{ m/s}) = 3.28 \times 10^7$ m/s

b) Eq.(46-15): $v = H_0 r$ (Hubble's law)

$$r = \frac{v}{H_0} = \frac{3.28 \times 10^7 \text{ m/s}}{20 \times 10^3 \text{ (m/s)/Mly}} = 1640 \text{ Mly}$$

46-29 **a)** $p + {}^2_1H \rightarrow {}^3_2He$ or can write as ${}^1_1H + {}^2_1H \rightarrow {}^3_2He$

If neutral atom masses are used then the masses of the two atomic electrons on each side of the reaction will cancel.

Taking the atomic masses from Table 45-2, the mass decrease is

$m({}^1_1H) + m({}^2_1H) - m({}^3_2He) = 1.007825 \text{ u} + 2.014102 \text{ u} - 3.016029 \text{ u} = 0.005898 \text{ u}$.

The energy released is the energy equivalent of this mass decrease:

$(0.005898 \text{ u})(931.5 \text{ MeV/u}) = 5.494 \text{ MeV}$

b) ${}^1_0n + {}^3_2He \rightarrow {}^4_2He$

If neutral helium masses are used then the masses of the two atomic electrons on each side of the reaction equation will cancel. The mass decrease is

$m({}^1_0n) + m({}^3_2He) - m({}^4_2He) = 1.008665 \text{ u} + 3.016029 \text{ u} - 4.002603 \text{ u} = 0.022091 \text{ u}$.

The energy released is the energy equivalent of this mass decrease:

$(0.022091 \text{ u})(931.5 \text{ MeV/u}) = 20.58 \text{ MeV}$

46-33 The Wien displacement law (Eq.40-28) says $\lambda_m T$ equals a constant.

$$\lambda_{m,1} T_1 = \lambda_{m,2} T_2$$

$$\lambda_{m,1} = \lambda_{m,2} \left(\frac{T_2}{T_1} \right) = 1.062 \times 10^{-3} \text{ m} \left(\frac{2.728 \text{ K}}{3000 \text{ K}} \right) = 966 \text{ nm}$$

Problems

46-35 **a)** $E_a = 2(7.0 \text{ TeV}) = 14.0 \text{ TeV}$

b) Need $E_a = 14.0 \text{ TeV} = 14.0 \times 10^6 \text{ MeV}$.

Since the target and projectile particles are both protons Eq.(46-10) can be used:

$E_a^2 = 2mc^2(E_m + mc^2)$

$E_m = \dfrac{E_a^2}{2mc^2} - mc^2 = \dfrac{(14.0 \times 10^6 \text{ MeV})^2}{2(938.3 \text{ MeV})} - 938.3 \text{ MeV} = 1.0 \times 10^{11} \text{ MeV} =$

1.0×10^5 TeV.

This shows the great advantage of colliding beams at relativistic energies.

46-37 $e^- + e^- \rightarrow e^- + e^- + \pi^0$

$Q = m_{\pi^0} c^2 = 135.0 \text{ MeV}$

The available energy E_a in the collision must be $E_a = Q + 2m_e c^2 = 135.0 \text{ MeV} + 2(0.511 \text{ MeV}) = 136.0 \text{ MeV}$.

Since the beam and target particles are the same we can use Eq.(46-10):

$E_a^2 = 2mc^2(E_m + mc^2)$, where m is the electron mass m_e.

$$E_m = \frac{E_a^2}{2mc^2} - mc^2 = \frac{(136.0 \text{ MeV})^2}{2(0.511 \text{ MeV})} - 0.511 \text{ MeV} = 1.81 \times 10^4 \text{ MeV}$$

$E_m = K + m_e c^2$, so $K = E_m - m_e c^2 = 1.81 \times 10^4 \text{ MeV} - 0.511 \text{ MeV} = 1.81 \times 10^4 \text{ MeV} = 18.1$ GeV

46-39 The total available energy must be at least the total rest energy of the product particles:

$$E_a = mc^2(\Lambda^0) + mc^2(K^+) + mc^2(K^-) = 1116 \text{ MeV} + 2(493.7 \text{ MeV}) = 2103.4 \text{ MeV}$$

Since the target and projectile particles are different we must use Eq.(46-9):

$E_a^2 = 2Mc^2 E_m + (Mc^2)^2 + (mc^2)^2$, with $M = m_p$ and $m = m_{K^-}$

$E_a^2 = 2m_p c^2 E_{K^-} + (m_p c^2)^2 + (m_{K^-} c^2)^2$

$$E_{K^-} = \frac{E_a^2 - (m_p c^2)^2 - (m_{K^-} c^2)^2}{2m_p c^2}$$

$$E_{K^-} = \frac{(2103.4 \text{ MeV})^2 - (938.3 \text{ MeV})^2 - (493.7 \text{ MeV})^2}{2(938.3 \text{ MeV})} = 1758.6 \text{ MeV}$$

This is the total energy of the K^- particle. Its kinetic energy is

$K = E_{K^-} - (m_{K^-})c^2 = 1758.6 \text{ MeV} - 493.7 \text{ MeV} = 1265 \text{ MeV}$

46-43 $\phi \rightarrow K^+ + K^-$

a) The mass decrease is $m(\phi) - m(K^+) - m(K^-)$. The energy equivalent of the mass decrease is $mc^2(\phi) - mc^2(K^+) - mc^2(K^-)$. The rest mass energy mc^2 for the ϕ meson is given in Problem 46-42, and the values for K^+ and K^- are given in Table 46-3. The energy released then is 1019.4 MeV $-$ 2(493.7 MeV) = 32.0 MeV. The K^+ gets half this, 16.0 MeV.

b) Does the decay $\phi \rightarrow K^+ + K^- + \pi^0$ occur?

The energy equivalent of the $K^+ + K^- + \pi^0$ mass is 493.7 MeV + 493.7 MeV + 135.0 MeV = 1122 MeV. This is greater than the energy equivalent of the ϕ mass. The mass of the decay products would be greater than the mass of the parent particle; the decay is energetically forbidden.

c) Does the decay $\phi \rightarrow K^+ + \pi^-$ occur?

The reaction $\phi \rightarrow K^+ + K^-$ is observed. K^+ has strangeness $+1$ and K^- has strangeness -1, so the total strangeness of the decay products is zero. But strangeness must conserved, so we deduce that the ϕ particle has strangeness zero.

π^- has strangeness 0, so the products $K^+ + \pi^-$ has strangeness -1. The decay $\phi \rightarrow K^+ + \pi^-$ violates consevation of strangeness and does not occur.

Does the decay $\phi \rightarrow K^+ + \mu^-$ occur?

μ^- has strangeness 0, so this decay would also violate conservation of strangeness and hence does not occur.

46-47 **a)** The energy equivalent of the mass decrease is

$mc^2(\Xi^-) - mc^2(\Lambda^0) - mc^2(\pi^-) = 1321$ MeV - 1116 MeV - 139.6 MeV = 65 MeV

b) The Ξ^- is at rest means that the linear momentum is zero. Conservation of linear momentum then says that the Λ^0 and π^- must have equal and opposite momenta:

$m_{\Lambda^0} v_{\Lambda^0} = m_{\pi^-} v_{\pi^-}$

$v_{\pi^-} = \left(\dfrac{m_{\Lambda^0}}{m_{\pi^-}} \right) v_{\Lambda^0}$

Also, the sum of the kinetic energies of the Λ^0 and π^- must equal the total kinetic energy $K_{tot} = 65$ MeV calculated in part (a):

$K_{tot} = K_{\Lambda^0} + K_{\pi^-}$

$K_{\Lambda^0} + \frac{1}{2} m_{\pi^-} v_{\pi^-}^2 = K_{tot}$

Use the momentum conservation result:

$K_{\Lambda^0} + \frac{1}{2} m_{\pi^-} \left(\dfrac{m_{\Lambda^0}}{m_{\pi^-}} \right)^2 v_{\Lambda^0}^2 = K_{tot}$

$K_{\Lambda^0} + \left(\dfrac{m_{\Lambda^0}}{m_{\pi^-}} \right) \left(\frac{1}{2} m_{\Lambda^0} v_{\Lambda^0}^2 \right) = K_{tot}$

$K_{\Lambda^0} \left(1 + \dfrac{m_{\Lambda^0}}{m_{\pi^-}} \right) = K_{tot}$

$K_{\Lambda^0} = \dfrac{K_{tot}}{1 + m_{\Lambda^0}/m_{\pi^-}} = \dfrac{65 \text{ MeV}}{1 + (1116 \text{ MeV})/(139.6 \text{ MeV}} = 7.2$ MeV

$K_{\Lambda^0} + K_{\pi^-} = K_{tot}$ so

$K_{\pi^-} = K_{tot} - K_{\Lambda^0} = 65$ MeV $- 7.2$ MeV $= 57.8$ MeV

The fraction for the Λ^0 is $\dfrac{7.2 \text{ MeV}}{65 \text{ MeV}} = 11\%$.

The fraction for the π^- is $\dfrac{57.8 \text{ MeV}}{65 \text{ MeV}} = 89\%$.

The lighter particle carries off more of the kinetic energy that is released in the decay.